Contemporary Ergonomics 1987

Contemporary Ergonomics 1987

Proceedings of the Ergonomics Society's
1987 Annual Conference
Swansea, Wales, 6-10 April 1987

'ERGONOMICS WORKING FOR SOCIETY'

Edited by

E.D. Megaw

Department of Engineering Production
The University of Birmingham

Taylor & Francis
London • New York • Philadelphia
1987

UK	Taylor & Francis Ltd, 4 John St, London WC1N 2ET
USA	Taylor & Francis Inc, 242 Cherry St, Philadelphia, PA 19106-1906

Copyright © Taylor & Francis Ltd 1987

All rights reserved. No part of this publication may be reproduced, stored in a retrieval system, or transmitted in any form or by any means, electronic, mechanical, photocopying, recording or otherwise, without permission in writing from the publisher.

British Library Cataloguing in Publication Data

Contemporary ergonomics — 1984 —
   1. Human engineering — periodicals
   620.8'2'05   TA166
   ISBN 0-85066-386-5
   ISSN 0267-4718

Library of Congress Cataloging in Publication Data is available

Printed in Great Britain by Taylor & Francis (Printers) Ltd, Basingstoke, Hants.

CONTENTS

KEYNOTE ADDRESSES

Lighting and Human Performance
P. Boyce 1

The Cognitive Bases of Predictable Human Error
J. Reason 21

HUMAN RELIABILITY

A Case Study Involving Simulation, Human Error Prediction and Quantification for Inclusion into a Probabilistic Risk Assessment
T. Waters 37

SHERPA: A Systematic Human Error Reduction and Prediction Approach
D.E. Embrey 39

A Guide to Reducing Human Error in Process Operation
P.W. Ball 46

Helping the Designer to Improve Human Reliability
S.P. Whalley 52

Human Reliability from a Nuclear Regulatory Viewpoint
D. Whitfield 58

An Integrated Approach to Improving the Performance of Process Operators
D. Visick 64

Applied Human Reliability: Quo Vadis?
B. Kirwan 70

HUMAN PERFORMANCE

Anxiety Prior to and During Decompression
M.H. Ussher and E.W. Farmer 77

Minor Illnesses and Performance Efficiency
A. Smith and K. Coyle 83

Effect of Visual-Lobe Size on Search Performance on Industrial Radiographs
P. O'Boyle and T. Gallwey 89

Compensation Strategies of Elderly Car Drivers
P. Van Wolffelaar, T. Rothengatter and
W. Brouwer 95

MENTAL WORKLOAD

Utilisation Focused Workload Evaluation in Systems Design
N. Mohindra, E. Spencer and R. Taylor 101

Fundamental Issues in Workload Assessment
R. Hockey 103

Measuring Subjective Mental Load
A. Craig 105

Mental Workload and the Organisation of Training
R.B. Stammers 106

Workload and Situation Awareness in Future Aircraft
T.J. Emerson, J.M. Reising and
H.G. Britten-Austin 108

DESIGN, SIMULATION AND EVALUATION

Military Training and Human Factors Contributions to the British Simulation Industry
T. Crampin 110

Contents

The Use of People to Simulate Machines: An Ergonomic Approach
M.A. Life and J. Long 117

Design Models and Design Practice: An Overview
S.E. Powrie 123

In Search of Methods of Prediction
J.R. Wilson, J.J. Ing, J.S. Cadman and
P.H. Barton 129

Hierarchical Task Analysis: Twenty Years On
R.B. Stammers and J.A. Astley 135

Ergonomic Implications in the Design of an Engine Assembly Line
W.L. Chan, A.J. Pethick and R.J. Graves 140

EQUIPMENT DESIGN

Defence Standard 00-25: Human Factors for Designers of Equipment
R.S. Harvey 146

Laboratory and Field Evaluation of Selected Fire Fighting Assemblies
F.D. Mawby, P.J. Street and C.J. Norman 150

Ergonomics for Designing Products
H. Kanis 156

Getting to Grips with Hand Tool Design
M.H. Mabey and R.G. Graves 162

Social and Economic Consequences of an International Ergonomic Standardization
I. Matzdorff 167

VEHICLE ERGONOMICS

Motorcycle Ergonomics: An Exploratory Study
S. Robertson and J.M. Porter 173

The Ergonomic London Bus
G.N. Davis and T.J. Lowe 179

Ergonomics of the Standard City Bus Cabin
M. Kompier, F. Van Noord, H. Mulders and
T. Meijman 185

The Effects of Posture and Seat Design on Lumbar Lordosis
J.M. Porter and B.J. Norris 191

WORKING POSTURE

Industrial Maintenance Tasks Involving Overhead Working
C.M. Haslegrave, M. Tracy and E.N. Corlett .. 197

An Evaluation of Office Seating
I.G. Kleberg and J.E. Ridd 203

The Effect on Nurses' Backs of a Switch from Beds to Chairs for Use by Kidney Dialysis Patients
M. Porter and A. Farrow 209

HUMAN-COMPUTER INTERFACE DESIGN

Software Ergonomics at the Sharp End: Development of a New Interface for a CAD Package
S.E. Powrie 215

What Is a Good CAD Dialogue?
M.A. Sinclair 221

Prescription, Description and Evolution: Design of User-Computer Interfaces for Changing Systems
H. David 227

Using Videotex to Order Goods from Home
S. Fenn and P. Buckley 233

Communication Failure in Dialogue: Non-
Verbal Behaviour at the User Interface
N.P. Sheehy, A.J. Chapman and M.A. Forrest   .. 239

A Framework for User Models
J. Long   .. 245

The Evaluation and Generation of Icons for
a Computer Drawing Package
R.M. Browne and R.B. Stammers   .. 251

KNOWLEDGE ACQUISITION

The Design, Development and Evaluation of a
Climatic Ergonomics Knowledge Based System
T.A. Smith and K.C. Parsons   .. 257

Problems of Knowledge Acquisition for
Expert Systems
N.K. Taylor, E.N. Corlett and M.R. Simpson   .. 263

Ergonomics and Mechanical Engineering
Design
P. John   .. 269

PHYSIOLOGICAL STRESS

A Model of Human Thermoregulation: The
Effects of Body Size on the Accuracy of
Prediction
S.L. Cooper, R.A. Haslam and K.C. Parsons   .. 275

Limiting Heat Strain: Can We Rely on the
Subject?
L.A. Gay and N.T. Thomas   .. 280

Predicting the Metabolic Cost of
Intermittent Load Carriage in the Arms
I.P.M. Randle   .. 286

A Comparison of Methods for Measuring Body
'Core' Temperature in Ergonomics
Applications
A.P. Payne and K.C. Parsons   .. 292

Reducing the Effects of Vibration Through the Control of Workplace Ambient Temperature
N.K. Akinmayowa 298

LIGHTING

Lighting for the Partially Sighted
P.T. Stone 303

Lighting for Control Rooms
J. Wood 309

Emergency Lighting and Movement Through Corridors and Stairways
G.M.B. Webber and P.J. Hallman 315

MANAGEMENT OF CHANGE

Managing Change in the Local Economy: Information Technology Support and the New Entrepreneurs
C. Brotherton, M. Aldridge, S. Charleton and P. Leather 321

Organisational Change and Management Training
A. Davis and T. Cox 328

Socio-Technical Change in Truck Manufacturing
P. Leather and C. Brown 333

Managing Individual Innovation
M. West and N. King 340

The Organisational Requirements of Modern, Automated Manufacturing Systems
S. Joyner 346

TRADE UNIONS AND ERGONOMICS

On the Till: Workers' Perceptions of Health and Safety in Supermarket Checkout Design
C. Thorne and D. Russell 352

The Record to Date: One Union's Experience
J. Church 356

Worker Designers: A Work Environment Act for Britain?
D. Feickert 359

New Technology and Redesigning Work
J. Winterton 365

Restructuring the Relationship: The Ergonomist as a Resource Person
K. Forrester 371

LATE PAPER

Capricious Behaviour and Human Reliability
W.W. Suojanen 376

Author Index 383

Subject Index 385

Keynote Addresses

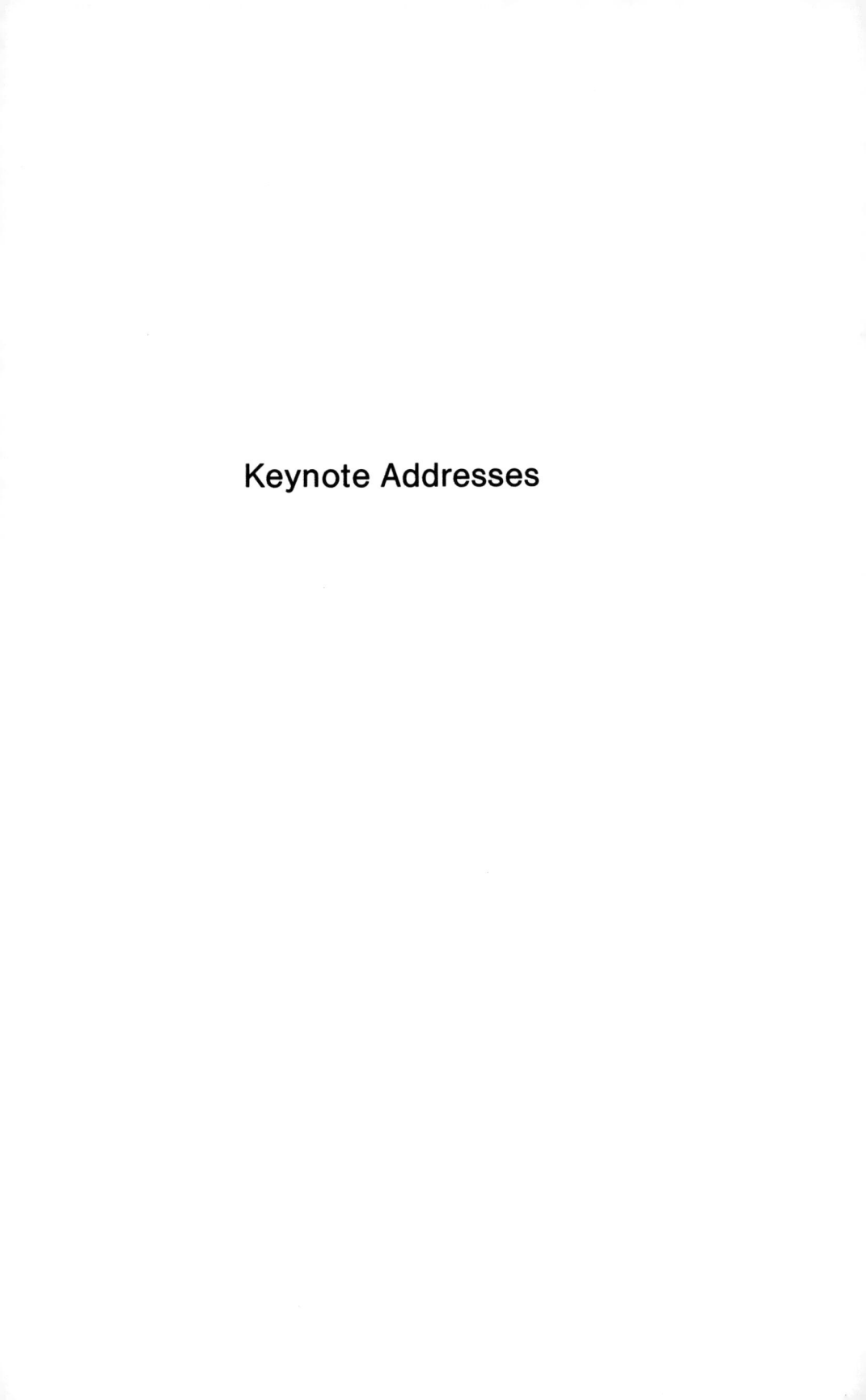

LIGHTING AND HUMAN PERFORMANCE

P.R. BOYCE

The Electricity Council Research Centre
Capenhurst, Chester CH1 6ES

The relationship between lighting and human performance is multi-faceted. The key to understanding this relationship, in all its complexity, lies in a knowledge of the characteristics of the human visual system. This paper attempts to paint a broad brush picture of the significant features of the visual system and how they interact with various aspects of lighting to influence human performance, whilst reading, searching, moving and judging colours.

INTRODUCTION
Nearly all forms of work involve vision. Light is the medium which allows the visual system to operate. Therefore it is reasonable to expect a link between the lighting conditions provided by a lighting installation and people's ability to perform work in the area lit by that lighting installation. There is no doubt that such a link does exist but the exact form it takes is likely to change markedly with the specific circumstances. There are four reasons for this. First, the visual component in work varies widely in magnitude. For example, copy typing is largely visual, whilst audio typing could almost be done with closed eyes. The implications of changes in lighting conditions for performance of copy and audio typing are, therefore, very different. Second, the significance of the visual component in work can vary greatly. For example, in proof reading, whilst the number of misprints missed will be influenced by the lighting conditions, the consequences of passing a misprint will depend on exactly what it is. A misprint which changed an author's royalties from 15 to 150 per cent would probably have graver consequences than simply passing a common mis-spelling. Third, the nature of the visual component can be expressed on a number of different

dimensions, such as size, contrast, colour, movement, etc. Different lighting conditions affect different dimensions in different ways. For example, changes in light distribution relative to the observer are unlikely to change the size of detail but may well change the contrast of the material being viewed. Fourth, the visual system is part of a living entity. Its capabilities will vary from individual to individual as well as showing the common effects of age and disease. Given these four factors, three of which relate to the form of the work being considered and one of which is determined by the nature of the visual system, it is hardly surprising that the various forms of relationship between lighting conditions and human performance are not well understood, in spite of a history of study that goes back at least 66 years (Elton, 1920).

The immediate aim of this paper is to present a broad brush picture of what is known of the relationship between lighting and human performance. The ultimate aim is to encourage more fruitful research and greater understanding of the way various factors intervene on the path from lighting conditions to human performance.

SIGNIFICANT FEATURES OF THE VISUAL SYSTEM
In order to understand how lighting conditions affect human performance, it is first necessary to know what the significant features of the visual system are. To appreciate the significant features it is not necessary to know how the visual system operates but only to be aware of its principal characteristics. The visual system can operate over a very wide range of luminances, from bright sunlight to starlight, but it cannot do so instantaneously. Rather, the visual system is continually adapting to the prevailing luminance. There are two mechanisms by which adaptation takes place, a pupillary/neural mechanism and a photochemical mechanism. The pupillary/neural mechanism is quick, adaptation is complete within a fraction of a second, but it has a limited range. The photochemical mechanism is slow, several minutes being required for adaptation to occur, but it has a very large range. The success of this combined system is shown by the fact that one is rarely aware of being misadapted, the most common experience of it being on entering a cinema. Initially, vision is very limited but after about 5 minutes much more detail can be distinguished. This phenomenon occurs because the sudden change of luminance from the luminance of the street or foyer to the luminance of the cinema itself is too great for the pupillary/neural system to cope and the

photochemical system takes several minutes to make the necessary change. Whether the visual system is fully adapted is an important consideration for the relationship between lighting and human performance.

Another important aspect of the visual system is the organisation of the retina. This can be considered in two dimensions; luminance range and spatial distribution of sensitivity. Fundamentally, man has two separate retinas, a night retina, in which only rod photoreceptors are used, and a day retina, in which the response is governed by the cone photoreceptors. When the night retina is operating, there is no colour vision and the ability to discriminate detail is sacrificed to the need for high sensitivity to low luminous flux. When the day retina is operating, colour vision is available and good discrimination of detail is possible. It should be noted that there is no abrupt transition between these two states of operation, rather a transition zone exists in which limited colour vision and discrimination of detail is possible. This feature ensures that the luminance to which the visual system is adapted has a definite effect on human performance. However, full night vision is used only in a few occupations, shipping for example, the vast majority of work being carried out using the day retina.

As for spatial distribution, the human retina has two different parts, a small central region, called the fovea, whose function is discrimination of detail; and a large surrounding peripheral region whose function is to detect the occurrence of events. Both detection of presence and recognition ability deteriorate with increasing deviation from the fovea but recognition deteriorates to a much greater extent (Johnson et al. 1978). This is why peripheral vision is mainly used to detect the occurrence of something that may be important, so that the central fovea can be turned towards it and its nature identified. Given this structure it should be apparent that an important characteristic of work, as far as the visual system is concerned, is the predictability of where relevant events will occur.

Yet another relevant feature of the visual system is the ability of the retina to discriminate detail. The most concise description of the sensitivity of the retina in this respect is the spatial modulation transfer function. This function shows contrast sensitivity plotted against spatial frequency for a sinewave grating target. Contrast sensitivity is the reciprocal of the threshold contrast. Spatial frequency is a measure of the size of detail in the grating, expressed in cycles per degree subtended at the eye. Measurements of the spatial modulation transfer function (Campbell & Robson, 1968; Daitch & Green, 1969) show that contrast sensitivity is a maximum around spatial frequencies of one to three cycles per degree and diminishes at both lower and higher spatial frequencies. This pattern is maintained as adaptation luminance is varied but there is an increase in absolute contrast sensitivity at all spatial frequencies and a shift in the maximum to higher spatial frequencies at higher luminances. This pattern occurs for both foveal vision and peripheral vision.

Sensitivity measures such as these reveal two important factors which affect the interaction of lighting and human performance. First that adaptation luminance affects the absolute sensitivity of the visual system. Second that there is always a minimum stimulus which cannot be seen; exactly what it is depending on the conditions of contrast, size and luminance that prevail. A consistent observation from studies of visual performance is that the closer the visual component of the work is to this limit the greater is the visual difficulty and the poorer is the performance, although this relationship is not a linear one. It should be apparent that the proximity of the stimulus to threshold conditions is an important factor in determining the relationship between lighting conditions and human performance.

Finally, there is the matter of colour judgement. The human visual system is very good at discriminating colours, particularly in the central fovea. However, it can only do this if the adaptation luminance is above about 3 cd m^{-2} i.e. the day retina is operating. Below this luminance colour vision gradually fails. This implies that the colour vision capabilities are influenced by adaptation luminance. Further, the stimulus received by the eye and from which it derives colour consists of a spectral distribution, determined by the spectral emission of the light source and the spectral reflectance of the material being examined. No one would pretend that this is all that is involved in colour vision but the physical nature of the stimulus cannot be ignored. For the purposes of this paper the important point to note is that different light sources can have very different spectral emissions, from the monochromatic low pressure sodium lamp, widely used for roadway lighting, to the continuous spectral emission of the incandescent lamp widely used for domestic lighting.

These properties of the visual system suggest a number of factors which are likely to affect the relationship between lighting and human performance. First there is the question of what the worker is being asked to do; simple detection, recognition or colour discrimination. Second there are the inherent features of the task; the size, contrast and colours of the detail, the predictability of where the required information can be found, the time for which the information is available, whether the material is moving or stationary. Third there is the state of adaptation of the visual system. And on top of all this is the organisation of the work and the place of the visual component in that organisation. It is hardly surprising that studies of the relationship between lighting and human performance have shown that different aspects of lighting are important in different circumstances. The rest of this paper is concerned with demonstrating the significance of different aspects of lighting for work categorised according to the above factors.

READING

Reading is one of the most widely studied visual tasks (Tinker 1963, Spencer 1968, Poulton 1969, Monty & Senders 1970). Reading is carried out by a series of fixations along a line of print. The reader operates by recognising groups of letters or words from the limited set that the message being read leads him to expect. As a general rule the more familiar the material being read the fewer the fixation points, the less familiar the material the more the fixation points. This method of operation is only possible because passages to be read are laid out following a convention so that the experienced reader can predict where the next word will occur. Further, reading is usually done under stable conditions where the observer has time to adapt to the prevailing luminance. In addition, reading material is usually printed in black or grey on a white background so colour judgements are not often required. Thus reading is an example of work which is done by fully adapted people and where the occurrence of information can be predicted. The important features of the task are the size, contrast and style of the print.

The most recent and most thorough study of the effect of lighting conditions on the performance of this type of task is that of Rea (1986a). The task used involved the comparison of two printed number lists for discrepancies. The viewing distance and hence the size of the task was fixed by locating the subject's head in a chin rest. The printed numbers subtended 13 min arc in width and 19 min arc in height. The contrast of the task was varied by changing the reflection of the ink used in the printing and by changing the position of the material relative to the position of the light source. Eight young subjects performed the comparison under four different adaptation luminances and at sixteen different contrasts. For each comparison the time taken to complete the comparison and the number of errors were recorded. The reciprocal of time taken to compare the lists plotted against contrast for each of the adaptation luminances is shown in Figure 1.

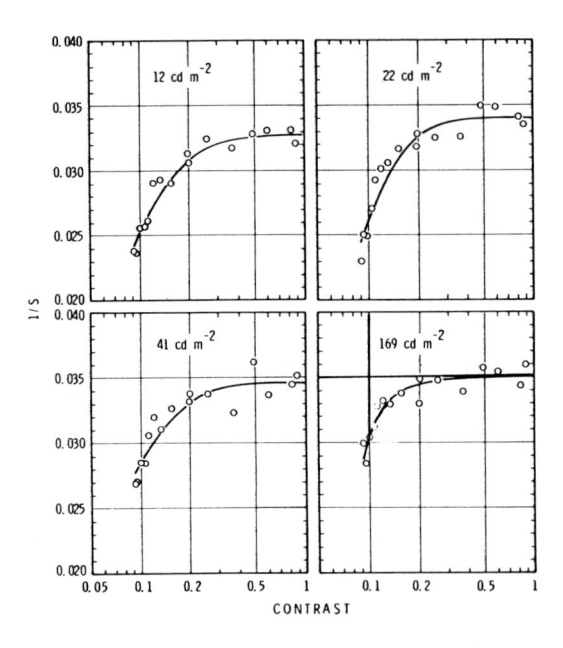

Figure 1. The reciprocal of time taken to complete the numerical verification task plotted as a function of contrast at four background luminances (after Rea 1986a).

Two points should be noted from Figure 1. The first is that as contrast is decreased from about 0.9, there is little change in speed until the contrast falls below about 0.3 but then speed is rapidly reduced. The second is that for the same contrast, performance tends to be higher at higher luminances and, as contrast increases from a low level, performance tends to saturate more quickly at high luminances. Taken together these results show that lighting which produces higher adaptation luminances not only increases the absolute level of task performance but also ensures that the performance is higher over a larger range of contrasts. Interestingly, the curves shown in Figure 1 have the same form as those obtained when plotting the strength of response measured electrophysiologically in the visual cortex of a monkey with a visual system very similar to man when exposed to stimulation of different contrast (Albrecht & Hamilton 1982).

Rea (1986a) has used these results to construct a model of visual performance, subject to two modifications and some interpolation and expansion. The modifications are the subtraction of the time taken to mark discrepancies in the list and the time taken to read the checklist from the total time taken to perform the complete task. The aim of these subtractions is to minimise the non-visual component in the measure of task performance. The interpolation occurs because the model covers all luminances between 12 and 169 cd m^{-2}. The expansion relates to the effects of age, expressed in terms of the change in contrast threshold and retinal illuminance with age. The model can be represented by a visual performance surface. With this model, and knowing the background luminance of the task and its contrast, it is possible to predict the level of visual performance that would be achieved by people of different ages relative to the maximum achieved under high luminance (169 cd m^{-2}) and high contrast (1.0) conditions. The background luminance of the task is determined by the illuminance on the task and the reflectance of the task material. The practical implications of this model have been considered in a second paper (Rea 1986b). Table 1 shows the levels of relative visual performance that are predicted at different illuminances for the numerical verification task where the contrast of the task is either 0.70 (conventional printing) or 0.15 (light pencil letters) and the reflectance of the paper is 0.8.

Table 1. Predicted relative visual performance by subjects of different ages reading materials of contrast 0.70 or 0.15 at three different illuminances (Rea 1986b).

Illuminance (lx)	Contrast	20 yr old	65 yr old
150	0.70	0.980	0.856
400	0.70	0.996	0.897
650	0.70	0.999	0.915
150	0.15	0.931	0.750
400	0.15	0.977	0.836
650	0.15	0.988	0.861

The illuminances recommended for offices in the U.K. range from 300 to 750 lx (CIBSE 1984), the lowest illuminance being recommended for areas where there is little fine detail to be seen (conference rooms) and the highest illuminance for areas where the work is visually demanding (drawing offices). Table 1 shows that these recommendations are not unreasonable given the need to cover the complete age range and low contrasts as well as high.

The question that now needs to be considered is how far can the model be extended to other types of work. The first point to make is that it applies to tasks where the position where information is presented is completely predictable and for which the observer is fully adapted to luminances usable by the day retina. The second point is that the model is based on data treated so as to maximise the effects of the visual component. In these conditions the effects of change of lighting conditions will also be maximised. Therefore for tasks with a smaller visual component it seems likely that the model will over-estimate the effects of any changes in lighting conditions. Unfortunately, it is not possible to predict what will happen with different types of work because the visual component can vary in both magnitude and significance. In order to estimate the likely departures from the model it is necessary to understand the organisation of each task and the place of the visual component in that organisation. From the model itself, all that can be said is that it is unlikely that changes in lighting conditions will have any greater effect than that predicted.

Two other factors may limit the application of the model. The first is the fact that the data on which it is based were obtained with only one size of print. Threshold measurements have shown that visual acuity varies with adaptation luminance so it is reasonable to suppose different sizes of print would lead to different levels of performance. However, some recent work suggests that the effect of size over the range of print sizes commonly found in commercial environments may be slight (Legge et al. 1985). The second is that the model makes no allowance for behavioural modification. The data used in the model were collected for a fixed viewing distance when in real life if something is difficult to see we tend to get closer to it. This in itself may carry penalties in terms of poor posture, back strain, etc.

In spite of these limitations, this model is a definite

advance in our understanding of the relationship between lighting and human performance. It quantifies the relative importance of contrast and luminance on the relative visual performance that can be achieved for an important type of activity done by people of different ages and it is broadly consistent with the electrophysiology of the visual system. Further exploration and development of the model would be well worth while. Areas which would be worth examining are the effects of print size and blur, caused either by poor printing or by inadequate refraction.

Finally, it should be noted that the model has implications for lighting conditions beyond the illuminance provided, because the major variable which affects task performance is contrast. Lighting installations which create veiling reflections on the task can change the contrast of the task.

Veiling reflections occur when a high luminance area is reflected specularly from the task material. What effects these have depend on the structure of the task. One of the most common situations where this occurs is with pencil writing. Pencil marks are specularly reflecting and are usually darker than the paper on which they are written. In this situation, high luminance reflections from the pencil marks will reduce the contrast of the task. The consequences of this for task performance can be predicted from the model. There are a number of simple guidelines for avoiding veiling reflections in interior lighting (Boyce & Slater 1981).

SEARCHING
The essential characteristic of this type of work is the peripheral detection and recognition of significant features. Examples of situations where this type of activity occur are visual inspection of products and driving. Studies of this type of activity frequently appear under the name visual search. Research in visual search, whilst extensive, has more often been concerned with the characteristics of the material being inspected rather than the lighting conditions. Nonetheless, sufficient understanding has been developed to outline the role of lighting in such work.

The measure commonly used to quantify the ease with which the target can be detected in the peripheral visual field is the visual detection lobe. The visual detection lobe is a surface centred on the visual axis which defines the probability of detecting the target at various deviations from the visual axis. Because of the characteristics of the visual system the visual detection lobe always has a maximum on the visual axis but the probability represented by that maximum and the fall off with increasing eccentricity depends on the size and contrast of the target; the smaller the size and the lower the contrast the lower the maximum probability of detection and the more rapid the fall off in detection.

The significance of the visual detection lobe for visual search lies in the eye movements associated with such search. Typically these show a pattern of fixation on a point followed by saccadic movement to a new fixation point. If a visual detection lobe is associated with each fixation point, it can be readily appreciated that the greater the detection lobe the fewer the number of fixations needed to cover a given search area and find the target. This is most evident when searching a very uniform field, such as a sheet of glass, for a flaw. Figure 2 shows the mean of the median search times for the inspection of glass sheets, each containing a single flaw (Drury 1975). As expected, as the size of the flaw decreases, there is a marked increase in search time because the visual detection lobe becomes smaller.

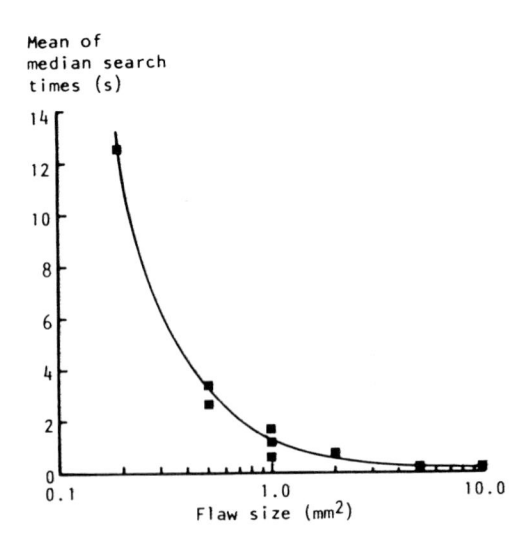

Figure 2. Mean of median search times for detecting a single flaw in a glass sheet, plotted against the flaw size (after Drury 1975).

Unfortunately, this relatively simple explanation becomes rather more complex as soon as we move to searching cluttered fields. In this situation, the task consists not just of detecting the presence of something but rather of detecting the defect amongst all the things that can be detected. Drury & Clement (1978) have shown that the search time in a cluttered field is strongly influenced by the amount of clutter in the search area. Other important factors are the density of the clutter and the extent to which the defects can be discriminated from the correct items. Howarth & Bloomfield (1969) have shown that mean search times for searching for a disc of a given size in an array of larger discs is closely related to a discriminability index which is based on the square of the difference in disc diameters. One interpretation of this dependence on discriminability is that the results are determined by an effective visual detection lobe; a lobe which is related not only to the target but also to the other items amongst which the target is seen. To further complicate matters Williams (1960) has pointed out that defects may differ from good products on several different dimensions. In this situation it is the dimensions which generate the largest effective visual detection lobes that are likely to determine the search time.

Given this understanding of what determines the efficiency of visual search, what role does lighting have to play? The most widely applicable factor is likely to be the illuminance in the search area. Increases in illuminance produce increases in luminance and hence increases in contrast sensitivity and visual acuity, both of which should increase the size of the visual detection lobe and the associated discriminability. Figure 3 shows the mean search times obtained in a task requiring the location of a specified number from a random array of 100 such similar numbers at different illuminances (Muck & Bodmann 1961). As expected, the mean search times increase as the illuminance falls but the increase is greater for the smaller, lower contrast numbers. This implies that the luminance gets more important as the size and contrast of the task decreases, a conclusion in broad agreement with that given by the model of visual performance described earlier.

Illuminance is effective as a lighting variable because it changes the operating state of the visual system, so it should be effective regardless of the type of defect. However, care is required with this conclusion. The essential role of lighting as an aid to visual search is to increase the discriminability of the target being sought.

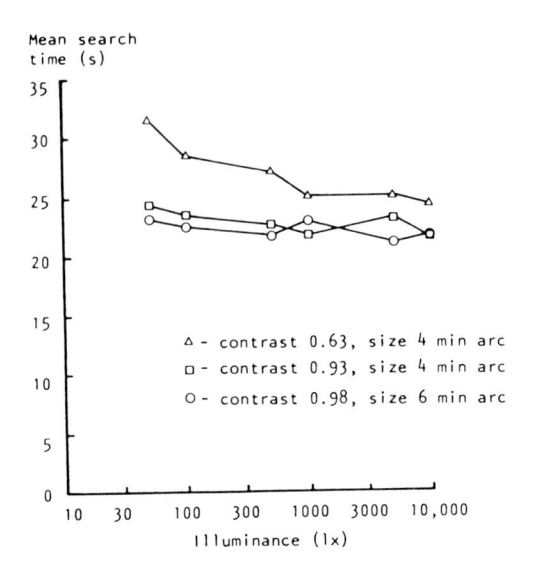

```
Mean search
time (s)
35
30
25
20
15
            △ - contrast 0.63, size 4 min arc
            □ - contrast 0.93, size 4 min arc
10
            ○ - contrast 0.98, size 6 min arc
 5
 0
    10   30  100  300  1000 3000 10,000
              Illuminance (lx)
```

Figure 3. Mean search times for locating a specified integer number from a random array of 100 such numbers, plotted against illuminance (after Muck & Bodmann 1961).

If the higher illuminances are provided in such a way that discriminability is reduced search times are likely to be increased rather than decreased. This is the negative side of lighting as an aid to visual search, but it has a positive role to play, particularly in three dimensional objects. By carefully controlling the distribution of a light it is possible to increase the visual detection lobe of scratches, cracks, dents and other defects in a wide range of materials, usually by improving the contrast (Boyce 1981). Advice is available on the use of lighting as an aid to visual inspection in practice (Faulkner & Murphy, 1973) but it should not be used without thought. In general, the feature of the defect which distinguishes it from the surrounding area and other objects will have to be identified in order for the appropriate form of lighting to be determined.

MOVING
    In order to catch the visual system seriously misadapted it is necessary for a large change of luminance to have occurred suddenly. This may well be the situation

following the failure of the normal room lighting in an interior where there is no available natural light. What happens here is that the normal room lighting, typically with an illuminance of 500 lx, is replaced, sometimes after a 5 second period of darkness, by emergency lighting which usually has an illuminance much lower, typically about 0.2 lx. The instructions given to people when such emergency lighting comes into operation usually suggest that they should leave the building immediately in an orderly manner. The illuminances that should be provided for evacuating a building have been the subject of several studies recently. What these studies do is to place people in an interior lit to the usual illuminances. Then the normal room lighting is extinguished and replaced by the escape route lighting providing a much lower illuminance. The subjects have to find their way out of the interior. The time they take to do this is recorded and sometimes their manner of movement is observed. Figure 4 shows the mean speed of movement over an escape route obtained in three different situations, by three different experimenters in three different experiments. Simmons (1975) used a convoluted, simulated corridor for his study. Jaschinski (1982) used a series of small interconnected offices which the subject had to find the way through. This escape route also involved

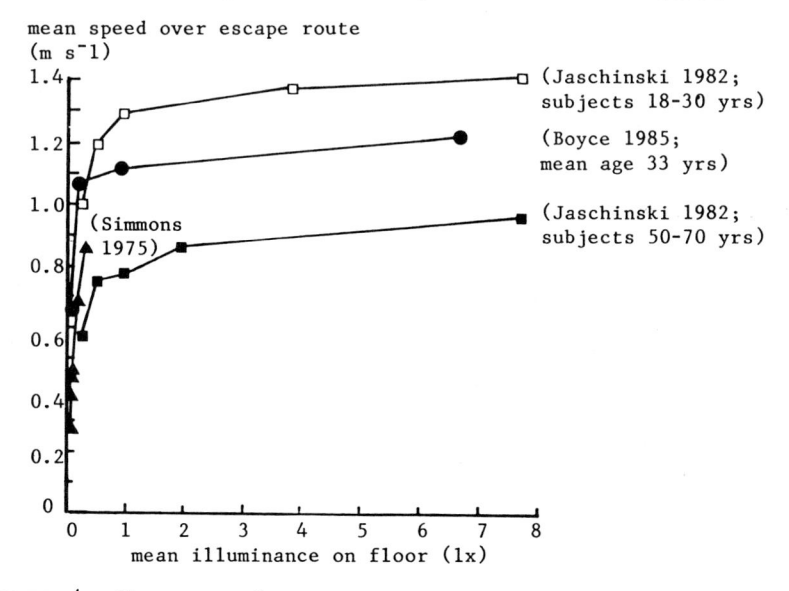

Figure 4. Mean speed over an escape route plotted against mean illuminance for different studies of people's performance under emergency lighting conditions.

changes of level. Boyce (1985) used a large open plan office where subjects were free to choose their own path through the array of office furniture to the exit door. In spite of these differences the similarity of results is encouraging. Above a mean illuminance of 1 lx there is little difference in speed of movement from what occurs under the normal room lighting but below 1 lx the speed of movement tends to decline rapidly. The actual illuminance at which the decline begins depends on the particular circumstances of each experiment.

The more obstructed and convoluted the initial escape route is the higher the illuminance at which the slowing of movement starts. In one experiment, the speed measurements are supported by observations of the manner of movement (Boyce 1985). As speed of movement decreases the subjects become more hesitant in their movements and ultimately confused, changing direction and backtracking as they try to find the way out. The reason for this hesitation lies in the range and time characteristics of the two mechanisms of adaptation used by the visual system. The change from normal room illuminance to escape room lighting above 1 lx can be dealt with by the pupillary/neural system but at lower illuminances photochemical adaptation is needed and this takes of the order of minutes, time which is not available in emergency conditions. This suggests that emergency lighting criteria should be determined by the need to keep the illuminance in the range of the pupillary/neural system so that the occupant can adapt to the new conditions rapidly.

JUDGING COLOURS

Colour judgement is involved in the manufacture and inspection of many different products. All such judgements involve discrimination of departures in colour from a norm. The norm can be a physical entity, such as a specimen of the product, or a memory, which is what happens when a variable product, such as fruit, is graded.

The perceived colour of the surface is determined by three factors; the state and organisation of the visual system used to observe it, the size and reflectance characteristics of its surroundings, and the light source used to illuminate it. The state and organisation of the visual system matters because colour vision fails at low luminances and the ability to make colour discriminations varies with age and with individuals. For accurate colour judgement work some form of visual screening is a wise

Lighting and Human Performance 15

precaution. The surround to the object being judged matters because if it is strongly coloured it will modify the spectral content of the light which reaches the object after reflection and it affects the chromatic adaptation of the visual system. A mid-reflectance neutral colour is recommended for surrounds where accurate colour judgement is required. The light source matters because different light sources produce very different spectral emissions.

Boyce & Simons (1977) examined the effect of light source on colour judgements using the Farnsworth-Munsell 100 hue test. This test requires the subject to arrange a series of coloured discs, which differ only in hue, into a consistent series. Mispositioning of a disc counts as an error, the error score being weighted to reflect the size of the mispositioning. Figure 5 shows the results obtained from a wide range of commonly used light sources plotted against their colour rendering index. This latter is a method of quantifying the ability of the light source to render colours in the same way as a reference light source.

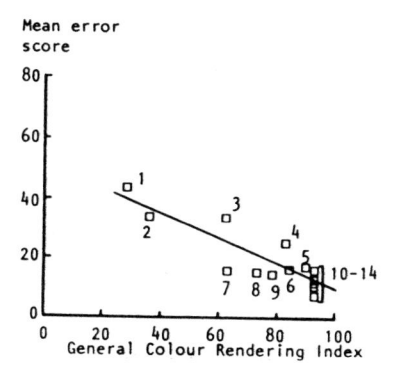

Figure 5. Mean error scores on the Farnsworth-Munsell 100 hue test for different light sources plotted against the general Colour Rendering Index. Light sources: 1, high pressure sodium discharge; 2, high pressure mercury discharge; 3, Homelite fluorescent; 4, Tri-band fluorescent; 5, Kolor-rite fluorescent; 6, Natural fluorescent; 7, Daylight fluorescent; 8, Plus-White fluorescent; 9, high pressure mercury discharge with metal halide; 10-14, Artificial Daylight fluorescent (after Boyce & Simons 1977).

A value of 100 implies perfect colour rendering (CIE 1974). It can be seen that there is a clear relationship between general colour rendering index and the mean error score with the artificial daylight fluorescent lamp, a lamp specifically designed for accurate colour judgement of work, giving the lowest mean error scores, and the high pressure sodium discharge lamp, a lamp widely used in industry because of its high efficiency in converting electricity to light, giving the highest mean error score. Closer examination of Figure 5 suggests that several different light sources allow good colour judgements to be made and the main thing to do is to avoid the use of lamps with low colour rendering indices, say below 60.

This result is supported by a field study of the inspection of freshly killed meat and poultry (Collins & Worthey 1984). In this study 18 poultry and 16 meat inspectors observed five different poultry or meat tissue samples under each of five different light sources. The light sources used were incandescent, cool white fluorescent, cool white de luxe fluorescent, high pressure sodium discharge and low pressure sodium discharge. The results indicated that more errors in inspection were made with the two sodium discharge light sources than the others. The general colour rendering index for these two light sources is below 60. For the other three light sources it is above 60.

There can be little doubt that avoiding lamps with colour rendering indices below 60 improves the accuracy of colour judgements but whether a higher index would improve it even more requires further investigation. It may well be that such an effect will only be found where high precision is required in the judgements and the stimulus material is very stable in colour.

Whilst the choice of light source is important the illuminance provided should not be neglected. Obviously there is a minimum illuminance because below a luminance of about 3 cd m^{-2} colour vision starts to fail. However, there may also be a need for higher illuminances because the rod photoreceptors have an identifiable effect at luminances above 3 cd m^{-2}. Boyce (1976) failed to find any significant change in mean errors for subjects doing the Farnsworth-Munsell 100 hue test at 300 and 1000 lx, provided by a number of different fluorescent lamps. The only situation in which illuminances above 300 lx have been found to be important is when older people (>55 years) were doing the test (Boyce & Simons 1977). Then increasing the illuminance from 400 to 800 to 1200 lx did lead to a significant decrease in the mean error score (Figure 6) although it was always much higher for the older subjects than for the younger ones.

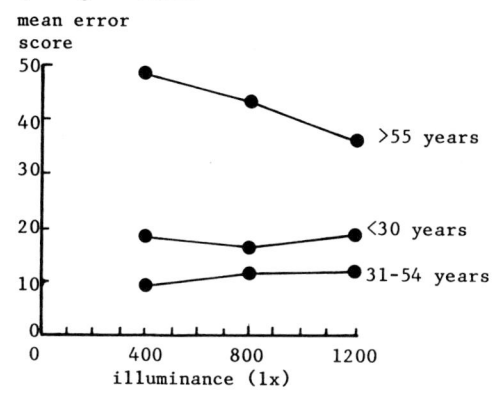

```
mean error
score
50┌
   •
40┤     •
         •  >55 years
30┤
20┤  •       <30 years
     •  •
10┤  •  •    • 31-54 years
 0└──┬───┬───┬───┬──
 0   400  800  1200
     illuminance (lx)
```

Figure 6   Mean error score obtained on the Farnsworth-Munsell 100 hue test lit to different illuminances, by people of different ages (after Boyce & Simons 1977).

It can be concluded that the most important aspects of lighting for colour judgement work are the selection of the light source and the illuminance to be provided. Of these two factors, there can be little doubt that within the illuminance range 300-1000 lx, which covers most industrial and commercial interiors, choice of light source is the more important of the two.

UNKNOWNS

Space has permitted only a broad brush picture of what is known of the relationship between lighting and human performance. Nonetheless there are two areas which are conspicuous by their absence. This is not because they are not important but rather because little is understood about their effects. They are the effect of lighting conditions on long term performance and the influence of discomfort. All the studies considered above have involved short term performance. For some, such as emergency lighting, that is all there is because longer times lead to full adaptation, but for the others there is a possibility that to maintain performance over a long period may involve some physiological cost which is not revealed in short term experiments. Attempts have been made to measure fatigue and other consequences of long term visual performance, often using physiological measures, but little by way of a consistent picture has emerged.

Similarly, attempts have been made to examine the indirect effects of lighting on human performance. By indirect effects, I mean an effect which operates, not by changing the visual difficulty of the task or the state of operation of the visual system, but through distraction, arousal or changes in mood and motivation. It is widely believed that such effects occur but it does seem difficult to prove that they do, at least in a laboratory context. What has been shown is that people will complain of visual discomfort in conditions when they are quite capable of maintaining performance (Muck & Bodmann 1961).

CONCLUSION

In this paper I have attempted to present a picture of the diversity of relationships between lighting and human performance. There is not one relationship, there are many and different aspects of lighting influence different tasks depending on what is being asked of the observer and the features of the task. I believe the key to understanding the influence of lighting on human behaviour lies in knowing the capabilities of the visual system. By understanding what aspects of a task are significant to the visual system we are more likely to be able to identify what features of lighting are important. This is surely a better way to proceed than what has often been done before, to simply take a lighting variable and examine its effect because it is widely used in design. By bringing together an understanding of task analysis and visual system capabilities it should be possible to develop a clearer picture of the relationship between lighting and human performance in all its complexity.

REFERENCES

Albrecht, D.G. & Hamilton, D.B., 1982, Striate cortex of monkey and cat: contrast response function. Journal of Neurophysiology, 48, 217-237.

Boyce, P.R.. 1976, Illuminance, lamp type and performance on a colour discrimination task. Lighting Research and Technology, 88, 195-199.

Boyce, P.R., 1981, The visual detection lobe and visual inspection. Proceedings 4th European Lighting Congress, Granada.

Boyce, P.R., 1985, Movement under emergency lighting: the effect of illuminance. Lighting Research & Technology, 17, 51-71.

Boyce, P.R. & Simons, R.H., 1977, Hue discrimination and light sources. Lighting Research and Technology, 9, 125-140.

Boyce, P.R. & Slater, A.I., 1981, The application of contrast rendering factor to office lighting design. Lighting Research & Technology, 13, 65-79.

Campbell, F.W. & Robson, J.G., 1968, Application of Fourier analysis to the visibility of gratings. Journal of Physiology, (London), 197, 551-566.

Chartered Institution of Building Services Engineers, 1984, CIBSE Code for Interior Lighting, CIBSE, London.

Collins, B.L. & Worthey, J.A., 1984, The role of colour in lighting for meat and poultry inspection. National Bureau of Standards, NBSIR 84-2829, National Bureau of Standards, Gaithersburg, Maryland, USA.

Commission Internationale de l'Eclairage, 1974, Method of measuring and specifying colour rendering properties of light sources. CIE Publication 13.2, CIE, Vienna.

Daitch, J.M. & Green, D.G., 1969, Contrast sensitivity and the human peripheral retina. Vision Research, 9, 947-952.

Drury, C.G., 1975, Inspection of sheet material - model and data. Human Factors, 17, 257-265.

Drury, C.G. & Clement, M.R., 1978, The effect of area density and number of background characters in visual search. Human Factors, 20, 597-602.

Elton, P.M., 1920, A study of the output of silk weavers during the winter months. Report No. 9, Industrial Fatigue Resarch Board, HMSO, London.

Faulkner, T.W. & Murphy, T.J., 1973, Lighting for difficult visual tasks. Human Factors, 15, 149-162.

Howarth, G.I. & Bloomfield, J.R., 1969, A rational equation for predicting search times in simple inspection tasks. Psychonomic Science, 17, 225-226.

Jaschinski, W., 1982, Conditions of emergency lighting. Ergonomics, 25, 363-372.

Johnson, G.A., Keltner, J.L. & Balestrery, F., 1978, Effect of target size and eccentricity on visual detection and recognition. Vision Research, 18, 1217-1222.

Legge, G.C., Pelli, D.G., Rubin, G.S. & Schleske, M.M., 1985, Psychophysics of reading - 1. Normal vision. Vision Research, 25, 239-252.

Monty, R.A. & Senders, J.W., 1970, Eye movements and physiological processes. Lawrence Emblaum Associates, Hillsdale, New Jersey, USA.

Muck, E. & Bodmann, H.W., 1961, Die bedeutung des beleuchtungsniveau bei praktischer sehtatigkeit, Lichttechnik, 13, 502-508.

Poulton, E.C., 1969, Skimming lists of food ingredients printed in different sizes. Journal of Applied Psychology, 53, 55-58.

Rea, M.S., 1986a, Toward a model of visual performance: foundations and data. *Journal of the Illuminating Engineering Society*, 15, 41-57.

Rea, M.S., 1986b, Some practical implications of a new visual performance model. *Lighting Research and Technology* (in press).

Simmons, R.C., 1975, Illuminance diversity and disability glare in emergency lighting. *Lighting Research & Technology*, 7, 121-132.

Spencer, H., 1968, *The visible word.* Lund Humphreys, London.

Tinker, M.A., 1963, *Legibility of print.* Iowa State University Press, USA.

Williams, L.C., 1960, The effect of target specification on objects fixated during visual search. *Perception and Psychophysics*, 1, 315-318.

THE COGNITIVE BASES OF PREDICTABLE HUMAN ERROR

J. REASON

Department of Psychology
University of Manchester
Manchester M13 9PL

This paper identifies two pervasive error-shaping factors: similarity and frequency. These biases are evident in a wide range of error types (mistakes, lapses and slips), involving many different cognitive activities. A 'dual-architecture' model of human cognition is outlined. This comprises a serial, restricted, but computationally-powerful 'workspace' interacting with an effectively unlimited, parallel, distributed knowledge base. It is argued that similarity and frequency effects are rooted in the simple but universal heuristics by which the outputs of stored knowledge units are identified and elicited.

ERRORS TAKE A LIMITED NUMBER OF FORMS
Human error is neither as abundant nor as varied as its vast potential might suggest. Not only are errors much rarer than correct actions, they also tend to take a suprisingly limited number of forms - surprising, that is, when measured against their possible variety. Moreover, errors appear in very similar guises across a wide range of mental activities. Thus, it is possible to identify comparable error forms in action, speech, perception, knowledge retrieval, judgement, problem solving, decision making, concept formation, and the like. The ubiquity of these systematic error forms forces us to formulate more global theories of cognitive control than are usually derived from laboratory experiments which, of necessity, focus upon very restricted aspects of mental function in artificial settings.
   The purpose of this paper is to sketch out some of the causal relationships between these pervasive error forms and the more fundamental properties of human information processing. It will be argued that systematic errors are inextricably bound up with those things at which the

cognitive system excels relative to other information-processing devices, and especially with the characteristic ways in which it simplifies complex information-handling tasks.

DISTINGUISHING ERROR TYPES AND ERROR FORMS
Error types

The term 'error type', as used here, relates to the presumed origin of an error within the stages involved in conceiving and then carrying out an action sequence. These stages can be described under three broad headings: planning, storage and execution. Planning refers to the processes concerned with identifying a goal and deciding upon the means to achieve it. Since plans are not usually acted upon immediately, it is likely that a storage phase of some variable duration will intervene between formulating the intended actions and running them off. The execution stage covers the processes involved in actually implementing the stored plan. The relationship between these three stages and the primary error types is shown in Table 1.

Table 1. The primary error types.

COGNITIVE STAGE	PRIMARY ERROR TYPE
Planning	Mistakes
Storage	Lapses
Execution	Slips

Actions may fail to achieve their desired consequences either because the plan was inadequate (mistakes), or because the actions did not proceed according to plan. In the latter case, it is useful to distinguish between those unintended actions which arise as a consequence of memory failures (lapses), and those due to the imperfections of attentional monitoring (slips).

For most practical purposes, however, the crucial distinction is between errors which occur at the level of intention (mistakes), and those which occur at some subsequent stage (lapses and slips). Mistakes can be further subdivided into (a) failures of expertise, where some pre-established plan or problem solution is applied inappropriately; and (b) a lack of expertise, where the individual, not having an appropriate 'off-the-shelf' routine, is forced to work out a plan of action from first principles, relying upon whatever relevant knowledge he or she currently possesses. These two types of mistakes correspond closely to the rule-based and knowledge-based levels of performance, as described by Rasmussen (1982).

Error forms
Whereas 'error types' are conceptually tied to underlying cognitive stages or mechanisms, 'error forms' are recurrent varieties of fallibility which appear in all kinds of cognitive activity, irrespective of error type. Thus, they are evident in mistakes, lapses and slips. So widespread are they that it is extremely unlikely that their occurrence is linked to the failure of any single cognitive entity. Rather, this omnipresence suggests that they are rooted in universal cognitive processes.

Two such error forms will be considered here: similarity-matching and frequency-gambling. Before describing them further, however, it is necessary to provide a preliminary sketch of the basic structural components of the cognitive system, and then to indicate how they might interact to specify a given action or thought sequence.

COGNITIVE STRUCTURES
The conscious workspace (Ws)
This 'sharp end' of the cognitive system receives input from both the outside world, via the senses, and from the knowledge base. It is related to - though not necessarily co-extensive with - conscious attention and working memory (Baddeley & Hitch, 1974; Mandler, 1985). Its primary concerns are with the setting of gaols, with selecting the means to achieve them, with detecting deviations from current intention, and with monitoring progress towards these desired outcomes. The products of this high-level planning and monitoring activity are accessible to awareness; but since consciousness is a severely restricted 'window', it is usually the case that only one such high level activity can be 'viewed' and worked upon at any one time. These resource limitations confer the important benefit of selectivity, since several high-level activities are potentially available to the Ws.

The conscious Ws has powerful, analytical, feedback-driven computational processes at its disposal, and is essential for coping with novel or changed circumstances, and for detecting and recovering errors. But it is severely resource-limited, slow, laborious, serial and difficult to sustain.

The knowledge base (Kb)

Human cognition is extremely proficient at modelling the useful regularities of its previous dealings with specific environments, and then using these stored representations as a basis for the automatic control of subsequent perception and action. The minutiae of mental life are governed by a vast community of specialised processors, each constituting a 'mini-theory' regarding some particular aspect of the world, and each being instantiated by highly specific triggers supplied by both the conscious workspace (intentional 'calling conditions') and by the environment (contextual or task-related 'calling conditions'). These 'knowledge packets' or 'schemata' thus possess two closely related elements: (a) they embody generic or prototypical knowledge concerning specific aspects of the world, and (b) they can generate pre-programmed instructions for eliciting particular actions, words, images, percepts, etc. Only these products of schema activity are available to the higher-level workspace; the processes themselves lie beyond the reach of consciousness.

In contrast to the conscious Ws, which can function over both long time spans and a wide range of circumstances, schemata within the Kb are tied to highly specific triggering conditions. The Kb, however, has no known limits on its capacity. It can process familiar information rapidly, in parallel, and without conscious involvement or effort. But it is relatively ineffective in the face of novel or unforeseen circumstances.

THE SPECIFICATION OF COGNITIVE ACTIVITY

Correct performance in any sphere of cognitive performance is achieved by activating the right schemata in the right order at the right time. Schemata may be brought into play (i.e. deliver their products either to the conscious Ws or to the outside world in the form of actions) by both specific and general activators. Specific activators are those which trigger targeted schemata at a particular time and place. Among these, intentional activity is likely to be paramount. Other specific activators include contextual cues and 'descriptions' (see Bobrow & Norman, 1975; Norman & Bobrow, 1979) passed on by other sequence-related schemata General activators provide background activation to schemata, regardless of the current intentional state. Of these, frequency of prior use is probably the most important.

The central thesis of this paper is that predictable error forms are rooted in a tendency to over-utilise what is probably the most conspicuous achievement of human cognition: its ability to simplify complex informational tasks by resorting to pre-established routines, heuristics and shortcuts. It is believed that the two most fundamental of these heuristics are (a) match like with like, and (b) resolve conflicts (between schema candidates) in favour of contextually-appropriate, high-frequency knowledge structures. Both of these tendencies are brought into particular prominence in conditions of <u>cognitive underspecification</u>.

Precisely what is missing from a sufficient specification, or which cognitive level fails to provide it, will vary with the nature of the task. But notwithstanding these possible <u>varieties</u> of under-specification, their <u>consequences</u> are remarkably uniform. The (often erroneuous) responses selected in conditions of under-specification tend to (a) show formal similarities to either the prevailing contextual features or to the currently intended (or normatively and evidentially appropriate) responses, or both; and (b) they are likely to be more familiar, more conventional, more typical, more frequent-in-context than those that would have been judged correct or appropriate.

SIMILIARITY AND FREQUENCY: COGNITIVE 'PRIMITIVES'

Such fundamental aspects of experience as the degree of likeness between events or objects and their frequency of prior occurrence have been termed <u>intuitive concepts</u>. Similarity and frequency information appear to be processed automatically without conscious effort or perhaps even awareness, regardless of age, ability, cultural background, motivation, or task instructions (see Wason & Johnson-Laird 1972; Shweder, 1977; Tulving, 1983; Hasher & Zacks, 1984). There is a strong case for regarding them as being pre-eminent among the computational 'primitives' of the cognitive system.

SIMILARITY EFFECTS IN MISTAKES, LAPSES AND SLIPS

Error forms can resemble the properties of both the current intentional specification and the prevailing environmental cues in varying degrees. The most obvious tendency in mistakes - particularly those in which the problem solver has been limited by an incomplete or incorrect knowledge base - is for the error forms to be shaped by salient features of the problem configuration. In the case of both lapses and slips, however, there can be matching to both intentional and contextualcues. The forms of these execution failures can show close similarities to the intended word or action, as well as, on occasions, being appropriately matched to the situation in which they occur. Exactly what is matched by the error form appears to be related to the extent to which the correct response is, or could be, specified at the outset of the thought or action sequence. Difficulties encountered at the level of formulating the intention or plan tend to create errors that are shaped primarily by immediate contextual considerations; those which occur at the level of storage or execution may reflect the influences of both intentional and environmental 'calling conditions' (specifiers). But irrespective of the precise nature of the matching, similarity effects are evident across all error types.

Similarity effects in mistakes

Similarity effects have been most clearly demonstrated in (a) laboratory studies of reasoning and inference; and (b) investigations of human judgement under conditions of uncertainty.

Wason and Johnson-Laird (1972) employed a variety of techniques (the Wason Card Test, syllogisms, the '2-4-6' rule-discovery task, etc.) to investigate the ways in which people draw explicit conclusions from evidence. In the course of this work, they identified a number of 'pathologies' of problem solving (the reluctance to utilise negative statements, the corresponding ease of handling affirmative statements, confirmation bias, the 'thirst for confirming redundancy', illicit conversion, and so on); but they concluded that all of these factors were the consequence of one general, overriding principle: "..whenever two different items, or classes, can be matched in a one-to-one fashion, then the process is readily made, whether it be logically valid or invalid".(Wason & Johnson-Laird, 1972, p.241). Thus, affirmatives are easier to handle than negatives because they involve just the single step of making a one-to-one relation between a statement and a state of the world. And the fact that people will naturally establish such a match when dealing with a problem will inevitably bias them toward affirming rather than falsifying their beliefs.

In a series of studies, Tversky, Kahneman and their associates (see Khaneman, Slovic & Tversky, 1982) have shown that when people are asked to judge whether object A belongs to class B, or whether A originates from process B, they typically over-utilise the representativeness heuristic. That is, their probability judgements are heavily determined by the extent to which A resembles B, regardless of such critical factors as sample size, base rates, and the like.

Similarity effects in lapses

Perhaps the commonest type of memory lapse is forgetting to remember to carry out intended actions at the appointed time and place (Reason & Mycielska, 1982). For the most part, these failures of prospective memory lead to the omission of isolated planned actions, and, as such, cannot readily manifest similarity effects. There are, however, other commonly occurring lapses, particularly failures to retrieve known items (names or words) from long-term memory, which show this bias in a variety of obvious ways. There is now a wealth of evidence (from Aristotle onwards) indicating that 'intermediate solutions' - wrong words dredged up in the course of an active memory search for a blocked word (in a tip-of-the-tongue or TOT state) - show close phonological, morphological and semantic similarities to the target item (see Reason & Lucas, 1984, for a discussion of the relevant literature).

Similarity effects in slips

As with TOT state 'intermediates', slips of the tongue show marked similarity effects between the actual and target (intended) utterances (see Fromkin, 1973, 1980). "The more similar a given unit is to an intended unit, the more likely the given unit or a part of it will replace the intended unit or a corresponding part of it" (Dell & Reich, 1980, p.281).

Different kinds of similarity are likely to operate at the various stages of formulating and executing the articulatory program. Fromkin (1971) has suggested that semantically-related substitutions may occur because of under-specification of the semantic features. At the more detailed level of phoneme specification, however, substitutions are facilitated by phonological similarity between word segments (Nooteboom, 1969; MacKay, 1970).

Although less widely investigated, slips of action reveal comparable similarity effects (see Reason & Mycielska, 1982). These are most obvious in slips involving wrong objects, recognition failures in which another item is substituted for the correct one during the execution of a highly routinised action sequence. A necessary condition for these and other action slips is some diminution of attentional monitoring, either through preoccupation or distraction. Just as some investment of the limited attentional resource is necessary for checking that the correct actions have been selected, so it is also involved in verifying the accuracy of perceptions, particularly during oft-repeated action sequences when the relevant recognition schemata seem ready to accept rough approximations to the expected object configuration (see Reason, 1979).

Absent-minded misrecognitions most commonly involve the unintended substitution of physically similar items, as in the following actual instances (Reason & Mycielska, 1972): "I intended to pick up the deodorant, but picked up the air freshener instead." "When seasoning meat, I sprinkled it with sugar instead of salt". "I filled the washing machine with oatmeal". "I put shaving cream on my toothbrush".

Similarity effects are also evident in strong habit intrusions. In one diary study (Reason & Mycielska, 1982), subjects were required to complete a standarised set of ratings in relation to each slip recorded. In particular, they were asked whether or not their intended actions were recognisable as 'belonging to' some other task or activity, not then intended. Such a relationship was identified in 77 (40 per cent) of the 192 action slips netted. In the case of these slips, the diarists were further asked to rate the extent to which the intended actions and the 'other activity' shared common features. Strong similarity effects were found in regard to locations, movements and objects, and somewhat weaker ones in relation to timing and purpose (see Reason & Mycielska, 1982, p.257 for the actual data).

FREQUENCY EFFECTS IN MISTAKES, LAPSES AND SLIPS

Errors in all types of cognitive activity tend to take the form of contextually-appropriate, high-frequency responses. The more often a sequence of perception, thought or action achieves a successful outcome, the more likely it is to appear unbidden in conditions of incomplete specification (see also the Law of Effect). The psychological literature is replete with terms to describe these high-frequency error forms: 'conventionalization' (Bartlett), 'sophisticated guessing' (Solomon & Postman), 'persistence forecasting' (Bruner, Goodnow & Austin), 'strong associate substitution' (Chapman & Chapman), 'inert stereotypes' (Luria), 'banalization' (Timpanaro), 'strong habit intrusions' (Reason), and 'capture errors' (Norman). Though it sometimes leads to incorrect responses, this tendency to gamble in favour of well-used knowledge structures is a highly adaptive strategy for dealing with a world that contains a great deal of regularity as well as a large measure of uncertainty.

Frequency effects in mistakes

Planning, decision making and problem solving all require the generation from the knowledge base of possible courses of action. The products of these cognitive activities emerge from a complex interaction between current perceptions of the world and the recall of previous states. But however these processes are initiated, the outcome tends to favour the selection of salient (vivid) or familiar (frequent) scenarios of future action (see Nisbett & Ross, 1980; Kahneman, Slovic & Tversky, 1982; Fischhoff, Lichtenstein, Slovic, Derby & Keeney, 1981).

Irrespective of what other kinds of under-specification may promote them, mistakes are almost always the result of incomplete or inaccurate knowledge. A recent study (Reason, Bailey & Horrocks, 1986), involving the identification of quotations from US presidents, demonstrated the greater effects of 'frequency-gambling' upon ignorant as opposed to educated guesses in the retrieval of declarative knowledge. The less subjects knew about US presidents (as measured by a recognition test comprising a jumbled mixture of both presidential names and those of their famous contemporaries) the more inclined they were to attribute quotations to contextually-appropriate, high-frequency presidents (frequency scores for all the 39 presidents were derived from ratings made by another group of comparable subjects). The results are summarised in Table 2.

Table 2. The relationship between domain knowledge and frequency-gambling.

	QUARTILE GROUPINGS BY KNOWLEDGE SCORES		
	Mn.Knowledge Scores	Mn.Frequency Scores	No.in group
Group 1	6.36 (1.37)	4.89 (0.71)	29
Group 2	8.75 (0.60)	4.84 (0.72)	28
Group 3	12.06 (1.31)	4.56 (0.70)	28
Group 4	26.20 (7.73)	3.80 (1.02)	29

NOTES:
(a) Knowledge scores run from 1-39 adjusted recognitions.
(b) Frequency scores run from 1 (lowest) to 7 (highest)
(c) Figures in parentheses are standard deviations
(d) $F(3,110) = 11.26$ (p .0001).

Frequency effects in lapses

In the memory block study, touched upon earlier (Reason & Lucas, 1984), 16 volunteers kept 'extended' diaries of their resolved TOT states over a period of one month. Of the 40 resolved TOTs recorded, 28 involved the presence of 'recurrent intruders' - recognisably wrong names or words that continued to block access to the target item during deliberate search periods. In 77 per cent of these TOTs, the recurrent intruder was ranked higher than the target for either contextual frequency or recency, or both.

These data are consistent with the view that recurrent intruders emerge in TOT states when the initial fragmentary retrieval cues are sufficient to locate the 'ball park' context of the sought-for item, but not to provide a unique specification for it. Recurrent intruders tend to be high-frequency items within this general context.

Frequency effects in slips

It has already been mentioned that a large proportion of action slips (40 per cent in one study) take the form of strong habit intrusions: well-organized action sequences that were judged as belonging to some other task or activity. In addition to assessing the similarity between this other activity and the intended actions, the subjects were asked to rate how often they engaged in the activity from which the erroneous sequences had apparently 'slipped'. Over 70 per cent of these unintended actions were judged as 'belonging to' a task that was performed very frequently indeed (see Reason & Mycielska, 1982, p.257).

There is also a considerable literature relating to the word frequency effect, the repeated finding that common words are more readily recognised than infrequent ones when their presentation is brief or otherwise attenuated (see Neisser, 1967). When a perceived word fragment is common to many words, and the subject is asked to guess the whole word from which it came, "..he will respond with the word of the greatest frequency of occurrence (response strength) which incorporates the fragment" (Newbigging, 1961, quoted by Neisser, 1967,p.117). An important corollary to this 'fragment theory' is that seen (or recalled) bits of relatively rare words will tend to be erroneously judged as belonging to common words sharing the same features (see also Gregg, 1976).

THE ROOTS OF PREDICTABLE ERROR: KNOWLEDGE RETRIEVAL

The first part of the paper made a distinction between error forms and error types, and between the properties of two parts of the cognitive system: the limited workspace (Ws) and the knowledge base (Kb). Evidence was then presented for the prevalence of similarity and frequency effects (error forms) across mistakes, lapses and slips (error types). In this concluding section, it will be argued that these two basic error forms have their origins in the processes by which knowledge items (schema products) are retrieved from the Kb.

It is postulated that the cognitive system has three mechanisms for bringing the products of stored knowledge into the conscious Ws and/or into action. Two of them - similarity-matching and frequency-gambling - constitute the computational primitives of the system as a whole, and operate in a parallel, distributed and automatic fashion within the Kb. The third retrieval mechanism - directed or inferential search - derives from the sophisticated processing capabilities of the Ws itself. Within the Ws, through which information must be processed slowly and sequentially, the speed, effortlessness and unlimited capacity characteristic of the Kb have been sacrificed in favour of selectivity, coherence and computational power.

Similarity-matching

The Ws has a cycle time of a few milliseconds, and each cycle contains between two and five discrete informational elements. During a run of consecutive cycles, these elements are transformed, extended or recombined by powerful operators which function only within this restricted conscious domain (see Mandler, 1985).

A useful image for the conscious Ws is that of a slicer. Information, comprising elements from both sense data and the Kb, is cut into slices which are then dropped into the buffer store of the Kb.

Once in the Kb buffer store, the informational features (the 'calling conditions') contained in these 'slices' are automatically matched to the attributes of stored knowledge items. No special fiat is required for this matching to ocur, and the process of relating calling conditions to corresponding attributes is both rapid and efficient.

All stored items possessing attributes which correspond to the elements of the most recent 'run' of Ws 'slices' (I.e. those held in the limited buffer) will increase their activation by an amount related to the goodness of match. The closer the match, the greater will be the received activation. When the activation level of a given schema exceeds a certain threshold value, its products may be delivered to the conscious Ws (images, words and feelings) and/or to the effectors (speech, actions). However, not all the knowledge units so matched will deliver their products to the conscious Ws. In some cases, only a partial correspondence will be achieved; in others, the products of activated schema may be pre-empted at the Ws by higher priority inputs.

Frequency-gambling

In many situations, the calling conditions emerging from the Ws are insufficient to match uniquely a single knowledge item. This can occur because either the calling conditions, or the stored item attributes, are incomplete. These two kinds of under-specification are functionally equivalent. In under-specified searches, a number of partially-matched 'candidates' are likely to receive an increase in their activation levels. Where these contenders are equally matched to the current calling conditions, the conflict is resolved in favour of the most frequently encountered item. This occurs because an oft-triggered knowledge unit (i.e. one with a proven utility) will have a higher 'background' activation level than one less frequently employed.

Notice that these two automatic search processes, similarity-matching and frequency-gambling, though exceedingly simple and involving no logical principles, can together elicit a reasonably adaptive response (i.e. one that is contextually well-matched and previously useful) in any situation. Thus, the cognitive system is sensitive to both the formal and the statistical properties of the world it inhabits. And it is just these characteristics which give predictable shpaes to a wide variety of error types.

Directed search

As is made apparent by such phenomena as TOT states, the conscious Ws has no direct access to the Kb. Its sole means of directing knowledge retrieval is through the manipulation of calling conditions. The actual search itself is performed automatically by the similarity-matching and frequency-gambling heuristics. All that the Ws can do, therefore, is to deliver the initial calling conditions, assess whether the search-product is appropriate, and, if not, to reinstate the search with revised retrieval cues. The Ws has the power to reject the high-frequency candidates thrown up by under-specified matches, but only when sufficient processing resources are available. In conditions of high workload, environmental stressors, preoccupation and distraction, the Ws has often little choice but to accept these 'default' options. While these may serve adequately enough in relatively routine situations, they can and do lead to predictable error forms when their the goals or the circumstances of action have changed.

So far, we have considered a system in which actions are set in train and knowledge products delivered to the Ws as an automatic consequence of the interaction between (a) prior conscious processing, and (b) knowledge unit activation. But these properties alone are not sufficient to guarantee the successful execution of goal-directed behaviour. What gives cognition its intentional character? How does it initiate <u>deliberate</u> actions or knowledge searches?

As with many other difficult questions, William James (1890) provided a possible answer: "The essential achievement of the will.. is to attend to a difficult object and hold it fast before the mind. The so doing is the fiat; and it is a mere physiological incident that when the object is thus attended to, immediate motor consequences should ensue" (James, 1890, p.561).

This statement maps readily onto the simple 'dual architecture' model of cognition, previously described. The "holding-fast-before-the-mind" translates into a sustained run of same-element Ws 'slices'. Once in the Kb buffer, the consistency of these 'slices' will generate a high level of focused activation within a restricted set of knowledge units. This will automatically and advantageously release their products to the Ws and/or the effectors.

But such an 'act of will' places heavy demands upon the limited attentional resources available to the Ws. The continuation of specific informational elements within consciousness has to be amintained in the face of other strong claimants to the Ws. Such an effort can only be sustained for brief periods. As James puts it: "When we are studying an uninteresting subject, if our mind tends to wander, we have to bring back out attention every now and then by using distinct pulses of effort, which revivify the topic for a moment, the mind then running on for a certain number of seconds or minutes with spontaneous interest, until again some intercurrent idea captures it and takes it off" (James, 1899, p.101).

CONCLUSION

It has been argued that the pervasive similarity and frequency effects, apparent in a wide variety of error types, are rooted in the parallel and automatic processes by which knowledge is retrieved from long-term memory. In particular, they are shaped by the cognitive system's remarkable ability to match stored represenations to current Ws 'calling conditions', and to resolve conflicts between partially matched items using a simple "most used, most likely" heuristic. This bias towards selecting the more frequent of the partially matched candidates is dependent upon an automatic facility for keeping a running tally of roughly how often an event or object has been encountered in the past. The picture presented of human cognition is that of an informational system which, though adept at internalising the complexity of the world around it, is driven by a restricted number of simple computational procedures (see also McClelland & Rumelhart, 1985; Norman, 1986).

REFERENCES

Baddeley, A.D. & Hitch, G. 1974 Working memory. In The Psychology of Learning and Motivation edited by G.H. Bower, Vol.8 (New York:Academic Press).

Battig, W.F. & Montague, W.E. 1969 Category norms for verbal items in 56 categories: A replication and extension of the Connecticut category norms, Journal of Experimental Psychology Monograph, 80, 1-46.

Bobrow, D.G. & Norman, D.A. 1975. Some principles of memory schemata. In Representation and Understanding: Studies in Cognitive Science, edited by D. Bobrow & A Collins (New York:Academic Press).

Dell, G.S. & Reich, P.A. 1980 Toward a unified model of slips of the tongue. In Errors and Linguistic Performance Slips of the Tongue, Ear, Pen, and Hand, edited by V. Fromkin (New York:Academic Press).

Fischhoff, B., et al. 1981, Acceptable Risk, (Cambridge: C.U.P.).

Fromkin, V. 1971 The non-anomolous nature of anomolous utterances, Language, 47, 27-52.

Fromkin, V. (Ed.) 1973 Speech Errors as Linguistic Evidence ,(The Haugue:Mouton).

Fromkin V. (Ed.) 1980, Errors in Linguistic Performance: Slips of the Tongue, Ear, Pen, and Hand. (New York: Academic Press).

Gregg, V. 1976, Word frequency, recognition and recall. In Recall and Recognition, edited by J. Brown (London: Wiley).

Hasher, L. & Zacks, R.T. 1984 Automatic processing of fundamental information: The case of frequency of occurrence, American Psychologist, 39, 1372-1388.

James, W. 1890 The Principles of Psychology, Vol.2. (New York: Holt).

James, W. 1899 Talks to Teachers on Psychology: And to Students on some of Life's Ideals. (London:Longmans)

Kahneman, D. et al. 1982 Judgement under Uncertainty: Heuristics and Biases (Cambridge:C.U.P.).

Mandler, G. 1985 Cognitive Psychology: An Essay in Cognitive Science (Hillsdale:Erlbaum).

MacKay, D.G. 1970 Spoonerisms: The structure of errors in the serial order of speech, Neuropsychologia, 8, 323-350.

Neisser, U. 1967 Cognitive Psychology, (New York:Appleton-Century-Crofts).

Nisbett, R. & Ross, L. 1980 Human Inference: Strategies and Shortcomings of Social Judgement (Englewood Cliffs: Prentice-Hall).

Norman, D.A. & Bobrow, D.G. 1979 Descriptions: An intermediate stage in memory retrieval, Cognitive Psychology, 11, 107-123.

Norman, D.A. 1985 New views of information processing: Implications for intelligent decision support systems. In Intelligent Decision Aids in Process Environments, edited by G. mancini & D. Woods. (San Miniato, Italy: NATO Advanced Study Institute).

Nooteboom, S.G. 1969 The tongue slips into patterns. In Leyden Studies in Linguistics and Phonetics, edited by A. Sciarone et al. (The Hague:Mouton).

McClelland, J. L. & Rumelhart, D.E. 1985 Distributed memory and the representation of general and specific information, Journal of Experimental Psychology:General, 114, 159-188.

Rasmussen, J. 1982 Human errors: A taxonomy for describing human malfunction in industrial installations, Journal of Occupational Accidents, 4, 311-335.

Reason, J.T. 1979 Actions not as planned: the price of automatization. In Aspects of Consciousness Vol.1. Psychological Issues, edited by G. Underwood & R.Stevens (London:Wiley).

Reason, J.T. et al. 1986 Multiple search processes in knowledge retrieval: Similarity-matching, frequency-gambling and inference (Unpublished report).

Reason, J.T. & Mycielska, K. 1982 Absent-Minded? The Psychology of Mental Lapses and Everyday Errors, (Englewood Cliffs, New Jersey:Prentice-Hall).

Reason, J.T. & Lucas, D. 1984 Using cognitive diaries to investigate naturally occurring memory blocks. In Everyday Memory, Actions and Absent-Mindedness, edited by J. Harris & P. Morris (London:Academic Press).

Shweder, R.A. 1977 Likeness and likelihood in everyday thought: magical thinking and everyday judgements about personality. In Thinking: Readings in Cognitive Science, edited by P. Johnson-Laird & P. Wason (Cambridge:C.U.P.).

Tulving, E. 1983 Elements of Episodic Memory (Oxford: Oxford University Press).

Wason, P.C. & Johnson-Laird, P.N. 1972 The Psychology of Reading (London:Batsford).

Human Reliability

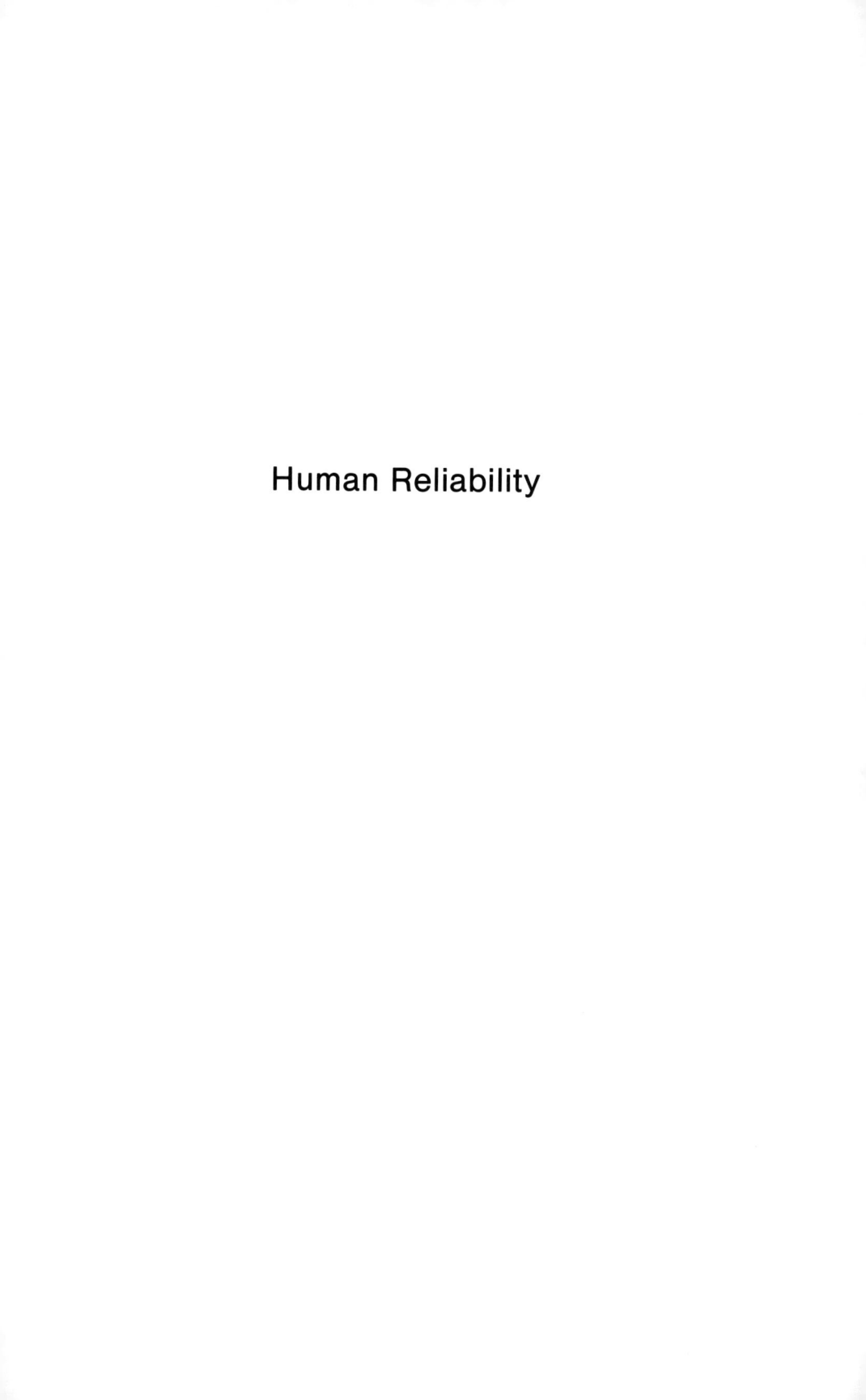

A CASE STUDY INVOLVING SIMULATION, HUMAN ERROR PREDICTION AND QUANTIFICATION FOR INCLUSION INTO A PROBABILISTIC RISK ASSESSMENT

T. WATERS

SRD, UKAEA
Wigshaw Lane, Culcheth
Warrington WA3 4NE

A fault tree has been produced as part of a safety analysis on one of the UKAEA's low power reactors. This revealed a number of human errors which could result in a hazardous situation if they are not recovered. It was decided to select one of the basic events which involved human error for more detailed investigation. There were two objectives to this investigation:
1. To increase operator reliability by traditional ergonomic measures of procedures, equipment and training improvements.
2. To assess quantitatively the probability of operator error for the optimum application of resources and for licensing purposes.

The event which was chosen occurred after a loss of coolant accident, when the automatic reactor protection systems operate and stop coolant circulation. In some circumstances it is necessary for the operator to restart circulation, and subsequently monitor coolant level and temperature. The event is of particular interest because the operator is faced with conflicting objectives, in that the maintenance of cooling results in a reduction of coolant level. Ideally the operator wants to maintain a high level of coolant while keeping the temperature low.

To investigate the performances of the operators in this situation a simple table-top paper and pencil simulation was performed. During the exercise operators were provided with information about alarms and coolant level and temperature, in as near as possible, real time. Eight operators took part in twelve simulations and although all of the operators succeeded in bringing the reactor to a safe and stable condition on each occasion, it appeared that two distinct strategies were being employed. One strategy involved the coolant being circulated intermittently whereas the other strategy involved constant circulation and the reliance upon automatic emergency cooling pumps.

In order to quantify the probability of operator error, the overall task was decomposed to its basic components of alarm detection, diagnosis, and operation of valve controls. The probability of error during operation of valve controls obviously depended upon the frequency of operation, which itself depended on the strategy which was employed and this was modelled using the results of the simulation exercise. For purposes of comparison, the human error probability was quantified using six techniques which are well known within the field of human reliability assessment. These techniques were: APJ (Absolute Probability Judgement), HEART, SLIM-MAUD, STAHR, TESEO and THERP. These methods vary greatly in the degree to which they consider the various factors which affect performance, which is one of the reasons why the overall human error probabilities varied by a factor of twenty. This degree of variability is on the upper limit of that which is generally acceptable within the reliability assessment community.

SHERPA: A SYSTEMATIC HUMAN ERROR REDUCTION AND PREDICTION APPROACH

D.E. EMBREY

Human Reliability Associates Ltd
1 School House, Higher Lane, Dalton
Parbold, Lancashire WN8 7RP

ABSTRACT

This paper describes a Systematic Human Error Reduction and Prediction Approach (SHERPA) which is intended to provide guidelines for human error reduction and quantification in a wide range of human-machine systems. The approach utilises as its basis current cognitive models of human performance.

THE OBJECTIVES AND STRUCTURE OF SHERPA

The overall function of SHERPA is to provide a framework within which human reliability can be analysed and assessed, both quantitatively and qualitatively. It also generates specific error reduction recommendations in the areas of procedures, training and equipment design. The optimal way in which these recommendations should be implemented is evaluated using the quantification module in SHERPA, which allows cost effectiveness assessments to be carried out. These essentially consist of sensitivity analyses which indicate the changes in the system which will have the greatest effect in enhancing human reliability. Ideally SHERPA should be applied at the design stage of a new system. In practice it will generally be used to modify an existing system.

Although, as described above, the separate modules within SHERPA are relatively independent, they are linked by a common theoretical orientation. This is that human reliability optimisation and quantification cannot be effectively achieved by considering only the surface aspects of human error. The underlying cognitive processes that give rise to errors must be addressed as part of the approach.

SHERPA consists of a series of sequentially applied modules which are described in the following sections.

Task Analysis

Any error prediction and reduction approach must begin with a task analysis to provide a detailed description of the situation being evaluated, and to identify the characteristics of the task likely to give rise to error. The recommended task analysis technique in SHERPA is Hierarchical Task Analysis (HTA) described in detail in Shepherd (1984).

HTA is a systematic method for identifying the various goals that have to be achieved within a task and the way in which these goals are combined to achieve the overall objectives. One of the major advantages of the HTA method is that it explicitly identifies the plans that are used by operators to achieve goals at various levels. Because HTA captures the operator perception of the goal structure of the task, this structure can be reflected in the procedures and training methods derived from the analysis.

Classification of Task Information Processing

The philosophy of SHERPA involves understanding the information processing involved in task performance, in order to facilitate error prediction and reduction. A number of different task classification schemes exist. For the purposes of analysing human reliability, SHERPA uses a classification scheme originally developed by Rasmussen, which has been adapted and extended for use in this context. The scheme is discussed in more detail in Rasmussen (1983). There are basically four types of mental processing represented in the classification.

Skill-Based (SB) Processing

This occurs in situations involving mainly simple physical operations of controls, or where equipment is manipulated from one position to another. In SB processing, the operator is sufficiently highly skilled or practised such that no conscious planning or monitoring is required to execute the actions.

Rule-Based Diagnostic (RBD) Processing

RBD processing involves diagnosis where a pattern of symptoms is associated with a cause by the operator using an explicit "production rule" of the form IF <condition X> THEN <cause Y>. For example, if a car fails to start, RBD processing could require invoking the rule IF <starter motor is OK> AND <lights dim when operating ignition key> THEN <cause is flat battery>.

Rule-Based Action (RBA) Processing

RBA is similar to RBD processing but the production rule directly connects a condition with an action or set of actions e.g. IF <situation X> THEN <do Y>. The situation that triggers the rule is often, but not invariably, made explicit as a result of prior RBD processing. Thus, RBD and RBA rules are often linked together in the following way:

    IF <Condition X> THEN <Situation Y>
    IF <Situation Y> THEN <do Z>

RBA processing generally occurs in situations where the operator is following an explicit procedure or rule of thumb, which may be written or memorised. SB processing is also associated in tasks with RBA processing, as it is required to actually execute the physical operations needed to carry out the task.

Knowledge-Based (KB) Processing

KB processing is required if the operator cannot refer to any existing procedure or rule of thumb, and therefore has to go back to first principles and utilise his overall technical understanding of the situation.

The type of processing which is required for a particular task depends on the level of training and experience of the individual. Thus, for a beginner, all tasks involve a high proportion of KB processing, because he or she will frequently have to refer back to first principles. With practice, most tasks enter the RBD or RBA domain, and some very frequently encountered operations, e.g. manipulations of commonly used controls, will become highly automatised (i.e. will involve mainly SB processing).

Application of the Information Processing Classification in SHERPA

The information processing classification is used in SHERPA in the following ways:

1. Defining the appropriate form of subsequent analyses. The qualitative error prediction technique used in the analysis module is applicable primarily to SB, RBD and RBA tasks. The KB dominated tasks must be dealt with using other approaches, which are currently under development.

2. Assisting in the identification of potential error modes during the error analysis. Characteristic error modes are associated with each of the information processing categories defined above. The prior classification of tasks therefore considerably facilitates the systematic postulation of possible error modes.

3. Defining the most effective error reduction strategy. Effective strategies can only be defined if the underlying root causes of errors are identified. These will depend on the types of information processing likely to be used in the task under consideration.

Other Information Produced by the Task Analysis

1. Definition of the main task elements to be executed during the performance of the task. This is basically the primary path via which the operator is expected to achieve the task goals. Alternative paths are analysed in the error analysis.

2. Definition of the plans and the conditions which determine transfer of control between the levels in the goal hierarchy and within task elements at a particular level. This has particular relevance for the definition of training content, and the identification of certain types of errors.

3. Commentary on (1) and (2) with regard to their implications for procedures, training and equipment design.

This information is subsequently supplemented by insights gained during the error analysis, and used in the error reduction module.

Human Error Analysis (HEA)

The HEA takes as its inputs the task elements identified in the HTA. For each task element the following analytical procedure is adopted:

1. Definition of external error modes.
   At task element it is assumed that one or more of 18 different error modes could occur. Examples of these are: action too early; right action on wrong object; information not obtained. These error modes are based on a taxonomy described in Pedersen (1985). A computer program is used to assist the analyst in deciding on which of the error modes is credible in the situation being examined.

2. Identification of psychological mechanisms.
   For each external error mode identified, the underlying psychological mechanism is identified. These mechanisms are also based on the error modelling work of Rasmussen. The derivation of psychological methods is important because it assists in prescribing appropriate error reduction strategies at later stages of the HEA. A computer program is again used to assist the analyst in this process.

3. Recovery Analysis.
   The ways in which error recovery can occur are analysed at this stage.

4. Consequence Analysis.
   The consequences of unrecovered errors are evaluated.

5. Development of error reduction strategies.
   Strategies to reduce the likelihood of the postulated errors are considered at this stage. These will include measures to reduce the initial likelihood of the error, together with methods for improving the probability of recovery.

The overall outcome of the HEA is a clearly documented evaluation of the errors likely to occur in the system, together with their expected consequences, and methods of control.

THE QUANTIFICATION MODULE

The quantification approach used in SHERPA is SLIM-MAUD. SLIM-MAUD (Success Likelihood Index Methodology using Multi-Attribute Utility Decomposition) is a computer-based technique developed under the sponsorship of the U.S. Nuclear Regulatory Commission (NRC) and Brookhaven National Laboratory, to quantify human error probabilities (HEPs) (see Embrey et al, 1984).

Quantification within SHERPA is primarily used to evaluate the degree of error reduction that will be produced by the different error reduction strategies postulated in the HEA. This allows cost-benefit analyses to be performed to identify the most effective error reduction approach within given cost constraints.

ERROR REDUCTION MODULE

The final module in SHERPA is concerned with applying the results of the previous two modules to obtain the most cost effective error reduction. As discussed previously, the outputs of the HTA and HEA provide very specific recommendations for the modifications to procedures, training and equipment design which will improve human reliability. After these recommendations are implemented, their effectiveness is monitored via a quality assurance phase.

CONCLUSIONS

The SHERPA approach is a comprehensive methodology for human reliability prediction and optimisation in NPP. It also has the advantage of an integrated quantification and cost-benefit analysis capability.

SHERPA has been applied to several critical systems areas in nuclear power plants and the results so far indicate that it produces auditable results, that are perceived by operators, engineers and managers to be valid, useful and cost effective. In a recent test of the technique SHERPA successfully predicted between 78% and 100% of errors that had actually occurred on a range of procedures used in a large plant.

We are continuing to develop SHERPA, with particular emphasis on a more systematic consideration of the organisational and other socio-technical variables which have a considerable influence on reliability at all levels in a system. Nevertheless, even at its present stage of development, we believe that SHERPA has wide applicability and considerable promise.

REFERENCES

1. D.E. EMBREY, P.C. HUMPHREYS, E.A. ROSA, B. KIRWAN and K. REA, "SLIM-MAUD. An Approach to Assessing Human Error Probabilities using Structured Expert Judgement", Vol. I & II. NUREG/CR-3518. Brookhaven National Laboratory, Upton, New York, USA (1984).

2. O.M. PEDERSEN, "Human Risk Contributions in Process Industry", Report no. Risø-M-2513. Risø National Laboratory, DK-4000, Roskilde, Denmark (1985).

3. J. RASMUSSEN (1983), Skills, Rules and Knowledge: Signals, Signs and Symbols and Other Distinctions in Human Performance Models. IEEE Transactions on Systems, Man and Cybernetics, SMC-13 (3) 257-266.

4. A. SHEPHERD, "Hierarchical Task Analysis and Training Decisions", Programmed Learning and Educational Technology 22.2 162 (1984).

A GUIDE TO REDUCING HUMAN ERROR IN PROCESS OPERATION

P.W. BALL

British Nuclear Fuels plc
Sellafield, Seascale
Cumbria CA20 1PG

INTRODUCTION

In February 1985 the Safety and Reliability Directorate published 'A Guide to Reducing Human Error in Process Operations' (Short Version) as SRD R 347. The Guide was prepared by a Sub Group of the Human Factors in Reliability Group. The principal author of the Guide was Lisanne Bainbridge.

The Short Version of the Guide was arranged as a check list of questions which were broken down into five sections:
   Operator/Process Interface
   Procedures
   Work Place and Work Environment
   Training
   Task Design and Job Organisation
Each of these sections were further sub divided into headings, each encapsulating an ergonomics principle.

Since that time an expanded group have been preparing a Longer Version of the Guide. The principal author of this version is David Whitfield.

The Longer Version will provide detailed advice in each of the above five areas. This paper indicates the type of information which will be given in that Guide.

OPERATOR/PROCESS INTERFACE

In this section common control panel deficiencies are highlighted and then recommendations are given on alarm annunciators and control instrumentation feedback. Eight such deficiencies (Hartley et al 1984) are listed:
   (a) Poorly designed scales and scale numerical progressions that are difficult to interpret.
   (b) Parallax problems in reading pointers to scale markings.

(c) Meters or recorders that can fail with the pointer reading in the normal operating band.
(d) Glare and reflections.
(e) Too many channels of information on chart recorders.
(f) No warning before a chart recorder pen runs out of ink.
(g) Lack of limit marks on meters used for check reading.
(h) Meters not arranged with 'normal' segments in the same relative positions (to facilitate check reading).

A number of common control panel deficiencies have been reported, eight are listed below:
(a) Lack of functional grouping.
(b) Controls with ambiguous pointers.
(c) Inconsistent component names from unit to unit.
(d) Ambiguous placement of labels.
(e) Temporary labels.
(f) Poor maintenance which caused missing labels.
(g) Unnecessarily detailed and obscure labels.
(h) Label lettering too small for the reading distance.

Where large numbers of annunciators are found in control rooms, priority should be applied so that operators can differentiate between 'serious' (and 'less important') alarms.

Three Levels of Priority are usually recommended. The first priority should include:
  Plant Shut-down, Noxious Release or Plant Condition which if not corrected immediately will result in noxious release, or require the plant to be shut-down.
The second priority should include:
  Technical Specification Violations (which are likely to lead to plant shut-down) and Plant Conditions which if not corrected quickly, may result in plant shut-down or noxious release.
The third priority is:
  Plant conditions which affect its operability.

These different priority annunciators could be colour coded Red, Yellow and White respectively.

Operators should be given information about the effects of their actions. Control and instrumentation systems sometimes give feedback to the operator based on the transmission of the control signal to the plant item, rather than on the actual response of the plant item.

This is undesirable since failure of the plant item might not be recognised.

PROCEDURES

In this section some common deficiencies are listed and then recommendations made to aid the preparation of procedures. It is not easy to write good procedures, a number of common deficiencies are listed:
- (a) Lack of consistency between nomenclature in procedures and on panel components.
- (b) Failue to indicate correct system response in control action instructions.
- (c) Excessive burden on operator's short term memory.
- (d) Failure to integrate charts/graphs with text.
- (e) Ambiguity about which procedures apply.
- (f) No formal method for updating procedures.
- (g) Little help for operators' diagnosis of problems in an incident.
- (h) Actions embedded near the end of complex paragraphs.

Well presented procedures have:
- (a) A title and a brief description of their purpose.
- (b) A clear indexing scheme; pages numbered 'y of x' to indicate missing pages.
- (c) Clear cross referencing *within* the document and *minimal* cross-referencing to other documents.
- (d) Basic information presented in short concise, identifiable steps.
- (e) Required actions must be specifically identified, control settings and parameter values *must* be expressed precisely.
- (f) Equipment or components must be identified exactly as on the plant itself.
- (g) Nomenclature and terminology must be familiar to the operators.
- (h) Precautions and explanations should be set out before the steps to which they apply.
- (i) Check list should be provided for tests and complex tasks, and *all* procedures should have space for recording task completion.

Prose can be improved by following a few simple rules:
- (a) Use simple words and phrases which are familiar to the operators.
- (b) Use brief and concise sentences.
- (c) Use simple sentence construction.
- (d) Use the active tense because the passive tense increases comprehension difficulties.

(e) Minimise the use of negatives.
(f) Only use upper case letters for isolated words thereby improving reading speed and understanding.

It should also be noted that alternatives to prose can be used to advantage in certain situations.

After procedures have been prepared, using prose or alternative forms, **do not** forget to check their usability by having 'walk-throughs' or 'talk-throughs' with operators. Finally check the consistency and compatibility of the procedures with the operator's training.

WORK SPACE DESIGN

There is much data on control room design, but work space design is equally important outside the control room. Seminara & Parson (1982) have noted some of the problems in maintenance operations:
  (a) Improper or inadequate equipment identification.
  (b) Limited access for maintenance.
  (c) Equipment not designed to facilitate maintenance.
  (f) Adverse environmental conditions.

For work where protective clothing may have to be worn, adequate allowance should be made for the limitations on vision, speech, reach, movement and manipulation.

WORKING ENVIRONMENT

In most control rooms it is not too difficult to achieve reasonable control of noise and temperature. Providing good lighting is much more difficult. In addition to choosing the levels of illumination which are appropriate to the activity (Kinkade & Anderson 1984) it is important to position light sources and choose working surfaces to prevent glare and reflections.

Glare can be minimised by using several low intensity rather than a few high intensity light sources.

Reflections can be effectively eliminated by the careful design and positioning of light sources in relation to work areas and by treating shiny surfaces. For visual tasks the size and contrast of materials can be as important as illumination.

TRAINING

There are a number of well established basic training requirements for operators:
  (a) Include all the operator's tasks and responsibilities in his training programme.

(b) The general task of operating the plant should be defined in terms of subordinate tasks, e.g.:
Startup, normal operation, fault detection and diagnosis, shutdown.
Normal operation should also be defined in terms of the various task components involved.
(c) Regular refresher training should be given, particularly for skills and knowledge which are infrequently used (e.g. fault diagnosis).
(d) For process operations which require the co-ordinated actions of a number of workers, the team should train together.

Team goals and interdependencies between team members <u>must</u> be included.

TASK DESIGN AND JOB ORGANISATION

Two important aspects of Task Design and Job Organisation are Team Responsibilities and Communications. Clear procedures must be established for communications, especially the transfer of responsibility between different shifts and workers with different responsibilities.

Status differences should not be allowed to interfere with efficient communication. Team members should be encouraged to draw attention to the errors of others <u>regardless</u> of status.

Job performance is dependent on job satisfaction. Herzberg (1968) identified two factors which affect job satisfaction:
(a) Intrinsic Factors which are the principle influences on satisfaction.
(b) Extrinsic Factors such as pay, which contribute only to levels of dissatisfaction.

Hackmann & Oldham (1976) tried to identify the 'job characteristics' which contribute to motivation and satisfaction. They listed five such factors:
Skill variety, task identity, task significance, autonomy, and feedback, direct information on his own performance.

It should also be noted that job performance will be improved if stress is optimised.

The intention of all the above recommendations is to reduce human error, but inevitably some errors will continue to occur and accidents will result. An effective incident reporting and information system will help to reduce future accidents.

The information contained in reports of incidents should:

(a) Be based only on genuine in-plant occurrences.
(b) Be based on events which are significant in terms of potential severity or frequency of occurrence.
(c) Be as free as possible from reporter bias.
(d) Be capable of easy extraction from reports.

CONCLUSIONS

In summary, I would recommend seven ways in which operator errors can be reduced:
(a) Improve control and display panels.
(b) Enhance communications system.
(c) Arrange regular refresher training.
(d) Improve the quality of procedures.
(e) Improve the work space design.
(f) Improve the operating environment.
(g) Upgrade maintenance and housekeeping, simply keep the place tidy.

ACKNOWLEDGEMENTS

The views expressed are my interpretation of some of the information in the draft Long Guide. They do not necessarily represent the views of BNFL.

REFERENCES

Hackmann, J R & Oldham, G R, 1976, Motivation through the design of work : test of a theory, Organisational Behaviour and Performance, 16, 250-279.

Hartley, C J et al, 1984, Potential Human Factors deficiencies in the design of local control stations and operator interfaces in nuclear power plants, V.S. Nuclear Regulatory Commission Report, NUREG/CR, 396.

Herzberg, G, 1968, One more time : how do you motivate employees?, Harvard Business Review, 46, 53-62.

Kinkade, R G & Anderson, J, 1984, Human factors guide for nuclear power plant control room developments, Electric Power Research Institute Report, NP 3659.

Seminara, J L & Parson, S O, 1982, Nuclear power plant maintainability, Applied Ergonomics, 13, 177-189.

**HELPING THE DESIGNER TO IMPROVE
HUMAN RELIABILITY**

S.P. WHALLEY

Applied Psychology Division
Aston University
Birmingham   B4 7ET

Emphasis has been placed on the role of Human Reliability Analysis (HRA) when assessing System Reliability with much effort going into quantification through objective and subjective methods. This paper considers a shift in emphasis towards assisting the engineer to *design* to reduce error likelihood and so ensure optimum human performance. A methodology for achieving this aim is briefly introduced. The remaining requirement, to use this as a system of interrogation for a comprehensive ergonomic data base, is discussed.

INTRODUCTION
It is now an accepted concept that human reliability, in the form of errors, affects system performance. What is only just becoming acceptable is that systems can affect human performance, this forces the issue into a regressive loop. It is necessary to identify a break point and this comes in the guise of Performance Shaping Factors (PSFs). By considering the notion of reliability in increasing detail it is possible to remove the overlaying issues ultimately revealing the PSFs.

THE POSITION OF PSFs
Figure 1 indicates that the first consideration of a reliability engineer or an HAZOP team (a HAZOP is a HAZard and OPerability study undertaken to identify all the possible process failure modes for each pipeline and each item of equipment) is "what can go wrong with the plant?" (Kletz1984). Certain identified failures can be directly caused by incorrect human performance. The next step is to move inwards to consider the possible types of human error that could have caused the particular process fault.

Proceeding towards the heart of the problem the underlying error causes can be examined in turn. Finally those aspects of the 'system' in the widest sense of the word which can enhance or degrade performance by affecting error likelihood are scrutinized. This simplified progression indicates the current position of ergonomic concern. The predominance of effort has been aimed at identifying potential errors and assessing their probability.

Figure 1. The steps towards providing design guidance in order to improve Human Reliability.

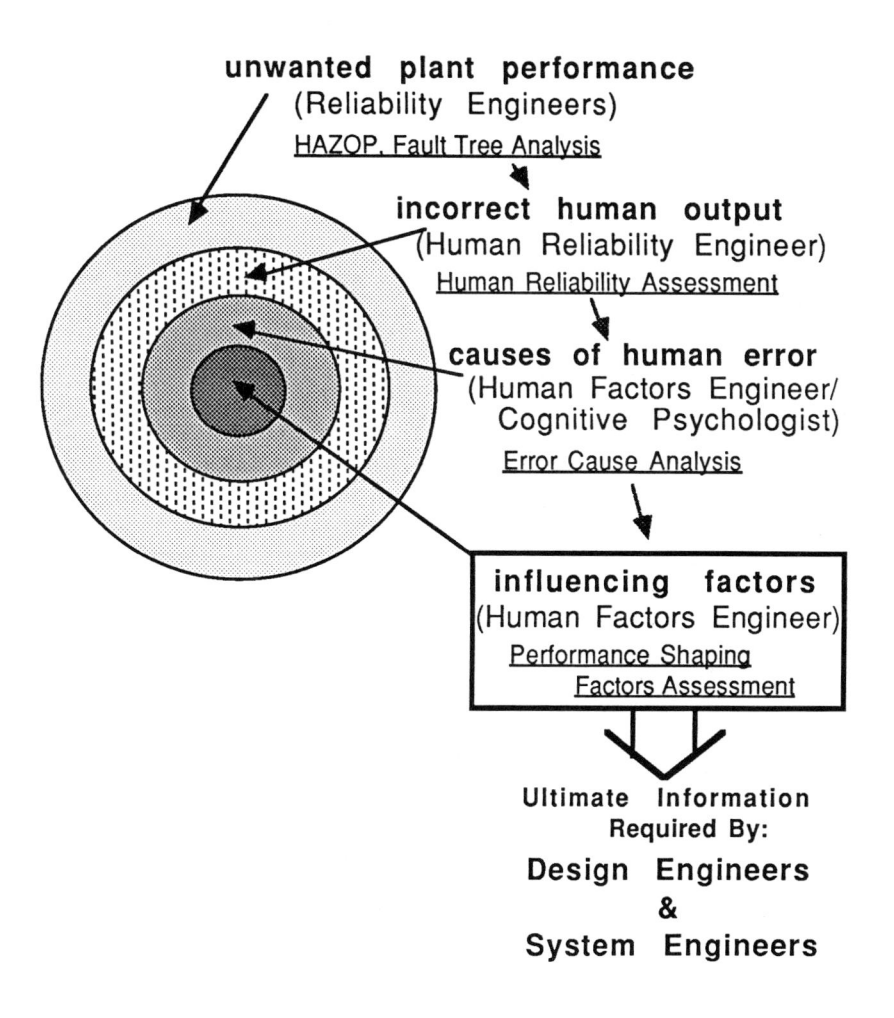

unwanted plant performance
(Reliability Engineers)
HAZOP, Fault Tree Analysis

incorrect human output
(Human Reliability Engineer)
Human Reliability Assessment

causes of human error
(Human Factors Engineer/ Cognitive Psychologist)
Error Cause Analysis

influencing factors
(Human Factors Engineer)
Performance Shaping Factors Assessment

Ultimate Information Required By:
Design Engineers
&
System Engineers

With the 1980's there has been an extension in the use of PSFs as HRA weighting factors (Embrey 1985), the originator of this practice was Swain (1964) for the assessment of nuclear reactor reliability onboard submarine.

HUMAN ERROR

As an interlocking issue human error has been considered by numerous psychologists over the years with a recent emphasis upon absent minded errors (Norman 1981, Reason 1985) and the importance of differentiating between 'slips', physical errors, and 'mistakes', mental errors. These theoretical notions have not as yet been linked into the human reliability equation nor been systematically considered within the work environment. Rasmussen (1986) has however recognised this distinction as an important qualitative aspect.

The role of error cause has however been substantial in defining one route towards identifying relevant PSFs for specific chemical plants and their processes (Whalley & Maund 1985).

DESIGN GUIDANCE

Once potential PSFs are identified for a specific workplace and task the remaining step is to provide associated ergonomic guidelines for inclusion by the designer. Many checklists exist that give ergonomic data, the most recent and specific for the process industries being the Short Guide to Reducing Human Error in Process Operation (HFRG 1986) with a long guide under development. In addition many traditional ergonomics and human factors engineeering text books are available providing basic design guidance. This information is however difficult to use by the non-expert and often requires extrapolation in order to fit the specific environment.

For process plant control rooms, particularly computer controlled processes, much pertinent research has and is still being undertaken in order to establish guidelines (Carey 1985, Umbers, White & Meijden 1986, Knauper & Kraiss 1986). Additionally, new methods of presenting this type of information to the designer are being explored; for example Whalley & Booth (1986).

Currently work is progressing towards a comprehensive compendium of basic human performance data, models of performance and ergonomic guidelines; the Engineering Data Compendium (Boff & Lincoln) of which two of the four volumes are currently available. This is by its very nature a long term project which has required careful consideration of data acquisition methods.

IDENTIFICATION OF RELEVANT PSFS

The author has been working towards a methodology for identifying likely error causes and ultimately specific PSFs. To achieve this specificity for a particular process plant, yet ensuring a generic method useable by all, three separate routes to PSFs have been established. The final PSF group includes individual PSF importance weightings obtained as a product of Set Theory. One of the three routes is illustrated in figure 2.

Figure 2. The system of mappings from generic task classification through to performance shaping factors and ultimate design guidelines.

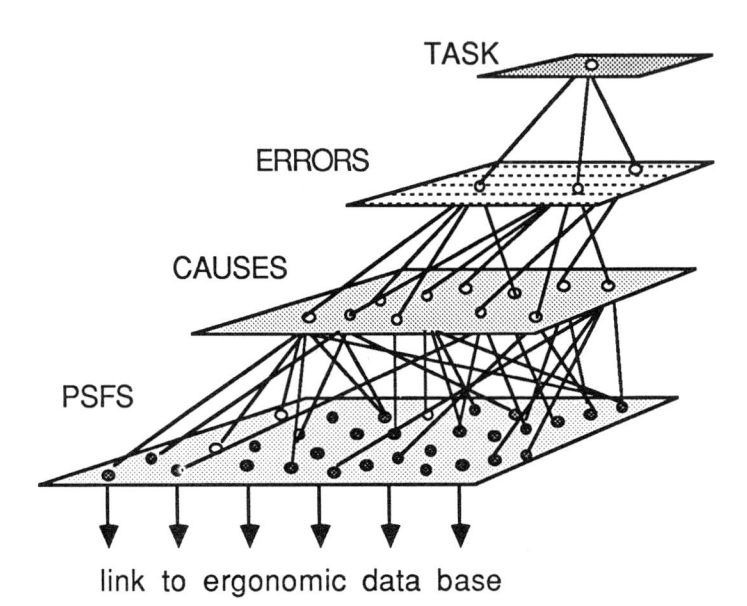

TO CONCLUDE

The trend has been to concentrate upon human error probabilities, obtained through expert judgement or data extrapolation, and reliability assessments using PSFs as weighting factors. One methodology; PHECA, the Potential Human Error Cause Assessment (Whalley & Maund 1986), considers PSFs in their own right as enhancers or degraders of human performance, though this is still under evaluation. Ultimately such a method requires integrating with a comprehensive ergonomic data base as an interrogation mechanism so that system engineers and design engineers can obtain realistic advice. Dedicated work is still required to achieve this aim and to establish the validity of positive and negative PSFs. Steady progress is however being made and the involvement of ergonomists will continue.

REFERENCES

Boff, K. & Lincoln, J., (1986 onwards) **Handbook of Perception & Human Performance** (New York: John Wiley & Sons)

Carey, M.(1985) **The selection of Computer Input Devices for Process Control** CR 2773(con) Applied Psychology Division (Warren Spring Lab, DTI)

Embrey, D., (1985)SLIM-MAUD a computer based technique for human reliability assessment. In **Ergonomics International 85**, edited by I.Brown et al (Taylor &Francis)

Human Factors in Reliability Group, (1985) **Guide to Reducing Human Error in Process Operation** SRD R 347 (Safety & Reliability Directorate, UKAEA)

Kletz, T.,(1984) **HAZOP & HAZAN- Notes on the Identification and Assessment of Hazards** (IChemE, Rugby)

Knauper, A. & Kraiss, K. F., (1986) An interactive graphic system for design and evaluation of man-machine interfaces. In **Human Decision Making and Manual Control**. edited by H.-P. Willumeit (North-Holland)

Norman, D.,(1981) Categorization of action slips. **Psychological Review** volume 88 no.1, 1-15

Rasmussen, J., (1986)**Information Processing and Human-Machine Interaction** (North-Holland)

Reason, J.,(1985) Slips and mistakes: two distinct classes of error? In **Contemporary Ergonomics 1985**, edited by D.Oborne (Taylor & Francis)

Swain, A., (1964) **THERP** SC-R-64-1338 (Sandia Laboratories, Albuquerque, N.Mex.)

Umbers, I., **An Experimental evaluation of layout, colour and Scale Design in some Process Control Displays**, LR 467 (con) M (Warren Spring Lab, DTI)

Whalley, S. & Booth, S.,(1986) A suggested structure for a computer based control room interface design aid. In **Contemporary Ergonomics 1986** edited by D. Oborne (Taylor & Francis)

Whalley, S. & Maund, J., (1986) Linking human error to performance shaping factors. In **Sixth European Annual Conference on Human Decision Making and Manual Control**, edited by J. Patrick & K. Duncan (Department of Applied Psychology, UWIST) p.19

Whalley, S. & Maund, J.,(1986) The use of performance shaping factors in process plant design aids. In **Contemporary Ergonomics 1986** edited by D. Oborne (Taylor & Francis)

White, T. & Meijden, P., (1986) VDT trend representations affecting human prediction accuracy. In **Sixth European AnnualConference on Human Decision Making and Manual Control**, edited by J. Patrick & K. Duncan (Department of Applied Psychology, UWIST) p.37

HUMAN RELIABILITY FROM A NUCLEAR REGULATORY VIEWPOINT

D. WHITFIELD

HM Nuclear Installations Inspectorate
Health and Safety Executive, Bootle
Merseyside L20 3LZ

The modus operandi of H.M. Nuclear Installations Inspectorate is described, and past and present concerns with human error are reviewed. In relation to the probabilistic safety goals for nuclear stations, comprehensive quantification of human reliability is not feasible, but it is necessary to have as much evidence as possible that human error will not be a limiting effect.

THE ROLE OF H. M. NUCLEAR INSTALLATIONS INSPECTORATE (NII)

NII, part of the Health and Safety Executive (HSE), is responsible for the regulation of the safety of all nuclear power plants and other nuclear installations in the UK, apart from those run by the UK Atomic Energy Authority and by the Ministry of Defence. NII does not set fixed regulations for design and operation, but rather requires each potential or existing licensee to make their own safety case. The safety case is assessed and negotiated by NII inspectors, and a licence is granted for a specific new installation. For example, with a new power station or nuclear processing plant, the licence is granted on the basis of the pre-construction safety report, and the licence contains conditions requiring further reports, tests and operational arrangements by the licensee. Operational reactors and other nuclear plant have to be shut down for regular annual or biennial maintenance, and require NII consent to start up again. The Inspectorate has the power to set any specific conditions or operating requirements in an individual licence, in the interests of safety.

Although the onus is on the licensee to make his own case, the Inspectorate has set out its safety assessment principles for power reactors (NII, 1979) and for nuclear chemical plant (NII, 1983a). These sets of principles are intended to promote consistency and uniformity in NII evaluations, and they do not serve as design guides. However, they give some indication of the NII's expectations of system characteristics, always bounded by the legal requirement on licensees to reduce risks "so far as is reasonably practicable".

NII AND HUMAN ERROR

Any evaluation of safety must take account of human error, and NII has recognised the importance of ergonomics for some time. Human error was discussed at length in the Public Inquiry into the proposed Pressurised Water Reactor for Sizewell 'B', and NII had convened previously a working group on ergonomics issues (NII, 1982). Ergonomics is specifically dealt with in part of the NII review of the Central Electricity Generating Board's pre-construction safety report (NII, 1983b). The major topics identified there were:

* <u>Design and construction</u>:

    - allocation of operator tasks
    - control room design
    - maintenance facilities and procedures
    - ergonomics of Quality Assurance procedures

* <u>Operational management</u>:

    - definition of operating goals
    - staffing levels
    - selection and training
    - operating procedures and documentation
    - monitoring of personnel performance

* Fault studies:

    - justification for omission of human actions from fault and event trees
    - operators as initiators of events
    - cognitive/conceptual errors
    - maintenance errors
    - task analysis for safety actions

An extensive programme of ergonomics development is now being pursued by CEGB, and this will be subject to the further NII assessment in the Sizewell 'B' design. The identification of possible human errors in operation and maintenance, and their reduction and mitigation in accordance with design safety goals, is an important component in that work. The general approach is to delineate the possible operator errors at each stage of a procedure, to assess the likely implications for the plant and safety systems, and to implement hardware and software modifications as appropriate.

Ergonomics issues were included in the recent NII audit of part of the Sellafield nuclear reprocessing plant, where specific attention was given to control room design, operating procedures, and plant operations (HSE, 1986).

For more general application within NII, an assessment guide on ergonomics is under discussion -- assessment guides develop in more detail from the principles documents referred to above. The guide will probably cover these facets:

* System design: Team structure, task analysis, roles of operators, procedural and knowledge based operation, decision support.

* Operator-plant interface: Information presentation on plant design and plant performance, support information, control facilities, maintenance and testing, communications.

* **Operating procedures**: Specification and presentation.

* **Operating environment**: Workspace, lighting, thermal, noise, radiological, and toxicological conditions in control rooms and on plant.

* **Selection and training**: Selection and allocation, initial training, regular simulator training, refresher training.

* **Organisation and management**: Management for safety, organisation of maintenance and testing, shift organisation and scheduling, human aspects of quality assurance.

NII AND HUMAN RELIABILITY

The safety targets for a station like Sizewell 'B' are that the frequency of a limited release of radioactivity to the environment should not exceed 1 in 10,000 per reactor year, and that the frequency of a large uncontrolled release should not exceed 1 in 1,000,000 per reactor year. The licensee constructs fault and event trees, to demonstrate that significant initiating events, and likely combinations thereof, do not breach these limits. The current policy of NII is that comprehensive quantification of human reliability is not feasible; nevertheless, it is necessary to show that all required human safety actions are within normal operator capabilities, and to have as much evidence as possible that operator error, passive or active, will not seriously qualify the outcomes of the fault analyses.

Of course, it is very doubtful that human error rates *per se* are compatible with the frequency goals set out above, particularly taking into account the varieties of human involvement in operation, supervision, maintenance, testing, engineering modifications, and management activities on the station. There has to be considerable dependence on highly reliable automatic safety systems, particularly for preserving reactor safety in abnormal conditions or after a major fault has developed. In addition, principle 124 (NII, 1979) requires that the safety case should not rely on any operator intervention for 30 min. after a major fault or reactor trip (automatic shut-down). The operator *is* expected to monitor the series of operations of the safety systems during this period, and to reinforce the series if required. Obviously, this implies exhaustive analysis, at the design stage, of his possible errors or misperceptions. It may be necessary to prevent any operator actions which threaten the success of the automatic systems. Progress is being made in analysing and solving such problems.

However, there are also the pervasive human influences through maintenance and testing of safety systems, and through the general management and supervisory climate of the plant. There are further problems of complex cognitive errors, which are difficult to predict and for which the interaction with complex plant and control systems is very intractable. Major nuclear accidents demonstrate these effects, and safety assessment should take account of them.

The resolution of modern safety goals and such human influences is a major concern. In the present state of the art, we can make only qualitative attempts. This qualitative assessment of human reliability is absolutely important in itself, and it should lay the foundations for developments in more quantified approaches more compatible with trends towards probabilistic analyses of system safety and reliability.

REFERENCES

HSE, 1986, HM Nuclear Installations Inspectorate: Safety Audit of BNFL Sellafield 1986, (HMSO).

NII, 1979, Safety assessment principles for nuclear power reactors. HSE - Report HA5 (HMSO).

NII, 1982, Ergonomics/human factors in the design and safe operation of pressurised water reactors. Report NII-ONSWG(82)P.1.

NII, 1983a, Safety assessment principles for nuclear chemical plant. Health and Safety Executive.

NII, 1983b, Sizewell 'B': A review by HM Nuclear Installations Inspectorate - Supplement 13: Human Factors. NII 01, SUPP 13.

Acknowledgement

I am grateful for discussions with Mr. J. L. Petrie. The views expressed herein are not necessarily those of the Inspectorate.

**AN INTEGRATED APPROACH TO IMPROVING
THE PERFORMANCE OF PROCESS OPERATORS**

D. VISICK

Warren Spring Laboratory
Stevenage, Hertfordshire SG1 2BX

Three Mile Island, Flixborough and more recently Bhopal and Chernobyl make the news, but many relatively safe process plants pay significant penalties for operator error each year. Work at Warren Spring Laboratory has shown that simply improving display design does not get to the root of the problem, which is a lack of real understanding of the process by operators. This problem is currently aggravated by technology, and will never be solved by it. Here, we present an integrated approach to designing the control room information environment covering display design, training and advisory systems.

INTRODUCTION

It has taken only a decade or so to reach the point where almost every new process plant built in the UK has at least a partly automatic control system, probably based on a central computer, and a man-machine interface built around visual display units. The latter of these innovations is the most obvious to the casual observer, and it is perhaps for this reason that display design has been researched exhaustively, and at the expense of other factors which clearly influence the performance of the man-machine system. These factors have more to do with the fundamental nature of the task than with the way in which process information is presented. This task has changed from one of direct manual process control, to supervision of an automatic system which itself controls the process. In the engineer's ideal world, this would be no task at all, since the process control computer would be infallible. However advances in control technology have been accompanied by advances in plant complexity, so that overall system unreliability has been preserved. What has changed is that the fraction of the process which designers do not fully understand - and therefore cannot automate - incorporates ever more complex problems. Yet it is precisely this fraction of the process which gives rise to unpredictable faults which the control room operator must cope with.

Most advances in interface design have been aimed at reducing operator error when extracting data from the system or changing its state. I would suggest that such errors are infrequent, and typically inconsequential relative to the far more fundamental conceptual errors which operators make when diagnosing process faults. In the nuclear industry, this fact is recognised and research is increasingly centred on training and advisory systems. Solutions, however, are typically costly (expert alarm analysis systems) and plant specific (high fidelity training simulators). Like research in the nuclear industry, the new research programme at WSL starts from the assumption that the human operator is there to perform occasional but important fault diagnosis and recovery tasks, and that there are many ways to help him perform at his best. The three major features of our approach are that we are aiming at general, practical operator aids; that all types of aid must be integrated so that they complement each other; and that the object is to <u>assist</u> the human operator rather than replace him.

WHY DISPLAYS ONLY GO SO FAR

When VDUs started to appear in control rooms, their quality was very low in comparison with today's high resolution colour monitors. Thus several problems associated with poor visual images came and went as available hardware improved. Another class of problem is associated with the flexible features of VDUs: Information content, layout and colour. System manufacturers, as well as the human factors community, have helped to increase awareness of good design techniques to the extent where truly bad design is the exception rather than the rule, and most display systems are merely sub-optimal in certain respects. This does not necessarily matter, since a display system once installed becomes a relatively fixed part of the information environment, and long term users - which most are - are able to adapt astonishingly well to quite major deficiencies.

Our own research and experience at WSL testifies to this fact. Experiments designed to investigate the effects of display density or 'clutter' indicated that varying visual or information density themselves had little effect as long as a few simple rules relating to display structure and layout were observed (Umbers & Visick, 1984). These studies used experienced operators as subjects, and together with other similar studies highlighted a basic problem: Any performance improvements that can be obtained by improved display design are invariably dwarfed by other factors which might be collectively termed 'individual differences'. One could appeal to individual differences in visual ability or other 'innate' factors, which would imply the use of selection tests as the best starting point for performance improvement. However while these studies were ostensibly based on display reading tasks, all relied on a reasonable grasp of the underlying process for successful performance. Later research at WSL indicated that deficiencies here, rather than in display reading, is the major obstacle for process operators during fault diagnosis.

STORING UP PROBLEMS

Very often in a new plant, there is not much new about the plant, but the control system is very different. Process operation is a comparatively low mobility job, so operators often go from the old plant to the new, or from plant maintenance to operation. In either case, they will already understand the plant and process, and will rely on the process information system only for current status information. However in time, a new generation of operators will be trained on automatic systems with VDUs. The restrictions imposed by the VDU based information system, being a limited window on the plant, are likely to become more apparent if it is the major means by which the operator learns about the plant. I have recently visited one plant where the operators admit that if they had not been trained on an old manual system, they would not be able to use the new automatic system.

This potential problem would be aggravated by the low investment in training in the UK. This means that the diagnostic skills of modern process operators are neither trained nor properly evaluated. On-the-job experience is the preferred method, but if not adequately controlled, this may reinforce misconceptions and poor practice. Simulated fault diagnosis experiments on a computer-controlled process plant (Visick et al, 1985) showed considerable deficiencies in operators' comprehension of the process, which was not correlated with the complexity of the problems. Various solutions have been proposed to ameliorate the deleterious effects on skill retention caused by the modern supervisory control task. These range from job enrichment or rotation (Engelstad, 1972) to computer-based game-playing intended to maintain a high level of fault diagnosis skill. Setting aside for the moment the difficulty of implementing some of these concepts, they do not address the major problem of how to train and advise operators for a task when you cannot define what that task will be.

TRAINING FOR FAULT DIAGNOSIS

Good fault diagnosis performance, by definition, cannot be trained and evaluated in the normal way. This is because the most intractable faults are not well enough understood to allow training in their specific solution. Given this fact, there are several possible approaches. One is to train 'general' fault diagnosis techniques, aimed at helping the operator think in an appropriate manner and avoid some of the more common pitfalls associated with human problem solving. Another is to instil a thorough knowledge of the plant and process, and all its cause-consequence relationships, in an effort to help the operator trace faults from first principles. Yet another method is to expand ones knowledge of rare and possible faults and simulate them in the hope that if they occur in practice, the operator will have learnt either by rote or by deeper understanding how to cope with the fault. These methods can be complementary, although I would argue that if a fault can be predicted someone will develop an automatic system to cope with it. The most practical approaches thus appear to involve teaching the operator <u>general</u> diagnosis skills and <u>particular</u> knowledge of normal plant operation, in the hope that he will infer the missing link - understanding abnormal plant operation.

Quite a lot is already known about the strengths and weaknesses of human problem solving. Particularly limited is the ability to assign weights and probabilities to multivariate evidence (Friedman et al, 1985). Research at WSL will initially concentrate on a detailed description of the fault diagnosis task in the modern computer controlled plant, highlighting differences and similarities with better researched areas such as electronic systems fault diagnosis. It is already clear that the range of skills required will vary considerably, depending on such factors as the level of automation and the nature of the control system. The emphasis will then turn to development of specific training tools aimed at teaching and reinforcing the individual skills. Where these are likely to be plant specific, the difficulty will lie in making the training system easily adaptable to local requirements. This has previously been a major obstacle, which we hope to overcome by decomposing the total domain of skill and knowledge into smaller units, only some of which will require plant specific training.

ADVISORY SYSTEMS

This term has been applied to almost any job aid which is used on-line, and which is directly relevant to a current problem.

Here, we are restricting the meaning to systems which provide assistance in the use of data, rather than the data itself. The latter is merely an information system. The advent of usable artificial intelligence techniques has rekindled interest in ambitious goals such as expert fault diagnosis systems (Moore, 1985) and alarm analysis systems (Hoenig et al, 1982). Except in certain applications where development of costly one-off systems is justified, these approaches are doomed to failure for the same reasons which make the original control systems fallible: There is not enough information to put in them. In a sense, an incomplete advisory system is worse than no advisory system, particularly if it cannot tell you the limits of its competence.

The new WSL research programme starts from the assumption that it is probably easier, and certainly better, to _assist_ rather than _replace_ the human operator. As a matter of principle, one should not attempt to mimic the best developed and most sophisticated features of human thinking, but to improve on the worst ones. A basic tenet of the programme is that all research streams should be complementary. Thus where it appears unlikely that a particular skill can be adequately trained, only then should the use of an advisory system be considered. For example, it should be better for an intelligent system to make probabilistic judgements about hypotheses generated by an operator, than for the operator to make probabilistic judgements about alternative scenarios put forward by a machine. The generation of plausible scenarios should be achievable by an adequately trained operator, and practice at doing so would be essential. Good human probability estimation is rare.

It is too early to say if the optimum level of operator advice would require machine intelligence or not, but this is one avenue which is being explored. Another possibility is to merely expand the capability of the plant information system so that it includes operating procedures as well as dynamic process data. Several users and manufacturers are actively exploring this idea.

SUMMARY

Human factors research in process control has been heavily slanted towards high risk industries, and results have not always been generally useful. Research is needed to take the majority of industry out of the knobs and dials era. The operator's task is now restricted largely to supervision of automatic systems, which periodically fail. There are far more influences on operator performance in this situation than just display design. The main issue facing plant owners now is how to generate and maintain adequate fault diagnostic skills in the operating team, and how to plug the gaps when this is not possible.

REFERENCES

Engelstad, P. (1972) 'A sociotechnical approach to problems of process control' in Davis & Taylor (eds) 'Design of Jobs' Penguin.

Freidman, L., Howell, W. & Jensen, C. (1985) 'Diagnostic judgement as a function of the preprocessing of evidence' Human Factors 27(6), pp665-673

Hoenig, G., Umbers, I. & Andow, P. (1982) 'Computer based alarm systems' IEE Conference on trends in on-line computer control systems, Warwick, 5-8 April. IEE, London. pp88-93

Moore, R. (1985) 'Adding real-time expert system capabilities to arge distributed control systems' Control Engineering, April 1985.

Umbers, I. & Visick, D. (1984) 'The effect of display density and related factors on the reading of VDU process displays' WSL report CR 2615 (CON)

Visick, D., Umbers, I. & Simpson, M. (1985) 'A study of current and potential methods of displaying sequence information in process plants' WSL report CR 2764 (CON).

APPLIED HUMAN RELIABILITY: QUO VADIS?

B. KIRWAN

Human Reliability Associates Ltd
1 School House, Higher Lane, Dalton
Parbold, Lancashire   WN8 7RP

This paper considers the current status and probable future directions of human reliability assessment, primarily in its role of assessing risk in large, complex systems. The future directions are postulated as probable responses to problem areas currently facing human reliability analysts. These areas require research and development because they are likely contributors to the risk of accidents, but are not yet addressed by current human reliability assessment methodology. Four areas are discussed: low technology risk; cognitive/diagnostic errors; management and organisational contributions to risk; and software reliability.

Introduction

The need for human reliability assessment

Recent accidents such as Bhopal and Chernobyl have shown the importance of human error in high risk systems. It is necessary to assess and prevent such accidents, and therefore to assess and reduce human errors in such systems. This is the primary application area of human reliability assessment (HRA).

The risk assessment of a large system (e.g. a nuclear power plant), requires the modelling of 'what can go wrong', usually in terms of equipment failure and human operator error. Although too detailed to present here, risk assessment essentially results in a set of probabilities of different accidents based partly on hardware failure probabilities, and partly on human error probabilities. It is then possible to quantify the likely human error contribution to an accident, and determine if means to reduce error (and hence risk) should be implemented. Thus, HRA has three parts: identification of errors; quantification of the probabilities of these errors; and specification of error reduction measures.

The current state of the art

Human Error Identification

HRA has been primarily involved with the analysis of procedural execution errors (e.g. failure to open a valve when required to do so, etc.), and the specification of what errors are likely to occur has been left largely up to the analyst. There is currently a move therefore both to develop more systematic methods, and to base them upon theoretical models of error causation, or underlying psychological theories of human action or error. One such approach is the Systematic Human Error Rate Prediction Approach (SHERPA: Embrey, 1986a).

Human Error Quantification

Since the early 1960's, much of human reliability assessment (HRA) has been concerned with the quantification of human error probabilities (HEPs), for a wide variety of tasks (e.g. opening a valve etc.) This trend continued in the years following the Three Mile Island accident in 1979, which spawned the development of several HRA techniques. Some of the more well-known techniques include the Technique for Human Error Rate Prediction (THERP: Swain & Guttmann, 1983), which uses a data base of HEPs for different tasks, and a set of rules for extrapolating from this data base to any nuclear power plant operator tasks. Another technique, SLIM-MAUD (Subjective Likelihood Index Method using Multi-Attribute Utility Decomposition: Embrey et al, 1983), uses a method of structuring expert judgement to generate HEPs. This expert judgement approach has been developed partly because of the lack of well-founded human reliability data, and also to quantify error rates in more complex scenarios, such as in emergency diagnosis in a power plant.

Another technique developed more recently is the Human Error and Reduction Technique (HEART: Williams, 1986) which assesses HEPs on the basis of the presence of a set of performance degrading factors. The quantified effects of these factors are based largely on research data available in the ergonomics literature, which have been interpreted and added to by the author. Also of note is the Time Reliability Correlation method (Hannaman et al, 1985) which quantifies human diagnostic performance almost solely as a function of time available for diagnosis. This technique is receiving much interest in the USA today. These techniques and others all calculate HEPs, using different assumptions and methods, and this has been their primary purpose.

Error Reduction

More recently HRA techniques are being used to assess the cost-effectiveness of various ergonomics improvements for a system, and this trend is likely to continue in the near future. This is achievable since the techniques allow the assessor to alter the human error probabilities, and thus the system risk, as a function of factors affecting performance. It is therefore possible, for example, to calculate how much safer an ergonomic layout of controls will make a system, or whether a higher level of safety could be obtained by improving the procedures, etc. This is an important development since it may hopefully allow more ergonomics into the design process, and hence into the design of complex systems, making them safer in general.

Future Directions

Each of the following sections discuss an area of human unreliability which can affect system risk, but which is not currently addressed by HRA techniques. These are therefore predicted to be probable and useful future directions for human reliability.

1. Low Technology Risk

The HRA field has mainly concerned itself with the high risk, high technology industry sector, including nuclear power plants, chemical plants, etc. There exist however, a large number of other high risk, lower technology sectors, e.g. mining. These industries often incur a high risk via a large number of 'small' accidents (e.g. 1 or 2 fatalities), rather than high risk technology industries where the high risk is caused by a very small probability of a very large accident. This is clearly an area where applied human reliability should be able to help reduce risk (see also Collier and Graves, 1986).

2. Cognitive Errors and Misdiagnosis

If the operating crew in a plant misdiagnose a situation, they may be slow to realise they have made a 'mistake' or incorrect diagnosis. They may reinterpret system feedback within their mental view of what they think the problem is, rather than what it actually is (also called 'mind set'). This is in fact what occurred during the Three Mile Island accident in 1979, and can result in the human operators actually defeating safety systems and making matters worse than if they did nothing at all. Also, evidence suggests that there is a low probability of operators recovering their mistakes at all (Woods, 1984).

The 'modelling' and analysis of cognitive errors is a difficult problem, and one that has not yet been resolved. Some work has been carried out on the psychology underlying cognitive errors (Reason and Embrey, 1985), in an attempt to predict the forms cognitive errors are likely to take. The Influence Modelling and Assessment System (IMAS: Embrey, 1986b) is a system which can be used to elicit mental maps of operators' perceptions of causal relationships in a nuclear power plant situation. These maps could be used to consider what misdiagnoses may occur. Also, in the Artificial Intelligence field, an attempt is being made to simulate the cognitive operator with an expert system, and estimate what errors will occur (Woods and Roth, 1986). These and other developments may solve the problem of cognitive errors.

3. Management, Organisational, and Sociotechnical Contributions to Risk

Three of the most recent and salient accidents are the Bhopal (1984), Chernobyl (1986), and Challenger Space Shuttle (1986) disasters. Each of these contained a significant human error contribution, without which these disasters would not have occurred. However, it is arguable that no current HRA techniques would have predicted them. This is because the types of error leading to the accidents were fundamentally neither procedural nor diagnostic in the conventional sense. Rather, all three involved a management and organisational error component. The Bhopal plant was possibly subject to economic pressures, resulting in management decisions which may have been more in favour of production than safety (since there is no official Bhopal Report the evidence is inconclusive, but see Bellamy, 1986). The Challenger Space Shuttle disaster also appeared to contain a significant element of management-type decision error, also influenced by economic considerations (Rogers et al, 1986). Lastly, the Chernobyl disaster (1986) contained such a bizarre series of events affected by low-level management that, had an analysis predicted this scenario prior to the accident, it probably would have been treated with derision.

It is clear that HRA must in the future consider not only the operators in the control room, but also the management running the plant, since the latter can have a profound effect on safety.

Several types of errors appear to occur when dealing with small groups, i.e. at a 'lower' management and organisational level than described above. This cause or group of causes may be categorised as individual, organisational, and sociotechnical factors (Bellamy, 1983), e.g. social pressures, personality conflicts, etc. These factors are rarely assessed in human reliability assessments, yet clearly can influence risks. Techniques should therefore be developed which identify, quantify, and specify how to reduce these types of error.

4. Software Reliability

Many complex high risk systems depend on computer control, with the operators in monitoring and supervisory positions. These systems are critically dependent on error-free or error-tolerant software, and since most software is still generated by programmers, human error in this area can constitute a significant source of risk. There is therefore a finite probability of a human error in software production or maintenance leading or contributing to a system accident. Examples of software-error induced accidents can be found in the development of programmable electronic systems. Software errors in these systems have been known to cause accidents, e.g. with robot-arms colliding with personnel.

A methodology should be developed to evaluate what software errors can occur, and how and why they do so, leading on to how they may be detected and prevented.

In summary, HRA has been mainly used to quantify human error probabilities for risk assessments. In the near future likely directions will be concerned with developing more robust and valid methods of error identification and quantification. HRA is also likely to play a larger role in the design process by virtue of its utility for making cost-benefit evaluations of ergonomic inputs into the design. In the medium term HRA should address the low technology sector, and in the longer term develop methods to assess cognitive errors, management, organisational, and sociotechnical errors, and software errors. If these goals are achieved, HRA will be a far more powerful tool for the assessment of risks and the prevention of accidents.

References

BELLAMY, L.J. (1983). Neglected individual, social and organisational factors in human reliability assessment. Proceedings of the 4th National Reliability Conference. Reliability 85, NEC, Birmingham, July.

BELLAMY, L.J. (1986). The Safety Management Factor: An Analysis of Human Aspects of the Bhopal Disaster. Paper presented at the 1986 Safety and Reliability Society Symposium, Southport, September 26.

COLLIER, S.G. and GRAVES, R.J. (1986). Improving Human Reliability: Practical Ergonomics for Design Engineers. In: Proceedings of the 9th Advances in Reliability Technology Symposium, University of Bradford, 2-4 April.

EMBREY, D.E., et al (1983). SLIM-MAUD, An Approach to Assessing Human Error Probabilities using Expert Judgment, USNRC, NUREG/CR-3518, Washington DC-20555.

EMBREY, D.E. (1986a). SHERPA: A Systematic Human Error Reduction and Prediction Approach. Paper presented at the International Topical Meeting on Advances in Human Factors in Nuclear Power Systems, Knoxville Tennessee, April 20-23.

EMBREY, D.E. (1986b). Approaches to aiding, and training operators diagnoses in abnormal situations. Chemistry and Industry, 7 July.

HANNAMAN, G.W. et al (1985). A model for assessing human cognitive reliability in PRA studies. In: Proceedings of the 3rd IEEE Conference on Human Factors and Power Plants. Institute of Electrical and Electronic Engineers, New York.

REASON, J. and EMBREY, D.E. (1985). Human Factors Principles Relevant to the Modelling of Human Errors in Abnormal Conditions of Nuclear and Major Hazardous Installations. Report for the European Atomic Energy Community, Human Reliability Associates Ltd., Lancs.

ROGERS, W.P. et al (1986). Report of the Presidential Commission on the Space Shuttle Challenger Accident, June 6.

SWAIN, A.D. and GUTTMANN, H.E. (1983). A Handbook of Human Reliability Analysis with Emphasis on Nuclear Power Plant Applications. USNRC, NUREG/CR-1278, Washington DC-20555.

USSR State Committee on the Utilisation of Atomic Energy (1986). The Accident at the Chernobyl Nuclear Power Plant and its Consequences. Information compiled for the IAEA Experts' Meeting, 25-29 August, Vienna (Part I).

WILLIAMS, J.C. (1986). HEART: A proposed method for assessing and reducing human error. In: Proceedings of the 9th Advances in Reliability Technology Symposium, University of Bradford, 2-4 April.

WOODS, D.D. (1984). Some Results on operator performance in emergency events. In: Proceedings of the Institute of Chemical Engineers Symposium on Ergonomics Problems in Process Operations, 1 ChemE Symposium Series No. 90, Rugby.

WOODS, D.D. and ROTH, E.M. (1986). Models of Cognitive Behaviour in Nuclear Power Plant Personnel: A Feasability Study. USNRC, NUREG/CR-4532, Washington DC-20555.

Human Performance

ANXIETY PRIOR TO AND DURING DECOMPRESSION

M.H. USSHER[*] and E.W. FARMER[**]

[*] Plessey Research Roke Manor Ltd
Romsey, Hampshire SO51 0ZN

[**] RAF Institute of Aviation Medicine,
Farnborough, Hampshire GU14 6SZ

Task performance, heart rate, and self-reported anxiety were recorded prior to, during, and after decompression. It was found that the emotionality component of anxiety was raised during the period 30 min prior to decompression, whereas the worry component was unaffected. Elevation of heart rate and degradation of performance on a logical reasoning task were noted immediately prior to and during decompression. These results are inconsistent with previous evidence that the effects of anxiety on cognitive task performance are solely attributable to worry. Implications of the present findings for research on decompression are discussed.

1. INTRODUCTION

Anxiety is an important determinant of performance in threatening situations. In some circumstances, such as deep sea diving or parachuting, anxiety-induced performance decrement may have hazardous consequences; in other circumstances, such as public speaking or examinations, in which physical danger is not involved, the consequences may nevertheless be significant.

Anxiety associated with the anticipation of a threatening situation might be expected to produce a performance decrement even before the stressor is experienced. In the present study, this notion was investigated by recording performance prior to decompression in a hypobaric chamber. Green and Morgan (1985) presented evidence that performance decrement under decompression may not be attributable solely to hypoxia. Performance on a logical reasoning task was poorer in a hypobaric chamber at approximately ground level than in a classroom. The authors speculated that anxiety associated with the anticipation of decompression was responsible for this effect.

Two major components of anxiety have been postulated: worry and emotionality. Previous research has suggested that the former is more closely associated with decrements in both cognitive and motor performance (Morris et al 1981). In the present study, measures of worry and emotionality, heart rate, and task performance were recorded prior to, during, and after decompression.

2. METHODS

Experimental Design

A 'time to a significant event' (T.S.E) paradigm was designed, involving a series of measurement sessions across the significant event period. An experimental group of 6 decompression-naive subjects completed a battery of tests on six occasions: 6 days before decompression (measure 1, baseline), 24 hours before decompression (measure 2), 30 mins before decompression (measure 3), in the hypobaric chamber at approximately ground level 10 mins prior to decompression (measure 4), at a decompression level of 9 000 ft (measure 5), and 24 hours after decompression (measure 6). A control group of 6 decompression-naive subjects was tested at equivalent times in the absence of any impending decompression.

Subjective Measures

State anxiety, was measured by the Revised Worry Emotionality Inventory (Morris et al 1981) which gives separate scores for the factors of worry and emotionality. The worry factor is a measure of the perception of negative thoughts in anticipation of a future event, whereas the emotionality factor is a measure of the perception of the physiological and affective condition associated with these thoughts. The experimental subjects responded with reference to the impending decompression. The control subjects responded without reference to any specific event.

Physiological Measure

Pulse rate, taken by hand, was used as a measure of autonomic arousal. This measure has previously been shown to rise significantly prior to the deep sea diving event (Baddelely and Idzikowski 1985).

Performance Measures

The performance task battery included both cognitive and motor tasks. Cognitive tasks with varying degrees of memory load were selected. The low memory-loaded version of the letter transformation task was used (Hamilton et al 1977). Subjects were presented with a single letter and were asked to make a forward transition through the alphabet. The length of the transition was either 1, 2, 3, or 4. For example, G+1 was transformed to H, W+2 to Y, B+3 to E, and L+4 to P. The logical reasoning task (Baddeley 1968) was used to measure grammatical transformation time. The subjects were presented with with a sentence describing the order of the letters AB or BA. They were required to tick whether the sentence gave a true or false description of the accompanying pair AB or BA. For example, the sentence "A follows B" is true for the letter pair BA, whereas the the sentence "A does not follow B" is false. The task included all 32 possible combinations of five binary variables: positive or negative, active or passive, precedes or follows, A or B mentioned first, and letter pair AB or BA following the sentence. For both the letter transformation and logical reasoning tasks, the order of the items was randomized independently for each measurement session.

Motor tasks with varying degrees of motor coordination were selected. Learned handgrip was used as a measure of 'motor intensity'. It has previously been demonstrated that subjects overestimate a criterion grip strength on a hand grip dynamometer during the period immediately prior to sports competition (Ussher & Hardy 1986). During the learning period, the subjects were required repeatedly to produce a criterion grip strength of 15 kg, the experimenter providing knowledge of results (K.R.). During actual testing, the subjects were asked to produce the criterion score over three successive trials in the absence of K.R.

Compensatory tracking was included as a measure of motor adjustment. whilst tick response was included as an unobtrusive index of motor production (Wing & Baddeley 1978).

3. RESULTS

Analysis of Variance revealed significant interactions between experimental/control group and time for emotionality ($F(4,40) = 5.56$, $p < 0.001$), pulse rate ($F(4,40) = 3.99$, $p < 0.008$), and learned handgrip ($F(4,40)$

= 6.26, p < 0.001). For each of these measures Dunnett's t-test was used to compare the three critical times (3,4,5) with the combined mean for the control times (2,6). There was a significant elevation of emotionality for the experimental group throughout times 3, 4, and 5, and a significant elevation of pulse rate at times 4 and 5. Although the interaction between experimental/control group and time for logical reasoning was not significant (F(4,40) = 1.55, p < 0.204), Dunnett's t-test revealed a significant depression in performance for the experimental group at times 4 and 5. Plots of these data are shown in figure 1.

Figure 1. Graphs showing interactions between experimental/control group and time for (a) emotionality, (b) pulse rate, (c) worry, and (d) logical reasoning.

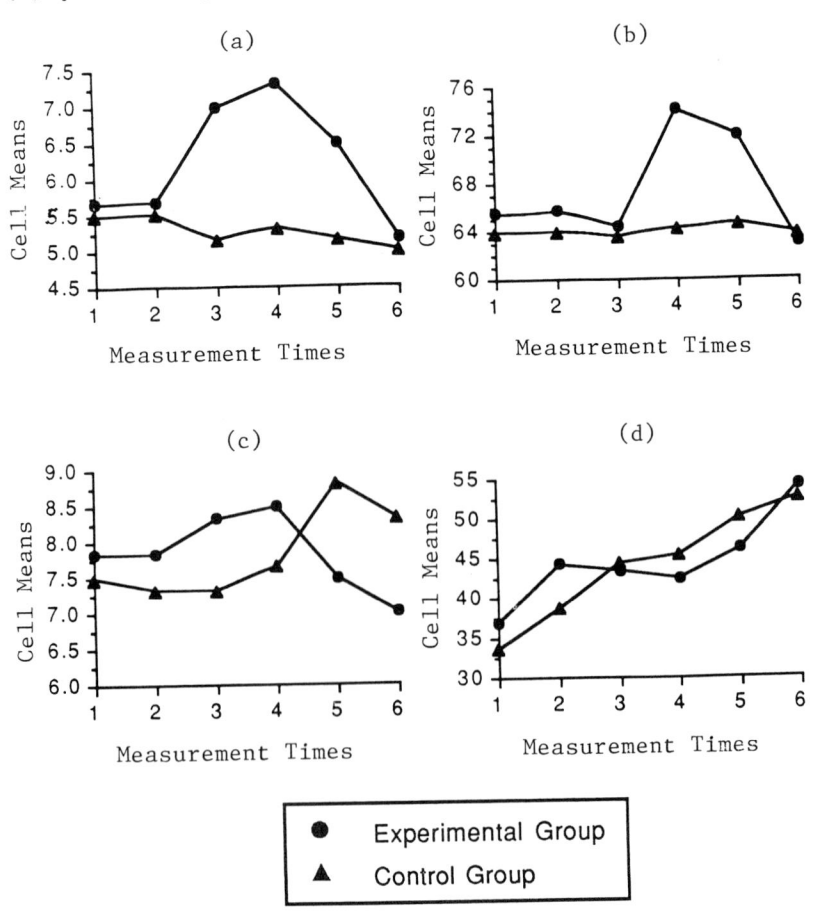

4. DISCUSSION AND CONCLUSIONS

There is evidence from subjective, physiological, and task performance measures for the presence of anxiety both prior to and during decompression. Baseline measures taken prior to decompression experiments may therefore be confounded by the effects of anxiety. Furthermore, the performance decrement found at high altitude may not be due solely to the effects of hypoxia.

These findings confirm the previous reports of a dissociation of anxiety (Spiegler et al 1968) by showing independent trends for worry and emotionality across the pre-decompression period. The T.S.E. paradigm could be used to demonstrate this distinction in other situations, and has a number of practical implications. For instance, the predominant mode of anxiety at specific times before a significant event (eg sports competition, dance performance, dramatic performance) could be matched with the most appropriate anxiety regulation procedure. In systematic desensitization, phobic anxiety could be measured across the time to the presentation of the object of the phobic's fear.

The finding of a decrement in cognitive task performance in association with subjective reports of elevated emotionality but an unchanged level of worry is a new finding. Previous research has suggested that the negative effect of anxiety on cognitive performance is due solely to the worry factor; the present study suggests otherwise.

The accuracy of hand grip tension showed a substantial decrement at the baseline measure relative to the learning session. A more extensive learning period was therefore evidently required. In the previous application of this task (Ussher & Hardy 1986), the oarsmen used as subjects may have learned the task very rapidly due to their familiarity with the control of muscle tension.

Although tick length did not vary between the experimental and control groups, observations of the ticks suggested that greater pressure had been exerted during the critical times (3,4,5). A measure of tick pressure using a response sheet overlaying a pressure sensitive pad may represent a more suitable variation of this task.

This study has indicated that anxiety may influence performance even prior to the application of a stressor. Researchers interested in the 'direct' effect of a stressor should therefore ensure that the confounding effect of anxiety has been removed.

ACKNOWLEDGEMENT

We would like to thank Dr. Paul Barber for his advice throughout the study.

REFERENCES

Baddeley, A.D., 1968, A 3-min reasoning test based on grammatical transformation, Psychonomic Science, 10, 341-342.

Baddeley, A.D. & Idzikowski, C., 1985, Anxiety, manual dexterity and diver performance, Ergonomics, 28, 1475-1482.

Green, R.G. & Morgan, M.B., 1985, The effects of mild hypoxia on a logical reasoning task, Aviation, Space and Environmental Medicine, 56, 1004-1006.

Hamilton, P., Hockey, G.R.J. & Rejman, M., 1977, The place of the concept of activation in human information processing theory: An integrative approach, in Attention and Performance, Vol. VI edited by S. Dornic (Erlbaum).

Idzikowski, C.I. & Baddeley, A.D., 1983, Waiting in the wings: apprehension public speaking and performance, Ergonomics, 26, 575-583.

Morris, L.W., David, M.A & Hutchings, C.H., 1981, Cognitive and emotional components of anxiety: Literature review and a revised worry emotionality scale, Journal of Educational Psychology, 73, 541-555.

Spiegler, M.D., Morris, L.W., and Liebert, R.M., 1968, Cognitive and emotional components of test anxiety: temporal factors, Psychological Reports, 22, 451-456.

Ussher, M.H & Hardy, L., 1986, The effect of competitive anxiety on a number of cognitive and motor sub-systems, Proceedings, 1st Annual Conference of the British Association of Sport Sciences.

Wing, A.M., & Baddeley, A.D., 1978, A simple measure of handwriting as an index to stress, Bulletin of the Psychronomic Society, 11, 245-246.

MINOR ILLNESSES AND PERFORMANCE EFFICIENCY

A. SMITH and K. COYLE

MRC Perceptual and Cognitive Performance Unit
University of Sussex, Brighton
Sussex BN1 9QG

Results from our research programme on the effects of experimentally-induced respiratory virus infections on performance have led to the following conclusions. First, these minor illnesses have signficant effects on performance efficiency, although the exact nature of the impairment will depend on the type of virus and the task being performed. For instance, influenza impairs the ability to attend and colds impair motor skills. Second, performance impairments have also been found with sub-clinical influenza infections and infections with certain cold viruses. These results have important implications for occupational safety and efficiency. Studies in progress are also outlined in the paper.

INTRODUCTION
Recent evaluations of the importance of respiratory disease show that acute infections and their consequences account for a substantial proportion of all consultations in general practice, and are the major cause of absence from work and education. Despite the frequency with which such illnesses occur there has been no research on their effects on the efficiency of performance. This has, in part, been due to the difficulties inherent in carrying out such studies. These difficulties have been overcome in the present project by studying the effects of experimentally-induced infection and illness at the MRC Common Cold Unit, Salisbury. The aims of the present paper are to describe the methodology used, report the results obtained so far, and to describe studies which are in progress or planned.
Studies of naturally-occurring illnesses could provide useful information but it is difficult to predict when they will occur,and it is often unclear which virus pro-

duced the illness (there are over 200 viruses that produce colds). Such studies would also only enable one to examine the effects of clinical illnesses. It is possible that sub-clinical infections may also influence behaviour and these can only be identified by using the appropriate virological techniques. The methodology employed by the Common Cold Unit overcomes these problems. The crucial features of the routine may be summarised as follows:

(1) Volunteers stay at the Unit for a period of ten days, being housed in groups of two or three and isolated from outside contacts.

(2) There is a quarantine period of about 3 days prior to virus challenge. Any volunteers who develop colds in this period are excluded from the trial.

(3) Each volunteer is assessed daily by the Unit's clinician who assigns a score based on objective measures such as temperature and number of tissues used, and on the presence of other symptoms (see Beare & Reed 1977 for full details of the scoring system).

(4) Nasal washings are taken so that virus shedding can be measured. Pre- and post-challenge antibody status can be measured from blood samples taken at these times.

The behavioural effects of challenge with the following viruses have been studied (1) Rhinoviruses, (2) Coronavirus, (3) Respiratory syncytial viruses (the symptoms produced by these are, in adults, indistinguishable from a common cold), and (4) influenza A and influenza B viruses. The clinical pattern and pathogenesis of the illnesses produced by the different viruses is quite variable. However, the major distinction is between colds and influenza, with colds producing mainly local symptoms (an increase in nasal secretion) whereas influenza also gives rise to systemic effects, such as fever, myalgia and malaise.

Many people feel that they already know about the behavioural effects of colds and influenza, and argue that it is not necessary to carry out empirical research on this topic. One view is that, almost by definition, if you are ill then you will perform less efficiently than normal. Another is that the illnesses are very minor and any behavioural effects are probably too small or transitory to deserve consideration. Recent research on stress has shown that different types of stress produce different effects on performance. Different illnesses might, therefore, affect performance in different ways.

We have carried out experimental studies to test which of these alternative views is correct. Two main experimental methods have been used. The first has consisted of the volunteers carrying out a battery of paper and pencil tests 4 times a day (at 8.00, 12.00, 17.00 and 22.00) on nearly all the days of the trial. The following tests have been used: (a) logical reasoning test, (b) search and memory tests with low and high memory loads, (c) a pegboard test, and (d) a semantic processing test. It was important to examine performance at several times of day for two reasons. First, it is well-established that performance varies over the day (see Colquhoun 1971). Second, we have found that there is diurnal variation in the severity of symptoms, with increases in nasal secretion and temperature being greatest in the early morning.

The second method of assessing performance consisted of administering computerised tests once during the pre-challenge period, once when symptoms were apparent in some volunteers, and sometimes during the incubation period. Subjects were always tested at the same time of day on all occasions, although some were tested in the morning and some in the afternoon. This meant that diurnal variation in the effects of the illnesses could be examined. These computerised tests have all been widely used to study the effects of stressors and drugs. They were selected to enable us to assess a range of functions such as attention, memory and motor skills. The tests which have been used are:-
(1) a variable fore-period simple reaction time task
(2) a five-choice serial reaction time task
(3) a numeric monitoring task (5's detection task)
(4) the Bakan vigilance task
(5) a pursuit tracking task
(6) a Sternberg memory scanning task
(7) stimulus and response set attention tasks
(8) a time estimation task
(9) a free recall task, involving both immediate and delayed recall.

RESULTS
(1) <u>Effects of clinical illnesses</u>

Smith <u>et al</u>. (in press) found that colds and influenza have selective effects on performance. Influenza slowed responses to stimuli appearing at irregular intervals but had no effect on a tracking task. In contrast to this, colds impaired tracking but had no effect on the attention tasks (variable fore-period reaction time task and number detection task). Smith <u>et al</u>. (submitted) confirmed these results using different performance tests (the motor task was the pegboard task and the attention task was the high memory load version of the search task) and different cold viruses.

Results from the five-choice serial reaction time task also show that activities involving movement of the hands are impaired by colds. In one study volunteers who developed colds following challenge with a coronavirus were significantly slower than uninfected volunteers. This result has been confirmed in a study using respiratory syncytial viruses.

Many types of behaviour involve both attentional and motor skills. The above results suggest that such activities may be vulnerable to the effects of colds and influenza. Studies using cognitive tasks have so far failed to show clear effects of colds or influenza. Further analyses are being carried out to examine individual differences in the effects of infection and illness on these tasks. Other moderating factors, such as the time of testing, are being taken into consideration. Indeed, we have some evidence that colds may amplify the diurnal changes in performance. In one study we examined the effects of changing the stimulus-response compatibility in a choice reaction time task (in the compatible condition stimuli presented on the left of the screen were responded to with the left hand, stimuli on the right with the right hand, whereas in the incompatible condition, stimuli on the left were responded to with the right hand and those on the right with the left hand). In the pre-challenge period subjects tested in the afternoon were faster but less accurate than those tested in the morning. In the post-challenge period, those with significant colds showed a much greater time of day effect than the uninfected subjects.

(2) Effects of sub-clinical illnesses

The effects of sub-clinical infection have been studied in two ways. First, performance has been examined during the incubation period, when by definition, no symptoms are apparent. Second, virological techniques have allowed us to sub-divide volunteers with no significant clinical illnesses into those who were infected with the virus and those who were not.

Smith et al. (submitted) found that volunteers with sub-clinical influenza infections were impaired on a search task. Performance was also impaired during the incubation period of this illness. However, we found no evidence of behavioural effects of sub-clinical colds on this task or on a pegboard task. However, a study using respiratory syncytial virus has shown impairments during the incubation period (the task was a five-choice serial reaction time task). Volunteers who had no significant clinical symptoms, but showed a significant antibody rise following virus challenge, were also slower on this task.

RESEARCH IN PROGRESS
(a) Mediators of the changes in performance
It is known that interferon is endogenously produced following infection with influenza virus. Interferon may produce systemic effects which could be responsible for the attentional deficits. If this is the case, then similar impairments should occur when volunteers are given an injection of interferon but no virus challenge. Preliminary studies have provided some support for this view.

(b) Drugs
A great deal of research is being carried out at the Common Cold Unit on prophylactic and therapeutic drugs. We are currently examining whether the drugs change the behavioural effects of the illnesses.

(c) Motor impairments
Studies are in progress to determine whether the motor impairments found with colds are due to peripheral effects, such as changes in muscle tone, or whether they reflect impairments in more central processes, such as response organisation.

(d) Prediction of illness from behavioural indicators
During the course of our research programme we have observed that volunteers who develop illnesses are sometimes worse in the pre-challenge period than those who are subsequently uninfected. This raises the interesting possibility that we may be able to predict illness from behavioural measures.

(e) New aspects of behaviour
We are currently examining the effects of infection and illness on cognitive functions such as comprehension and decision making.

PLANNED RESEARCH
(1) After-effects of illness
We intend to examine performance after the clinical symptoms have gone to see whether any impairments still exist, and if so, for how long. Changes in immunological indicators will also be taken and it will be possible to examine the relationship between performance and changes

in immune system function.

CONCLUSIONS

Our research programme has shown that minor illnesses do have significant effects on performance efficiency. The exact effect will depend on the activity being performed and the type of virus. Viral infection, unaccompanied by clinical symptoms, can impair performance although these effects appear to be largely restricted to influenza infections and attention tasks. These results have strong implications for occupational safety and efficiency, and suggest that detailed studies of the impact of these illnesses in real-life should also be carried out.

REFERENCES

Beare, A. S. & Reed, S. E., 1977, The study of antiviral compounds in volunteers. In Chemoprophylaxis and Virus Infections, Vol. 2, edited by J. S. Oxford (Cleveland: CRC Press), p. 27 - 47.

Colquhoun, W. P., 1971, Circadian variation in mental efficiency. In Biological Rhythms and Human Performance, edited by W. P. Colquhoun (London: Academic Press), p. 39 - 107.

Smith, A. P., Tyrrell, D. A. J., Coyle, K. B. & Willman, J. S., Selective effects of minor illnesses on human performance. British Journal of Psychology (in the press).

Smith, A. P., Tyrrell, D. A. J., Al-Nakib, W., Coyle, K.B., Donovan, C. B., Higgins, P. G. & Willman, J. S. The effects of experimentally-induced respiratory virus infections on performance (paper submitted).

EFFECT OF VISUAL-LOBE SIZE ON SEARCH PERFORMANCE ON INDUSTRIAL RADIOGRAPHS

P. O'BOYLE and T. GALLWEY

Department of Mechanical and Production Engineering
National Institute for Higher Education
Limerick, Ireland

A set of 96 radiographs of castings was assembled in three equivalent blocks. Six student subjects performed a tachistoscopic test of visual lobe size and then received training on 68 radiographs. Response times on the test set were measured for three types of radiograph within each block and an error score was calculated from wrong classifications in each block. Multivariate ANOVA showed no effect of blocks and multivariate regression showed no effect of lobe size. But simple regression gave a high correlation between lobe size and two responses for five of the subjects.

INTRODUCTION
Visual inspection has been shown over more than thirty years to be an error prone operation. On wiring and soldering Jacobson (1952) found that only 80% of faults were detected, whereas Fox (1964) found that coin inspectors only detected 55% of the defects. Later Mills and Sinclair (1976) examined knitwear inspectors who detected 50% of defects and more recently, on the inspection of saucepans, Thornton and Matthews (1981) obtained 65% detection. Obviously such performance levels are just not good enough. One type of inspection task which is growing is the use of X-rays to examine welds in pipelines and pressure vessels, for example. On humans it was shown by Yerushalmy (1969) that over 20% of the faults in 15,000 X-ray plates were missed by radiologists and Gale and Worthington (1985) reported that 20% to 30% of abnormal X-rays are reported incorrectly.
A local manufacturer of castings in a special alloy X-rays each one for voids, cracks, and so on. Because of

the critical environment of use a very cautious approach is taken. Any radiograph which suggests that there may be a doubtful area results in the casting being sent for sectioning and metallurgical examination. The net result is that the decision component of the inspection is relatively uncomplicated and the emphasis has to be placed on the search component. So it was decided to examine factors affecting this aspect.

One factor of interest is that of the visual lobe i.e. the area around the line of sight in which there is more than a chance probability of detecting a defect. Early work on this by Erickson (1964) and Johnston (1965) looked at peripheral visual acuity and search performance. Later Engel (1977) measured the visual lobe and obtained a relationship to search and Bellamy and Courtney (1981) found that search speed was correlated with visual lobe area. Gallwey (1982) obtained a significant effect of lobe size on search errors. So the effect is well established. But all these investigations used laboratory search tasks and the question remains open with regard to performance on a real task so we decided to examine it.

METHOD
Stimuli

Radiographs are retained for a period of six months before disposal. One particular product was chosen by company personnel as being typical in terms of volume and complexity. In this case twenty appear on one X-ray plate. From a study of the stock of plates it appeared that there were three main types - fault free (called Accepts), acceptable imperfects (called Seconds), and definitely unacceptable (Rejects). Individual examples of each were cut from a plate, taking care to ensure that only one imperfect or reject condition was present, if at all. A total set of 96 radiographs was assembled for search tests and consisted of three blocks, each of which contained 16 Accepts, 8 Seconds, and 8 Rejects. Another set was assembled for training.

Lobe Size Test

A computer plotter was used to produce sheets of paper containing two square brackets (e.g. [and]) at various distances apart with two Lefts, or two Rights, or one of each. These were photographed to produce positive black and white slides which were presented by means of a Lafayette Tachistoscope (Model No. 43011) set at 200 msec exposure time. The bracket subtended 11 x 4 minutes of

arc and eccentricities used were 1.8, 3.6, 5.4, 7.2
and 9.0 degrees of visual angle. A fixation slide
containing a "+" at the centre alternated with these.
There were 100 slides in completely random order with
equal numbers of all four possible combinations and
subjects had to decide whether the brackets were Same
or Different. As training for this a set of 20 exactly
analagous slides was used, the first 5 of which had a 500
msec exposure time.

Apparatus

For presenting the stimuli the company provided a
viewer which has a frosted glass vertical front backlit by
fluorescent light (S & S Medical Products, Brooklyn, N.Y.).
It was modified by putting two guide rails across the
front so that stimuli could be slid into position for
viewing. A response panel switched on the power by a
relay when the "NEXT" button was pressed and switched it
off again by means of a button for each of the product
classifications. A Lafayette clock/counter (Model No.
54035) was connected to each of the latter.

Subjects

All were student volunteers obtained from notice board
calls for subjects. They numbered six and ages ranged
from 18 to 22 (mean = 19.8). In an attempt to equalise
motivation they were paid £7.50 for attending plus a bonus
of 10p for each radiograph correctly classified and a
penalty of the same amount if wrong. They also received
a bonus of £2 if they completed the test in not more than
3/4 hour. None had previous experience of this type of
work.

Procedure

Initially subjects were tested by Orthorator to ensure
that vision was at least 6/6 (corrected if necessary).
They then did a number of tests which included that for
lobe size and this was followed by training. It used
an active progressive part approach as advocated by Czaja
and Drury (1981). It consisted of 68 radiographs of which
60 required search and feedback was given on times and
errors, except for the last 24 on which times only were
given followed by the error score at the end of each lot of
12. It took about 30 minutes to complete. After a
5 minute break they inspected the test set of radiographs.

RESULTS
Times and Errors

Actual values are given in Table 1. Each block contained 32 trials which were used to calculate the proportion of wrong classifications. A pilot experiment had shown that the three response times differed and that each was not Normally distributed. Therefore for ANOVA purposes they were logarithmically transformed and the 2 arc sin x transform was used on the error scores. Multivariate ANOVA of SPSS (Nie et al, 1975) was used to test for the effect of blocks. Three runs were used with each response time paired with the error score. Blocks were not significant so we concluded that there was no order effect.

Table 1. Summary of raw data by subjects and blocks.

Subject	Block	Lobe Score	Error Score	Accept	Second	Reject
1	1	0.56	0.125	11.055	10.117	8.332
1	2	0.56	0.2187	10.338	8.876	5.062
1	3	0.56	0.25	9.785	7.900	6.995
2	1	0.72	0.25	5.980	4.870	3.416
2	2	0.72	0.185	6.099	4.613	3.012
2	3	0.72	0.3438	5.460	4.863	3.478
3	1	0.695	0.185	10.530	6.931	3.969
3	2	0.695	0.1563	8.567	5.228	3.490
3	3	0.695	0.2187	7.752	6.610	3.884
4	1	0.73	0.3125	6.769	5.546	4.600
4	2	0.73	0.25	5.489	4.547	2.503
4	3	0.73	0.3438	5.918	5.422	3.651
5	1	0.785	0.185	12.696	11.883	9.902
5	2	0.785	0.25	16.822	12.475	9.645
5	3	0.785	0.2187	20.108	18.076	9.319
6	1	0.67	0.125	7.342	5.932	4.056
6	2	0.67	0.2187	7.794	5.115	4.051
6	3	0.67	0.25	8.009	8.275	6.458

(Response Times in sec)

Lobe Size

The average proportion correctly detected was calculated over the five eccentricities and included a value of 1.0 at zero eccentricity. Values obtained are given in Table 1. This value, and the 2 arc sin x transform of it, were used in a multivariate regression with SPSS using the same pairs as before as dependent variables. There was no significant effect. However there was doubt about a trade-off between these times and these errors, especially due to the behaviour of subject 5. So simple regressions were conducted on the data of the other subjects for the response times against lobe score. For Accepts the relationship was not significant (r = 0.868, $p > 0.05$) but it was significant for Seconds (r = 0.980, $p < 0.01$) and for Rejects (r = 0.983, $p < 0.01$). All three are plotted in Figure 1.

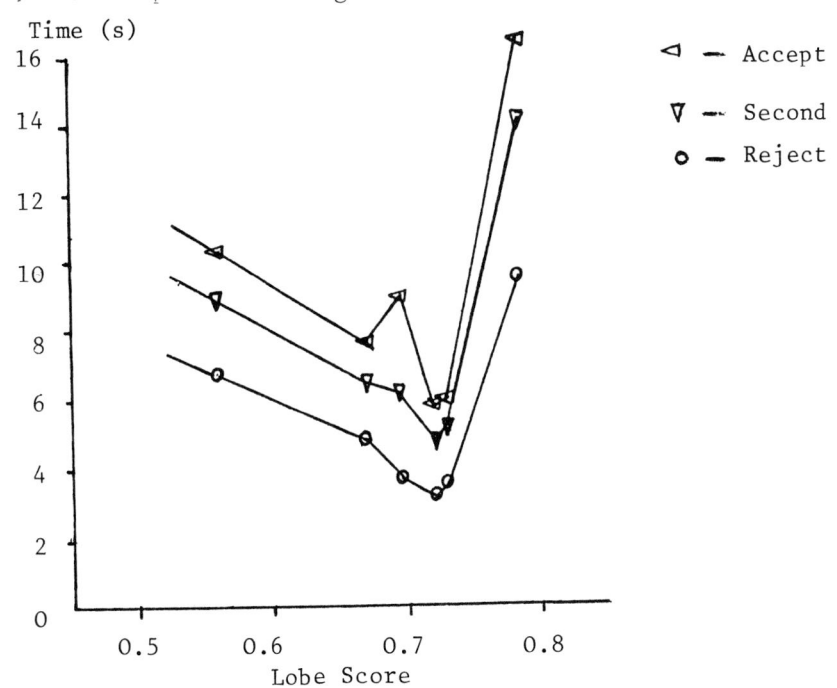

Figure 1. Mean Response times vs Lobe Score.

DISCUSSION

Consideration of the errors indicates that they were largely due to decisions and not search and so are not likely to be related to lobe size. Support for this is provided by subject 5 who could be seen to take a long time over decisions. A pure search task would eliminate this confounding factor.

REFERENCES

Bellamy, L.J. and Courtney, A.J., 1981, Development of a search task for the measurement of peripheral acuity, Ergonomics, 24, 497 - 509.

Czaja, S.J. and Drury, C.G., 1981, Training programs for inspection, Human Factors, 23, 473-484.

Engel, F.L., 1977, Visual conspicuity, visual search and fixation tendencies of the eye, Vision Research, 17, 95-108.

Erickson, R.A., 1964, Relation between visual search time and peripheral visual acuity, Human Factors, 6, 165-177.

Fox, J.G., 1964, The ergonomics of coin inspection, Quality Engineer, 28, 165-169.

Gale, A.G. and Worthington, B.S., 1985, Visual inspection of the chest radiograph, Contemporary Ergonomics, Taylor and Francis, 42-47.

Gallwey, T.J., 1982, Selection tests for visual inspection on a multiple fault type task, Ergonomics, 25, 1077-1092.

Jacobson, H.J., 1952, A study of inspector accuracy, Industrial Quality Control, 9, 16-25.

Johnston, D.M., 1965, Search performance as a function of peripheral acuity, Human Factors, 7, 527-535.

Mills, R. and Sinclair, M.A., 1976, Aspects of inspection in a knitwear company, Applied Ergonomics, 7, 97-107.

Nie, N.H., Hull, C.D., Jenkins, J.G., Steinbrenner, K. and Bent, D.H., 1975, Statistical Package for the Social Sciences (2nd Ed), McGraw-Hill, N.Y.

Thornton, D.C. and Matthews, M.L., 1981, An examination of the effects of a simple task alteration in the quality control inspection of cookware, Proceedings 15th Annual Meeting of Human Factors Association of Canada, 134-137.

Yerushalmy, J., 1969, The statistical assessment of the variability in observer perception and description of roentgenographic pulmonary shadows, Radiologic Clinics of North America, 7, 381-392.

COMPENSATION STRATEGIES OF ELDERLY CAR DRIVERS

P. VAN WOLFFELAAR, T. ROTHENGATTER and W. BROUWER

Department of Neuropsychology and Traffic
Research Centre, University of Groningen
Groningen, The Netherlands

Elderly drivers form an accident risk which is only slightly higher than that of adult drivers despite their decreased cognitive functioning. The differences in accident patterns suggest that elderly adapt their driving strategies to compensate for their decreased capabilities. The reported experiment tested this hypothesis both in the laboratory and on the road. In both conditions compensatory strategies are found.

INTRODUCTION

Elderly road users are more likely to get involved in a fatal traffic accidents than adults. This applies in particular to elderly pedestrians and cyclists, but elderly (65 yr and over) car drivers also have a fatality risk (deaths per distance driven) which is 2 to 5 times as high as that of younger adults (25 to 64 yr of age) (OECD, 1985). This increased fatality risk must be largely contributed to the increased physical fragility of the elderly, since the accident involvement risk (accidents per distance driven) itself is only slightly higher than that of other car drivers (Brouwer, 1987). The fact that elderly are not disproportionately involved in traffic accidents seems surprising considering that one of the aspects of aging is an impairment, and in particular a decrease in the speed of perceptuo-motor and attentional processes relevant for the driving task (Planek, 1981;

Salthouse, 1982). More detailed analysis of the accidents in which elderly are involved demonstrates that the accident patterns of elderly drivers differ from those of younger adult drivers. Elderly drivers are more likely to get involved in accidents in situations with high and unavoidable time-dependent demands on the driver, e.g. when crossing an intersection with high traffic intensities. On the other hand, single vehicle accidents caused by excessive speeding or alcohol usage occur significantly less often with elderly drivers. This suggests that elderly drivers adapt different driving habits or styles to compensate for their increased risk due to their perceptual and attentional impairments. Several studies report differences in general driving habits such as lower mean speeds, avoidance of rush hours, shorter trip distances and a marked reduction in night driving, but these differences may well be attributed to the changes in the professional and socio-economic roles as a result of aging and to general cultural factors (the cohort effect) (OECD, 1986). Whether elderly road users actively adapt their strategic and tactic driving behavior to compensate for their decreased capacities, is as yet unclear. In the present study the perceptuo-motor and attentional capabilities and the risk compensatory behavior of elderly were studied both in the laboratory and on the road.

METHOD

The subjects in this study were 60 drivers above 60 years of age, who had recently applied for driving licence renewal. In addition, 20 drivers who were aged under 60 were used as a control group in some of the tests. The subjects were administered a **visual choice reaction time** test both under non-distracting and distracting conditions to measure the degree of age related decrease in speed of cognitive information processing and a **compensatory potential** test (i.e. "Tower of London" test) to measure the degree to which subjects were able to adapt their strategy to their level of competence (see Van Zomeren, 1981). In addition to these laboratory tests, two on-the-road tests were administered. The first consisted of a **traffic merging task**, in which the subjects had to select a safe gap for merging from a side road into the main road. This task was considered as representative for controlled information processing during driving. The second road test consisted of **lane tracking task** while driving a 60 km long segment of a four lane divided highway at a constant speed of 90 kmh with an instrumented vehicle. During the test the lateral position of the vehicle was registered at a sample rate of 1 Hz. The ability to keep the lateral position of the vehicle under control was considered as a measure for automated information processing during driving.

RESULTS
Laboratory tasks

The results of the visual reaction time task indicate that motor component of the task becomes impaired with increasing age ($F = 7.68$, $p < .001$). The time required to reach a decision is independent of age if no distraction is presented, but significantly age dependent when the task has to be performed under distracting circumstances ($F = 4.76$, $p < .02$) (see figure 1).

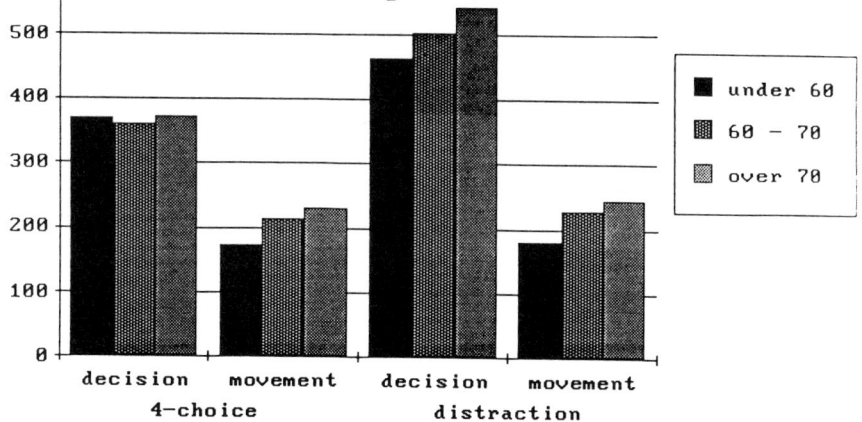

Figure 1. Reaction time in msec for the decision and movement components without and with distraction (first bar = under 60 yrs, second bar = 60 - 70 yrs, and third bar = over 70 yrs).

These results indicate that in particular under distraction the speed of controlled information processing does indeed decrease with increasing age, which can be attributed to attentional deficits.

The results of the compensatory potential test indicate that compensatory behavior is indeed found with increasing age. The mean response latencies increase with increasing age (F = 8.60, p < .001) as does the number of item restarts (F = 8.30, p < .001). Notwithstanding this compensatory behavior, the number of correct responses decreases with increasing age (F= 5.60, p < .01). These results indicate that elderly do indeed compensate for their decreasing capabilities, but not sufficiently to reach a same level of performance as adults.

Driving tasks

In the traffic merging task several striking differences were found between elderly and adult drivers. Firstly, the response time, necessary to make a decision to merge or not, appears to be markedly increased amongst elderly drivers. Adult drivers under 60 yrs of age require on average a response time of 1.6 secs, whereas drivers above 60 yrs of age require a response time close to 2.4 secs. This difference is highly significant (F = 15.09, p < .001). The standard deviation in the response times also increases with increasing age (F = 11.04, p < .001). The gap judged as being safe to merge, which can be considered as the criterion (β), also increases with increasing age (F = 6.69, p < .01). However, the sensitivity (δ') to the presented gaps decreases with age (F = 6.68, p < .01), despite the increased response time. These results indicate that elderly drivers do indeed perform worse than adult drivers on this task, even when they are not under time pressure to make a decision. The elderly drivers compensate on two accounts for their decreased performance. Firstly, they take more time for coming to a decision to merge, and secondly, they are more inclined to reject gaps than adult drivers.

Lane tracking task

In order to determine the performance on the lane tracking task the standard deviation of the lateral position of the vehicle was calculated for every 10 km segment of the test ride. For 50% of the segments it was found that elderly drivers significantly differed from the adult group of drivers in the s.d. of the lateral position of the vehicle. No difference was found between the age group 60 to 70 yrs and above 70 yrs on any segments but one (see figure 2).

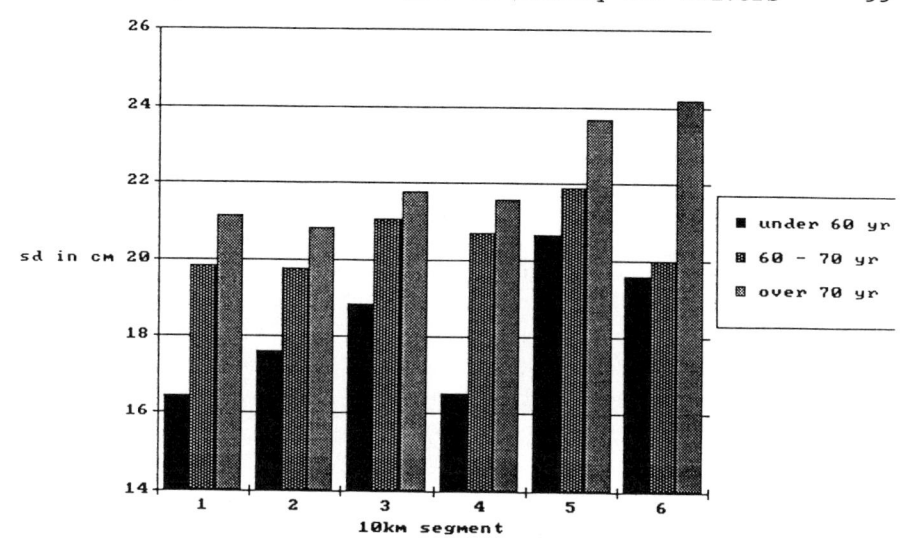

Figure 2. Standard deviation of the lateral position of the vehicle in cm during the test ride over six 10 km segments (first bar = under 60 yrs, second bar = 60 to 70 yrs, third bar = over 70 yrs).

The possibility to compensate for increased swerving on the highway lane would under normal conditions entail either reducing speed or keeping an increased distance to the fast lane. However, since the subjects in this study were instructed to keep a constant speed of 90 kmh, compensation for deteriorating performance on this task can be measured on the basis of the mean deviation from the lane midpoint. The difference between the mean deviation from midpoint between elderly and adult drivers appeared to be considerable. The young adult drivers average around 10 cm right of the midpoint, whereas the elderly drivers consistently score above 30 cm from midpoint, and at times reach level close to 50 cm from midpoint. These differences are significant for all segments (see figure 3).

DISCUSSION

The results clearly indicate that elderly drivers have a decreased performance on laboratory and driving tasks that require high level controlled information processing, even when they are not under time pressure. The decrease is much less in driving tasks requiring automatic processes, but still demonstrable. Elderly do use strategies to compensate for their decreased performance by using more time before reaching decisions and by using larger safety margins in the traffic merging task. In the lateral position control task elderly adapt a position right of the lane midpoint, which allows them to swerve without entering the fast lane.

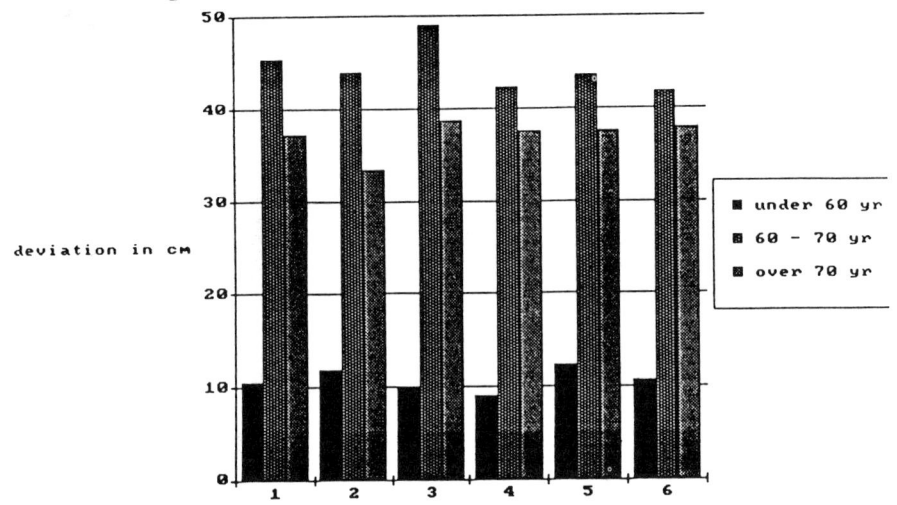

Figure 3. Mean deviation from midpoint in cm during the test ride over six 10 km segments (first bar = under 60 yrs, second bar = 60 to 70 yrs, third bar = over 70 yrs).

These results provide an explanation for the accident data, which indicate that elderly are involved in different types of accidents than younger adult drivers. They also account for the fact that the accident risk of elderly is only slightly higher than that of younger adult drivers. Further research should investigate the cues used for deciding to adopt a safer driving strategy. This could provide important indications for driver improvement courses aimed at elderly drivers.

ACKNOWLEDGEMENT

This study was carried out with a grant from the Netherlands Foundation for the Advancement of Science (ZWO).

REFERENCES

Brouwer W.H., 1987, Bejaarden in het verkeer. In: Sociale Verkeerskunde edited by C.W.F. van Knippenberg, J.A. Rothengatter & J.A. Michon (Assen: Van Gorkum, in press).
OECD, 1985, Traffic safety of elderly road users. (Paris: OECD).
Planek, T.W., 1981, The effects of ageing on driver abilities, accident experience, and licencing. In Road safety edited by H.C. Foot, A.J. Chapman (Eastbourne: Praeger), p. 171-180.
Salthouse, T.A., 1982, Adult cognition. (New York: Springer).

Mental Workload

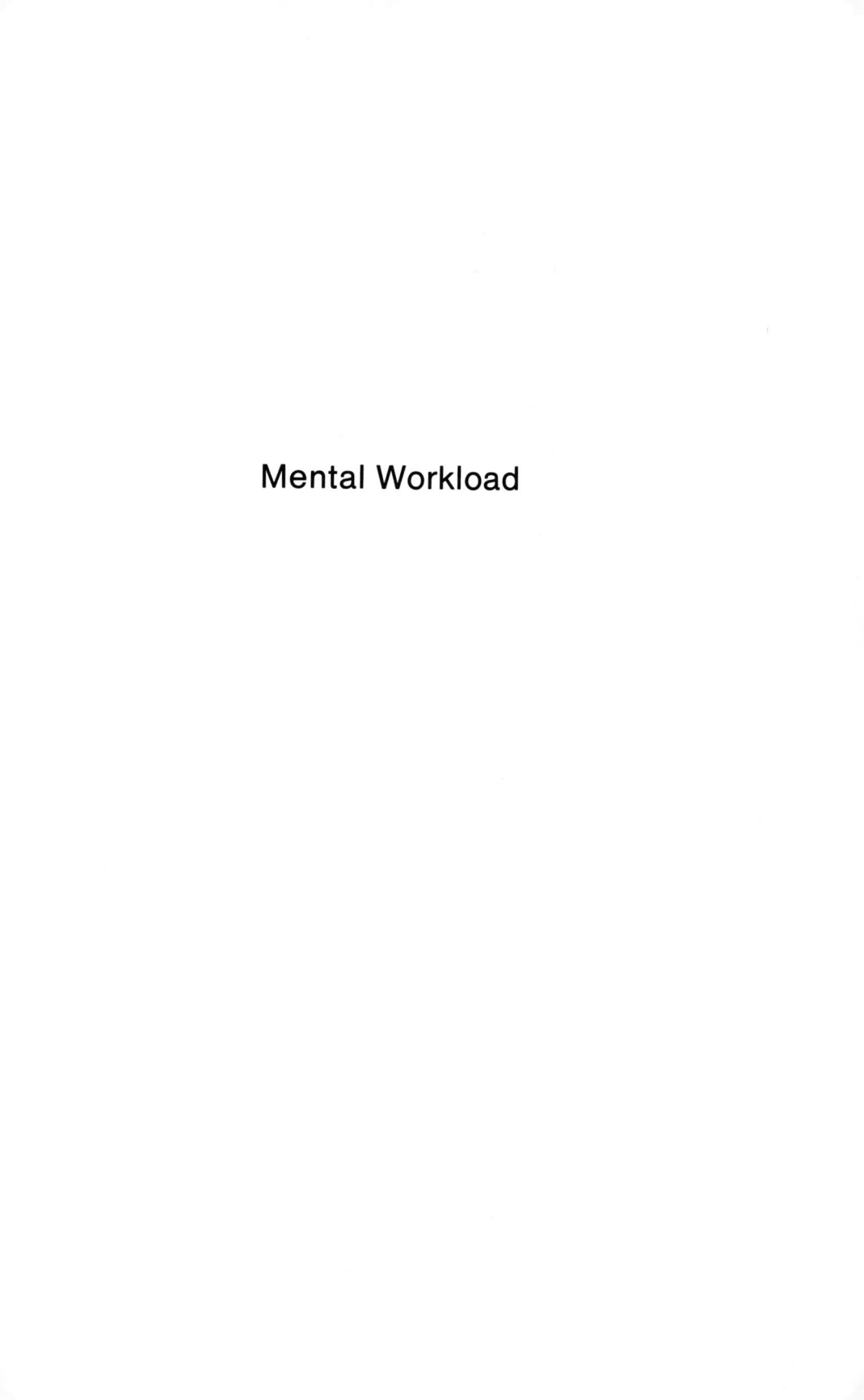

**UTILISATION FOCUSED WORKLOAD EVALUATION
IN SYSTEMS DESIGN**

N. MOHINDRA[*], E. SPENCER[**] and R. TAYLOR[***]

[*] ARE, Queens Road, Teddington, Middlesex TW11 0LN

[**] Easams Ltd, Frimley, Camberley, Surrey GU16 5EX

[***] RAF Institute of Aviation Medicine, Farnborough
Hampshire GU14 6SZ

The efficacy of human performance in any complex man-machine system is in many ways an unknown quantity to the system designer. The variable manner in which humans respond further compounds the problems of predicting the effectiveness of the total system within an operational environment. In recent years this problem has grown even more critical with the introduction of new computing technology which easily allows the design and construction of complex systems that can present and update information at rates which are far beyond the instantemous capacity of the human operator.

Assessments of the limits of human performance under strictly controlled but realistic conditions may be one means of ensuring that the information management capabilities of an operator are not exceeded and of guaranteeing successful man-machine relations. Knowledge about, and use of techniques for measuring, mental workload may offer some hope to the systems designer, as in principle, they provide the facility to indicate the demands imposed upon the operator by the system.

In 1977 the NATO Special Panel on Human Factors sponsored a workshop on 'Mental Workload' the proceedings of which have become a classic text on the theme. The workshop was initiated with the aim of synthesizing the enormous body of available knowledge on the topic and of producing a summary which would be both theoretically sound and practically useful. In summarising the problems faced by researchers of mental workload, Hopkin remarked: "As of this point in time theory has not produced an umbrella technique to partition, predict or measure the multitude of variables (affecting mental workload) in an operational field problem." In addition he also made a number of recommendations for future research such as the requirement for extending the knowledge base by carrying out more complex experiments in the laboratory to minimize extrapolation and the need for more field work to improve and test techniques.

Although over the last ten years in the area of mental workload assessment has grown substantially, there is little evidence to indicate that research has produced a solution to the problems faced by the systems designer. Discussion in the literature on the availability of yet more instruments or methods for workload assessment abound, but the application of these techniques to the evaluation of workload within the context of real systems are almost non existent.

The aims of the current symposium/workshop are to address the issues of workload measurement ten years on from the afore mentioned workshop and to establish, within the context of recent or ongoing studies, the methods being employed and how successful these have been, how researchers are tackling the problem of assessing mental workload in different applied settings, and finally to ask how useful the concept has been for the system designer in helping to predict *total* man-machine system effectiveness.

REFERENCES

Hopkin, V.D., 1979, General Status and Goals for workload applications. In Moray, N (Ed.) *Mental workload: Its theory and measurement,* Plenum Press, N.Y.

FUNDAMENTAL ISSUES IN WORKLOAD ASSESSMENT

R. HOCKEY

MRC/ESRC Social and Applied Psychology Unit
Department of Psychology, University of Sheffield
Sheffield S10 2TN

Issues will be considered under three headings:

A. Conceptual and Theoretical Issues

   *General capacity vs specific resources
   *Relation between workload and performance
   *Effort and control - management of resources
      may involve costs even when workload is
      normal
   *Dimensions of workload - relation between
      subjective, performance and physiological
      domains depends on task factors, skill
      level, emotional involvement etc.

B. Methodological Issues

   *Comparison of assessment techniques -
      each measurement domain may relate to a
      different workload dimension and interact
      with the performance/effort distinction
      in a different way
   *Reliability and validity/sensitivity and
      diagnosticity - what can measures tell us?
      how well? about what aspects of system
      function?
   *Changes in workload - practice, stress
   *Individual differences - strategies, control
      skills, expectation of effort etc

C. Application Issues

*Value of realistic simulation in order to ensure adequate operator-task coupling
*Problems of intrusiveness of workload measures

MEASURING SUBJECTIVE MENTAL LOAD

A. CRAIG

MRC Perceptual and Cognitive Performance Unit
University of Sussex, Brighton
Sussex BN1 9QG

In brief, this contribution asks: What is subjective mental load (SML)?, how can we measure it?, and what use is it?

The contribution looks at the progress that has been made in recent years in evaluating SML, and asks how far we have advanced to meet the needs identified by Moray in his 1982 review. The paper focuses on the links between SML, objective task load and performance: it asks how successful SML measurement techniques have been in predicting failures to cope on re-configured tasks; and it examines the role of SML in modelling human information processing.

A number of different measurement techniques are considered, including SWAT, NASA-bipolar and modified Cooper-Harper scales, as well as approaches based on techniques of psychophysical scaling. Differences between the techniques prompt questions about the dimensionality of SML - whether it is unidimensional or multidimensional, and in the latter case, how many dimensions and which particular dimensions should be measured - and the questions are linked to current theorising about information processing, to Multiple Resource Theory in particular, which is seen to hold promise in resolving some of the dissociations that have been reported between performance and SML. The role of individual differences and of performance criteria in SML evaluation are also discussed in this context.

The paper concludes with some practical recommendations and with a look to future needs.

MENTAL WORKLOAD AND THE ORGANISATION OF TRAINING

R.B. STAMMERS

Applied Psychology Division
Aston University
Birmingham B4 7ET

The concept of mental workload is not one that is very explicitly used within the occupational training area. However, a closer examination of the organisation of many everyday training schemes suggests an implicit use of mental workload ideas. The most obvious area where this occurs is that of increasing the complexity of training tasks as the learner progresses. This would seem to suggest a recognition that the capacity for handling task demands, arising from training exercises, increases with the development of knowledge and skill.

A focus on learning in relation to mental workload raises a number of interesting questions on the nature of knowledge and skill. The development of long-term memory structures for knowledge, and the development of procedures for retrieval information from the memory is a relevant topic. In addition the development of automatization of physical and mental skills is clearly something that has to be related to the concept of workload.

All too often workload is examined just in terms of cognitive capacity. But this capacity must be related to the adequacy of the knowledge base of the performer and to the development of skills for dealing with the task demands. These skills or routines are needed both for accessing items in memory store and for organizing the complex temporal patterning of performance.

Examples will be given in the paper of where these ideas have application in the design of training schedules. Firstly traditional topics in training will be addressed. For example the part or learning area and the topic of transfer of training, with particular reference to the transfer from difficult to easy tasks. These topics will be examined in terms of the way training material is scheduled to provide tasks appropriate for the learner at different stages of skill development. More contemporary issues that will be addressed will include the idea of adaptive training. This emerged in the early days of computer assisted learning, and has had particular application in physical and cognitive skill trainers. basic idea here is that the task demands increase as the learner's skill develops, so that a constant level of mental workload is achieved. Skill is developed by using task demands just beyond the current state of competence. Finally, attention will be given to training techniques drawing on the development of knowledge and cognitive skills in the area of intelligent tutoring. The basic philosophy here is that the instructional system should be able to diagnose the current state of learner's knowledge and skill from performance measures. It then presents appropriate tasks for the learner to carry out in order to move him or her towards the state of knowledge and skill closer to the model of the learner the system has.

WORKLOAD AND SITUATION AWARENESS
IN FUTURE AIRCRAFT

T.J. EMERSON, J.M. REISING, H.G. BRITTEN-AUSTIN

Air Force Human Resources Laboratory
Wright-Patterson Air Force Base, Dayton
Ohio 45433-5000, USA

Since 1980 aircraft crewstation design has progressed very rapidly into the electro-optical age. This has required a significant change in aircraft avionics architectures, which when coupled with the employment of new and improved sensors and artificial intelligence, will have a profound effect on both the role of the pilot and the nature of his workload problems. In older aircraft the problems centered around physical workload and placing the controls and displays where the pilot could reach them easily; however, digital fly-by-wire systems virtually eliminate the physical problems involved in "manhandling" the aircraft, and multifunction controls and displays, centrally located, minimize reach envelope considerations. What will become increasingly important, however, is the information processing load placed upon the crew.

One key issue relative to workload in future crewstations is to predict, during the design phase, workload imposed on the crew by the future system. Another aspect deals with the workload the pilot is encountering while performing the mission; the inflight problems further breakdown into 1) uncertainly caused by corrupted data, 2) stress caused by rules of engagement which are often very restrictive, and 3) time-compressed information overload and real time mission priority changes which require inflight mission re-planning.

Prediction will center around the use of algorithms which take into account the equipment performance mission scenarios, crew size, and levels of automation. The USAF Cockpit Automation Technology (CAT) Program has, as one of its major goals, the accomplishment of this end.

Real-time measurement will be gathered through physiological and performance monitoring of critical pilot functions relating to information processing workload. Fatigue, a critical component in long missions, will also be measured through physiological techniques.

The potential information overload caused by extremely dynamic increases in decision making and processing will be overcome through the inclusion of an electronic crew-member (EC) to aid the pilot. The workload shedding from the pilot to the EC will be accomplished through the functioning of a close knit team composed of both crew-members. If the human crewmember becomes overloaded, or short-term prediction indicates that he will soon become overloaded, the EC will pick up the slack. The crucial aspect of this relationship are the dynamic rules which describe when the human becomes overloaded and what level of responsibility the EC is allowed to assume.

Workload will also be intimately related to the crewmember's situation awareness (SA), which can be defined as knowledge of the internal and external states of the aircraft and the environment. The EC will aid the pilot's SA by fusing data from various sensors and communicating the product to the crewmember through pictorial formats and voice; the EC will be concerned with degrees of uncertainty in the data and will not present the fused data until a present level of certainty is reached (as a function of the rules of engagement, for example.) The EC will also monitor the crewmember's situation awareness primarily by performance but supplemented by physiological measures in order to assist or even take control, for example during blackout. The goal is to have the entire cockpit "play together" so that a consistent, intuitive pattern of situation awareness results; the pattern, in turn, will lessen information processing workload by easing decision making, thereby giving the crewmembers the performance edge.

Design, Simulation and Evaluation

**MILITARY TRAINING AND HUMAN FACTORS
CONTRIBUTIONS TO THE
BRITISH SIMULATION INDUSTRY**

T. CRAMPIN

Liveware Human Factors Consultants
160 The Street, Clapham Village
West Sussex   BN13 3UU

This paper describes, by example, how Human Factors has made a valuable contribution to simulation and military training. The appropriate application of task analysis techniques deployed in a commercial environment is explained. Examples include flight refuelling and helicopter low-level operations. A brief explanation is provided on the importance of matching the technology to the human skills to be trained. The paper concludes by highlighting the important role played by Universities and Research Establishments. The value of marketing Human Factors in a manner understood by the non-expert is emphasised.

KEY WORDS
Training;   Simulation;   Task analysis;   Skills

SUGGESTED FILE REF
Military Simulation and Training Analysis

1.   INTRODUCTION
Military training requirements are very demanding. Unlike the training needs of the commercial airline operators, military scenarios provide a formidable repertoire of widely varying tasks for aircrew. Military aircraft fly at varying heights and speeds in hostile environments shrouded with uncertainty. Military customers for flight simulation equipment are now very aware of the need to satisfy specific training requirements. Consequently, customers are intent on selecting the appropriate technology to meet those demanding training needs. The corollary of this argument is that training requirements must be defined first.

2. SIMULATION AND TRAINING REQUIREMENTS

Simulate - a word meaning 'facsimile' or 'very much like' or 'to assume the appearance of'. In military parlance, the word has come to mean a model or programme similar to or exactly replicating equipment used during combat or combat support operations or the environment surrounding that equipment. The word also suggests an annual world wide expenditure of around $900 million for government contracts to manufacturers of simulation equipment and applicable software.

This paper is concerned principally with those features of training relevant to the visual display. The information fed into the visual system arises from a thorough analysis of the tasks being carried out by the crew at any one moment during the mission. Only in this way can the visual system support the training function throughout that mission.

This paper reviews helicopter operations and flight refuelling.

The significance of defining user training requirements is to achieve optimum distribution of the available computer generated surfaces throughout the visual scene. That means, putting the information where it is needed and not displaying information which is irrelevant to the task undertaken by the crew. For example, in air-to-air refuelling, most of the important information resides on the underside of the tanker. For ground attack, targets and significant ground features take precedence in the allocation of surfaces to the scene.

In summary it can be seen that the tasks being performed by the crew dictate distribution of scene detail.

3. TRAINING ANALYSIS

This section discusses scene content, human skills and task analysis.

Scene Content

Training effectiveness relies on the provision of appropriate information in the visual. Scene content can broadly be described in terms of a collection of cues. These cues are defined in terms of two parameters: quality and quantity.

```
        ┌─────────────────────────┐
        │  TRAINING EFFECTIVENESS │
        └─────────────┬───────────┘
                      ▼
              ┌───────────────┐
              │ SCENE CONTENT │
              └───┬───────┬───┘
         ┌───────┘       └────────┐
         ▼                        ▼
  ┌─────────────┐          ┌──────────────┐
  │ CUE QUALITY │          │ CUE QUANTITY │
  └─────────────┘          └──────────────┘
```

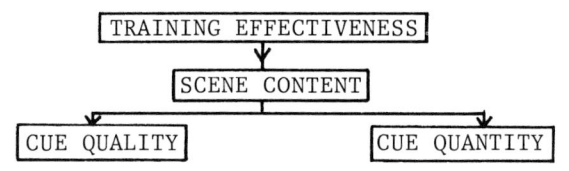

A procedure is discussed here, which lists a series of events leading up to a definition of the scene content. The scene content is directly related to training objectives, and involves an appreciation of user requirements.

Human Skills
Four important, and markedly different, human skills require consideration:

1   Psycho-motor
2   Perceptual
3   Procedural
4   Cognitive

Over the years the skills required to fly aircraft have changed. Dexterity required in mastering the psycho-motor skills has been diluted by advanced computer systems such as auto-pilots and auto-landing systems. The pilot of the future will be more of a decision maker and will require more training in mastery of cognitive skills.

Task Analysis Techniques
There are a variety of human factors techniques used to carry out task analysis. In a commercial environment many of these approaches, with respect, are impractical unless large resources of time, money and manpower are available. In fact, Singer Link-Miles make use of extensive literature provided by research personnel who do have the above resources. An effective approach is to take an overview to the training requirement at scenario level. Four techniques are mentioned briefly, here, which help to break the scenario down into a sequence of mission phases. These mission phases can be further broken down until the individual tasks of the crew are revealed and understood.

- Strategic Analysis
It is important to understand the overall strategy of the military scenario in question. That scenario can be considered in systems terms. It is appropriate, first, to identify the individual system elements, for example, the friendly and hostile platforms whether land, sea or air.

Having identified these elements the next process is to
establish the communication links between those elements.
If possible, draw a picture of this scenario of inter-
related dependencies on one sheet of paper so that the whole
scenario is visible at a glance.

- Analysis of Training Curricula
Almost without exclusion, military customers provide
thorough manuals which set high standards of user
capability. These manuals usually take the form of
Operating Procedures. From this data, information can be
extracted which determines the tasks of each individual.

- Flight Experience
Gathering data and talking to users provides invaluable
information. However, there is no substitute for
experiencing and getting involved in the actual task for
real. Details emerge which are not easily put into words.
Information becomes apparent which the user may not have
considered important to convey. Procedures are revealed
which may vary slightly from the procedures manual or
which may have arisen from recent changes in operational
methods. Singer Link-Miles go to great lengths to arrange
flight experience where availability and restrictions
permit. Investment in sophisticated video equipment
suitable for varying visual conditions has been made. By
this means it is possible to record visual impressions
and operational procedures and, further, analyse them in a
laboratory environment.

- Knowledge of Future Strategies
Over a period of time operational requirements can change.
During that period, new training techniques and equipment
emerge. If it is at all possible, flexibility must be
designed into the visual system to accommodate these
changes. For example, target designation by laser sighting
systems during low and high speed ground attack. Equally
important is the need to be aware of future techniques
in simulation which, for example, provide considerable
improvements in fields of view and resolution.

4. MILITARY SCENARIOS
This section reviews some of the visual cues found to be
important for helicopter Nap-of-the-Earth and flight
refuelling operations.

Helicopter Nap-of-the-Earth Operations

In respect of the huge amount of information available during low-level helicopter flying, visual simulation has been termed the last frontier in Computer Generated Graphics. Under an MOD 'Visual Studies' program Singer Link-Miles were permitted to film NOE manoeuvres from a Gazelle helicopter at RAE Bedford.

Analysis of the film revealed some subtle cue requirements.

- Terrain Profile

The pilot makes use of small depressions in the terrain to maintain cover. These depressions are estimated to be of the order of 10 feet from dip to peak. Such information is useful in defining to what accuracy CGI must represent terrain contours.

- Power Lines

When lines are flown over, the pilot crosses at the posts. It is considered unnecessary, in these circumstances, to model the actual wires. Only when flying under power lines need the actual lines themselves be modelled.

- Firing Position

The video film highlighted the minor changes necessary in aircraft attitude to adopt the best firing position. The parallax between the screen of trees and the background is striking. The importance of selecting a dip in the screen is revealed since, in this position, the helicopter is lower down and therefore less vulnerable.

Overall, the filming demonstrated the pilot's concern for visual information close to him in contrast to the gunner's interest in objects several thousand metres away. This is an example of the latest capability of the Singer Link-Miles Computer Generated Graphics system:

Flight Refuelling

In contrast to the MOD funded visual research programme this section reports on a commercially driven need for Human factors. In a much shorter space of time the above task analysis techniques were used to define the visual cues necessary for flight refuelling training for American Airlines. Of special relevance was the participation in flight training operations in the air, carried out at Barksdale Air Force Base, USA. Extensive pilot interviews and witnessing of many tens of refuelling connections consolidated by filming enabled the training requirements to be defined.

The training requirements were oriented around the many subtle visual cues which a pilot uses when refuelling behind a KC-10 tanker. These cues were reproduced in the simulator visual system from the analysis. American Airlines were satisfied with the integrity of the training medium and awarded Singer Link-Miles a $70m contract.

5. UNIVERSITIES AND RESEARCH ESTABLISHMENTS

Severe competition precludes anything other than a succinct and realistic approach to a client's problem. This demands a speedy response from all contributers, including the Human Factors engineer, but not at the expense of thoroughness. It is here that Universities and Research Establishments can help. Human Factors in industry must rely on a comprehensive library of research information at its finger tips. A computer data base with search facility is ideal for this purpose. However, a key issue is educating the above establishments to produce documents for industry which are themselves easy to find and easy to comprehend. It is felt that many basic Human Factors principles could be applied to the vast array of available literature.

6. MARKETING AND MANAGEMENT OF HUMAN FACTORS

A Human Factors Group in industry is unlikely to make an effective contribution to the company's product or service if it remains isolated. Human Factors is no more important than all of the other contributions; in simulation this equates to the hardware and software. Above all, it is the job of the Human Factors specialist to market his or her services in a manner which is understandable, especially to the company decision-makers in upper management. A Human Factors Group should strive for upper management backing where possible. Management of human factors at a high level has been a key discriminator in the ability of the Singer Link-Miles Human Factors group to contribute and add value to the product.

7. CONCLUSIONS

This paper has reported on a small but significant contribution that Human Factors has made to British Industry. However, the main thrust has been to offer some key issues for further discussion on what Human Factors specialists could be doing to further this contribution. The practising of Human Factors or Ergonomics, in many ways, relies on a number of activities separate from our main discipline. For example: marketing and management. All of these disciplines must be drawn together in the pursuit of Human Factors excellence.

THE USE OF PEOPLE TO SIMULATE MACHINES:
AN ERGONOMIC APPROACH

M.A. LIFE and J. LONG

Ergonomics Unit
University College London, 26 Bedford Way
London WC1H 0AP

People have been used successfully to simulate certain types of future machine in the absence of the technology required to produce them. This paper describes a simulation which demands a particularly close adherence by a human to a technological specification; it will be necessary to identify any limitations in the human's ability to simulate the target system and to provide aids to minimize them. An approach is described for applying ergonomic methods in optimizing performance of the task of the human simulator.

SIMULATION IN THE DEVELOPMENT OF TECHNOLOGY

Experiments using simulations permit designers to evaluate products at early stages in development. Product development typically involves novel utilization of existing technologies, and so development simulations can normally be implemented using existing technologies. For example, a simulation for the development of the human interface for a manually guided missile system could be constructed using current computer technology (Evans & Scully, 1986).

In the case of technology development, simulation offers a means of determining user requirements so that development is led by these, rather than by the evolution of an engineering concept. This feed-forward function is important because a failure to establish user requirements is likely to result in the technology producing unusable products. Unfortunately, it may be impossible or too expensive to simulate future technology using extant technology; however, for certain types of future system a solution lies in the use of people to simulate system behaviour. People are capable of adapting their behaviour to match a predefined specification, i.e. role-playing, and this paper describes an approach to assessing the demands of simulation tasks and for identifying the need for modifications to ensure adequate fidelity.

USING PEOPLE TO SIMULATE MACHINES: PREVIOUS APPLICATIONS

Human simulation experiments typically take the form shown in Figure 1. A simulation is developed to study the performance of users, a sample of which population will be represented in an experiment by "user subjects" (USs). USs operate the simulated system through a communication interface, A, equivalent to that in the target system; for example, A might consist of a keyboard terminal or a speech interface. However, rather than operating a computer system, the US is actually communicating with a person whose task is to imitate the system (a "system subject", SS) via a second interface, B. This paper is particularly concerned with the requirements for interface B in supporting the task of SS.

Figure 1. A generalised human simulation.

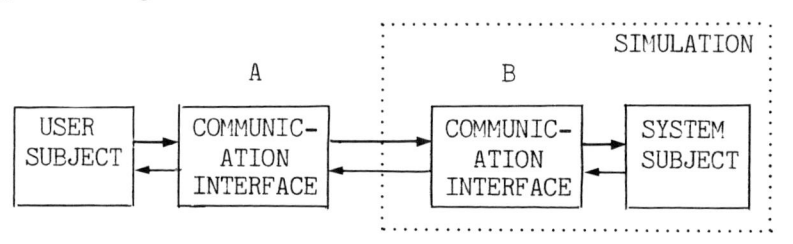

```
              A            :           B        SIMULATION :
  ┌─────────┐   ┌──────────┐ : ┌──────────┐   ┌─────────┐
  │  USER   │──▶│ COMMUNIC-│─:▶│ COMMUNIC-│──▶│ SYSTEM  │
  │ SUBJECT │◀──│   ATION  │◀─:│   ATION  │◀──│ SUBJECT │
  │         │   │ INTERFACE│ : │ INTERFACE│   │         │
  └─────────┘   └──────────┘ : └──────────┘   └─────────┘
```

Human simulation studies may be compared in terms of the complexity of behaviour demanded of SS in order to represent the target system. An example of a simulation requiring relatively simple perceptual-motor behaviour of a person is in the imitation of an isolated word speech recognizer. Pullinger (1980) explored the user interface requirements for a videotex system (Prestel), comparing user performance with a simulated speech interface and with a keypad. In the former condition, SS listened for voiced commands from USs and keyed them on their behalf. In this case, the behaviour required to mimic the machine was recognition of spoken numbers and translation of these to key-presses on a numeric keypad. There was no physical instantiation of interface A in this study, as the SS was hidden behind a screen close to US, listening to commands directly. Interface B was a keypad.

In contrast, the simulation of an expert system (ES) and its user interface involves a number of complex cognitive activities. Warren (1985) explored the interaction of users with a simulated ES, and compared their performance using natural language and command language front-ends. The target ES provided a consultancy service relating to computer equipment, and the USs used the simulated machine to advise

them on a suitable system within a limited budget. The SS and USs communicated via keyboard terminals, and the SS simulated not only the language interface, but also the expert system itself. To perform the task, the SS therefore had to:
(a) interpret US communications in terms of the constraints imposed by the type of interface being simulated, (i.e. natural or command language)
(b) request further information to determine US requirements
(c) consult a knowledge base
(d) maintain a record of USs' intended purchases and budgets
(e) compose and key in replies constrained appropriately by the rules of the interface being simulated.

These two examples illustrate the range of applications in which human simulators have been utilized successfully, but there are many examples of implementations involving behaviour of intermediate levels of complexity; e.g. in speech transcription (Gould 1982); and speech interfaces for public information systems (Richards & Underwood, 1984).

In general, experiments using human simulation have been employed in studying the user behaviour elicited by general classes of technology and have not intended a close representation of specific technology or actual products. Given these aims, fidelity is not a major issue, as claims are not made about performance with systems other than those which, coincidentally, have specifications generally similar to those presented in the simulations. However, fidelity becomes very important if the aim is to produce feed-forward information by determining the relationship between different implementations of future technology and user behaviour. The next section describes a study which imposes the requirement for such high fidelity human simulations.

DETERMINING USER REQUIREMENTS FOR FUTURE SPEECH INTERFACES

Although speech interfaces have been exploited in military applications (e.g. on aircraft flight-decks; Simpson et al., 1982) they have not been employed by the British Army for use on the land battlefield. One reason for this is that currently available devices do not meet the demands of operators. However, speech can offer unique benefits over other input/output methods (e.g. portability, operation in low ambient light and operation while the hands are otherwise occupied), and it is therefore likely that, with appropriate development, speech I/O could enhance the performance of certain classes of battlefield data communication task.

Research currently in progress at University College London (UCL) aims to develop methods for predicting the user requirements for speech interface devices in the context of specific tasks. This will enable those developing the technology to direct their efforts most effectively. However, to achieve this it will be necessary to simulate a range of speech I/O devices in order to determine how various parameters of the specification, such as recognition error rate, vocabulary size and syntax constraints, influence performance. Experimentation using human simulation is a potentially useful approach, but the project demands high fidelity in the representation of target systems. It will therefore require a closer adherence of the human simulator to a technological specification than has been necessary in previous studies.

As the fidelity of the simulation will depend critically upon the performance of a human-machine system (the SS and interface B in Figure 1), it is appropriate that ergonomic methods be applied in its optimization. The next section describes an ergonomic approach to human simulation which will enable it to meet the rigorous demands of the study.

AN ERGONOMIC APPROACH TO HUMAN SIMULATION

Ergonomics seeks to optimize the relationship between people and work, and the ergonomist would typically achieve this by modification of tasks or by the provision of task aids: alternative measures might be modification of training or modification of personnel selection. Such intervention is initiated as a consequence of system evaluation against criteria such as operator peformance (e.g. speed or errors) and measures of operator comfort (both physical and social).

Ergonomic approaches have been applied in the optimization of human-computer interfaces. For example, Buckley & Long (1985) identified difficulties that users experienced in using a system, by observing the errors that they made. Errors were assumed to be an indication of incompatibility between users' existing knowledge representations and the "ideal knowledge" representations necessary to operate the system correctly. Buckley & Long showed that observations of errors may be used diagnostically (to identify the causes of incompatibility) and prescriptively (to suggest ways in which incompatibility might be reduced and hence performance improved). Ways to improve compatibility might be to change the system such that the "ideal knowledge" for error-free performance coincided with that held by the user, or to change the user's knowledge by means of training or documentation.

A similar approach may be applied to the case of a human simulating a machine: mismatches between the performance necessary for an adequate simulation and that actually achievable indicate a requirement for performance aids, training or selection of SSs against tighter criteria. In this instance, it would not be sufficient only to consider compatibility between the ideal <u>knowledge</u> required to simulate the target system and that held by the user: incompatibility might also be manifested in the dynamic behaviour of the simulation and in the ability of the SS to reproduce the physical signals (audible and visible) emitted by the target system.

To apply the approach would initially require specification of the parameters of the target system which are to be included in the simulation, and their values. A study would then be conducted to determine the difficulties encountered by the SS in meeting this specification unaided. These difficulties would be used to develop a model of the SS, expressed in terms of incompatibility with respect to knowledge, dynamic and physical representations. This would provide a basis for identifying appropriate interventions to achieve the performance demanded by the specification; these might take the form of aiding devices for SS, and/or training to reduce the incompatibility of SS representations, or alternatively procedures might be developed for selecting SSs with more compatible representations. The effect of changes would then be evaluated empirically to ensure adequate fidelity in the simulation.

Warren's study of an ES will be used to illustrate how SS's performance might be analysed using this model. In his simulation, incompatibility concerning knowledge might have occurred as a consequence of interference between the SS's knowledge of natural language and that necessary to simulate a command language interface. This might have been manifested as consistent errors in which SS generated responses appropriate to a natural language interaction rather than responses which were legal within the target interface. Incompatibility might have been reduced by means of training or by a representational aid such as a display of legal commands.

Incompatibility concerning dynamics might have occurred as a consequence of the SS's inability to respond at the same speed as the target expert system, perhaps due to inadequate keyboard operating skill. Again, incompatibility might have been reduced by means of training, by selecting SSs with fast performance or by providing the SS with a representational aid such as a high speed typist to transcribe SS spoken responses.

Incompatibility concerning physical signals was trivially present in the Warren study: the subjects had no means of communicating unaided in a way comparable with the target system, so the simulation had to be implemented with representative text displays and keyboards. A better, albeit hypothetical, illustration of incompatible physical representation would be a system intended to simulate a speech recognizer: incompatibility might occur due to the SS being a more effective decoder of speech signals than the target system, causing him not to generate representative word recognition errors. In this case the incompatibility could be reduced by introducing errors into the SS's transcription according to an appropriate set of rules of type and frequency.

DEVELOPMENT OF THE ERGONOMIC APPROACH

The ergonomic approach described here is currently being used at UCL in the development of simulations of speech interface devices, but, given this approach, the use of human simulations might be extended to new application domains. For example, many human tasks currently performed manually might in future be performed by autonomous or semi-autonomous robots, in which case the human operator will become a supervisor with responsibility for occasional manual intervention. Human simulation of autonomous elements in such systems provides a means of investigating experimentally appropriate allocations of function.

The title of this Conference is "Ergonomics Working for Society". If ergonomics is working for the society of the future it must provide feed-forward information to ensure that developing technologies produce products which are usable by people. Human simulation is a tool which could help ergonomics to achieve this goal.

ACNOWLEDGEMENT

This research is carried out for the Royal Signals and Radar Establishment under Contract No. 2047/127 (RSRE). Any views expressed in this paper are those of the authors and do not necessarily reflect those of the Ministry of Defence.

REFERENCES

Buckley, P. & Long, J.B., 1985. Identifying usability variables for teleshopping. In D. Oborne (ed.), Contemporary Ergonomics 1985. London: Taylor & Francis.

Evans, J.L. & Scully, D.C., 1986. The simulation of a ground to air guided weapon system: the requirement and a solution. IEE Second International Conference on Simulators, IEE Conference Publication no. 267, 192-196.

Gould, J.D., Conti, J. & Hovanyecz, T., 1982. Composing letters with a simulated listening typewriter. Proceedings: Human Factors in Computer Systems, Gaithersburg, Maryland 367-370.

Pullinger, D.J., 1980. Voice as a mode of instruction entry to viewdata systems. MSc dissertation, University of London.

Richards, M.A. & Underwood, K., Talking to machines: how are people naturally inclined to speak? In E.D. Megaw (ed.), Contemporary Ergonomics 1984. London: Taylor & Francis.

Simpson, C.A., Coler, C.R. & Huff, E.M., 1986. Human factors of voice I/O for aircraft cockpit controls and displays. Proceedings: Workshop on Standardization for Speech I/O Technology, Gaithersburg, Maryland. 159-166.

Warren, C.P., 1985. The Wizard of Oz Technique: a comparison between natural and command languages for communicating with expert systems. MSc dissertation, University of London.

DESIGN MODELS AND DESIGN PRACTICE:
AN OVERVIEW

S.E. POWRIE

PAFEC Ltd
Strelley Hall, Strelley
Nottinghamshire NG8 6PE

Several approaches to the design process are examined in the context of computer aided design and related to evidence from studies of designers at work. It is concluded that theoretical models cannot be relied upon as a guide to system development.

INTRODUCTION

The review of design theory and design practice summarised here has been undertaken in the context of a wider study. This aims both to establish the potential for the extension of computer support to the design process and to review the usefulness in this field of the new generation of intelligent or quasi-intelligent techniques. The context of support for the design process needs to be considered as carefully as its content: the information available to the support process and the way in which its output should be framed will differ with the state of the solution process. The working model used, adapted from Hykin (1972), is similar to many found throughout the literature:

 1. Identification of need
 2. Specification of problem
 3. Generation of alternative solutions
 4. Selection of solution
 5. Development of selected solution
 6. Detail design
 7. Final evaluation of design
 8. Final design

The diversity between models centres on their analysis of activities underlying the higher order tasks, and particularly in stages 2-4. We will assume here that the activities belonging to stages 1, 6 and 8 are reasonably transparent and uncontroversial, although it is recognised that this may be an unduly optimistic assertion.

The progress of the designer through this route is not unidirectional: any design exercise is iterative to some degree. The system is an open one, and interactions with the outside world occur not only at the beginning and end of the cycle, but additionally take the form of interventions, expected or otherwise, at any point in between. The case studies used in Fuchs and Steidel (1973) and Hykin provide ample evidence of such events. A design support system should therefore provide for unplanned backtracking without losing data derived during the interrupted process.

SPECIFICATION OF PROBLEM

Approaches from different sources diverge in the importance attached to the specification and the manner of its derivation. Theorists concerned mainly with the engineering environment e.g Pahl and Beitz (1977) may treat the original specification as a fixed entity, save for absolutely unavoidable modifications, once its content has been fully detailed and agreed. Contrast this with the approach of Jones (1983), who suggests that the inevitable modifications which occur during the course of design imply that far less emphasis should be placed on the specification, and of Rittel and Webber (1973) who contend that design problems in the social fields and in some cases in engineering are fundamentally undefinable or 'wicked'. The observational studies reviewed in Lera (1983) report designers in a variety of fields as evolving the specification while the design progresses, several sources noting that some requirements were not recognised until highlighted by consideration of a tentative solution. Evidence from the Fuchs and Hykin case studies shows similar strategies. Darke (1979), in a study of architects characterises the subset of objectives from which such a solution is derived as the 'primary generator', noting that the choice of subset is a subjective rather than a logical one.

Thus while theorists divide between the evolutionary and the fixed views of the specification, such practical work as exists appears to support the former. It would seem that the specification data held in a design support system needs to be readily accessible for modification and amplification. 'Soft' specifications may not however be legitimate in all applications and more work is necessary before the role of the system as guardian of the specification's integrity can be determined.

GENERATION OF ALTERNATIVE SOLUTIONS

It is convenient to discuss here the meta-process which parallels the design task: the development of a control strategy. An explicit treatment is to be found in Archer (1965), where it is argued that before the problem solving itself can begin the problem must be decomposed into sub-problems, links between these established, and a solution order derived. Note that this can be considered as a formalisation of the decisions implicit in the 'primary generator' approach. Luckman (1967) describes a technique (AIDA) for the systematic analysis of decision relationships and the compatibility of their associated solution options. Formal treatments place this phase between the specification and the initiation of the solution process, but it is difficult to find direct evidence of this. Important problems in the case studies frequently do not emerge until further down the solution process. Implicit examples of identification of the main problem abound, but recognition of the effects on the rest of the decision structure seems not to occur except in post-hoc analysis.

For the 'primary generator' or solution-focussed approach, generation of multiple solutions is not an issue at this point: a proto-solution has already been derived and attention will turn to alternatives only when this has been found wanting. Other theorists emphasize the need for the production of as wide a range of solutions as possible, some espousing brainstorming and kindred techniques; others the systematic consideration of a design element's essential function, ways of providing this function in terms of physical principles, and all feasible means of embodying these principles. While one of these solution styles is preferred, elements of the other are usually incorporated. It is unsurprising that a mixture of styles is reported within many cases. What is notable is the almost complete absence of an exhaustive pursuit of alternatives, except where the range of possible alternatives is small. Tovey (1986) quotes a rare example of near-exhaustive search, but this was an exercise undertaken as a student project. Nonetheless, from this author's preliminary fieldwork with engineering designers, it is evident that some practitioners at least see thorough and systematic generation of alternative solutions (and the construction of decision trees) to be the only sound basis for a good, rather than an adequate, solution.

In many cases the generation process stops after a small number of alternatives. In some the stopping mechanism is clear: generation stops when the first alternative meeting the solution criteria is found. The designer is satisficing, not optimising. Otherwise, the mechanism is more obscure. The role of an automated design support system can only be uncertain in the state of present knowledge. It is possible to prod the imagination with simulated brainstorming techniques - Nevill and Crowe (1974) provide an example. It is possible, given an intelligently constructed database of parts and functions and expert systems techniques, to build a generator of alternatives, although for this to work economically a greater understanding of stopping mechanisms is needed. It is possible to conceive of a machine for checking the implications of partial solutions for the rest of the problem. Whether such techniques should act as a back-up to intuition or whether a designer in any circumstances could or should be constrained into a systematic approach is an open question for the present.

SELECTION OF SOLUTION

For the solution-focussed designer, the main elements of his choice have been made. The more rigorous of the problem-focused methodologies prescribe a systematic evaluation of alternative solutions against each aspect of the specification. Fine tuning is achieved by assigning weightings to the specification criteria. A complex numerical analysis can be constructed on this basis - for a recent example, see Kuppuraju et al. (1985). This includes an interesting recognition of the intuitive approach: if the outcome of the analysis is unexpected, the designer is recommended to change the original weightings.

In practice designers commonly use only those criteria which they find most compelling. These may or may not be those emphasised in the original brief, some parts of which may be ignored altogether, while considerations outside the narrow context of the problem may be introduced. As well as, for example, general engineering principles, these last may include criteria such as emotional appeal to the designer and the likely opinion of the peer group (Foz, 1972), or a general preference for certainty over innovation (Hykin). Nonetheless, there is also evidence that within the scope of the chosen criteria, evaluation proceeds in a reasonably systematic fashion (Lera, 1980). Finally, there are some cases where either because of a lack of information or because the optimisation of the solution is exceptionally important, alternative designs are developed in parallel. System design implications at this stage seem to be that there is a role for aids to the quantitative assessment of alternatives, provided that the parameters of this operation can be determined by the designer.

DEVELOPMENT OF SELECTED DESIGN

By this stage in the design process the steps to be taken are superficially straightforward. Theorists do not differ

greatly (or have much material) on the method of embodying a
design solution into a concrete form, but vary in their
treatment of difficulties encountered here according to the
degree of immutability attached to the original specification.
In practice also the course of action adopted varies with the
tightness of the design context. A major source of difficulty
found in the case studies arises from the fact that it is often
necessary for parts of the design to become crystallised at
different times, or be developed by different parts of the
team. Thus decisions in one sub-area can become fixed before
implications for other parts of the design are fully
appreciated. What may be needed here is an extension of the
type of decision network analysis system mentioned in the
consideration of the solution generation stage above. There are
of course well-established techniques for the modelling and
analysis of the physical properties of designs as they near
completion: their appropriateness is considered briefly below.

FINAL EVALUATION OF DESIGN
   It should be stressed that all sources are agreed that
evaluation against design goals is not a once-for-all exercise
but is undertaken as part of each of the preceding steps. Any
system for storing the design specification must therefore
permit easy access to the updated details. The task now is
both an assessment of the overall design in its final form,
and an analysis of those areas which can only readily be
evaluated once details have been determined. There are a wide
range of automated analysis tools available here. However both
established techniques such as stress analysis, and more recent
development of rule based systems assessing, for example, ease
of manufacture, share a major potential defect if used without
discretion: it becomes much easier to ignore such
considerations in the early stages of design and rely on the
system to trap out the worst of the errors. This is not the
same as finding the best solution, and it may cause much
avoidable backtracking. The extent to which this happens in
practice, and the form of more appropriate support mechanisms
are at present undefined but it seems likely that techniques
which are able to utilise design information in an imperfectly
defined form will be advantageous. Any system supporting and
documenting design evaluation must also be able to incorporate
the post-release performance feedback which, while desirable
in all cases, is essential in the many applications whose scale
is too large to permit prototyping.

CONCLUSIONS
   It is clear that design techniques vary considerably with
the designer and the field of design, and that design theory

does not always accurately reflect the nature of design activity. It is necessary to determine how much of this variation can be accommodated within a flexible design support system, and where the operation of such a system is fundamentally context-dependent. A detailed task analysis of the design activities in question is essential. As to the role of the system, views of design have moved away from the early systematic approaches, but the following extract from Archer states an aim which remains both valid and largely unfulfilled:

> ...to reduce the dull, imagination-suppressing chores which the designer now has to undertake, releasing him to devote more of his time to equipping himself for his crucial task - that of making the creative leap.

REFERENCES

Archer, L.B., 1965, Systematic Method for Designers, in Cross N., Ed., Developments in Design Methodology, (Wiley), 1984.

Darke, J., 1979, The Primary Generator and the Design Process, Design Studies, 1 (1), 36-44.

Foz, A.T.K., 1972, Some Observations of Design Behaviour in the Parti, MA thesis, (MIT press).

Fuchs, H.O. & Steidel, R.F., Eds., 1973, 10 Cases in Engineering Design, (Longman).

Hykin, D.H.W., 1972, Design Methods and Design Practice, PhD thesis, (Imperial College London).

Jones, J.C., 1983, Continuous Design and Redesign, Design Studies, 4 (1), 53-60.

Kuppuraju, N., Ittimakin, P., & Mistree, F., 1985, Design through Selection: a Method that Works, Design Studies, 6 (2), 91-106.

Lera, S.G., 1980, Designers' Values and the Evaluation of Designs, PhD thesis, (Royal College of Art, London).

Lera, S.G., 1983, Synopses of Some Recent Published Studies of the Design Process and Designer Behaviour, Design Studies, 4 (2), 133-40.

Luckman, J., An Approach to the Management of Design, Operational Research Quarterly, 18 (4), 345-58.

Nevill, G.E. & Crowe, R.A., 1975, Computer Aided Conceptual Design, in Spillers, W.F., Ed., Basic questions of Design Theory, (North-Holland).

Pahl, G., & Beitz, W., 1984, Engineering Design, (The Design Council/Springer Verlag).

Rittel, W.J., & Webber, M.M., 1973, Planning Problems are Wicked Problems, Policy Sciences, 4, 155-69

Tovey, M., Thinking Styles and Modelling Systems, Design Studies, 7 (1).

IN SEARCH OF METHODS OF PREDICTION

J.R. WILSON[*], J.J. ING[*], J.S. CADMAN[*] and P.H. BARTON[**]

[*] Department of Production Engineering and Production Management, University of Nottingham Nottingham NG7 2RD

[**] Post Office Research, Wheatstone Road, Dorcan Swindon SN3 4RD

Standard and split QUERTY keyboards are compared for performance on a post coding simulation, with little real difference. Methods have been tested which may allow an assessment of physical workload and hence of long-term physical effects.

INTRODUCTION

All practitioners of ergonomics must have received comments along the lines of "you´re good at criticising a completed system/product, and at analysis, but where is the evidence for ergonomics success in constructive development and synthesis?" Our response I suppose is along the lines of "If only we were brought in early enough ...." However, much of ergonomics is still searching for predictive tools, models and techniques, at a stage when we have prototype systems to play with. Where performance (time, error etc.) is concerned prediction of what will eventually be found in the field should be <u>relatively</u> simple (hence the plethora of performance measures in human-computer interaction research?). When we come to workload, mental or physical, things get somewhat tricky. How do we predict, from sometimes limited laboratory or controlled field trials, what will happen to the real operators under real conditions?

This paper is concerned with the examination of methods for assessment of a split keyboard in a specific task, Post Office letter coding. An attempt is made to predict possible benefits in terms of physical workload as well as of performance. Some points are made concerning keyboard design but the main area of interest is in the methods used.

PERFORMANCE WITH THE SPLIT KEYBOARD - EXPERIMENT I
The keyboards and experimental design
Recent work has re-opened the debate on advantages of using a split keyboard for continuous data or text entry (Nakeseko, et al, 1985). Claimed advantages rest around the premise that a design which allows keying to be performed with the fingers, hands and arms in a "natural" position, where ulnar deviations of the hands and pronation of the wrists are reduced, will reduce static postural loads. Since Nakeseko et al's work drew upon earlier experiences, as well as upon biomechanical considerations, this was taken as the basis upon which to build an experimental keyboard. The main characteristics of this were a split separation of 95mm (mid-"G" to mid-"H"), an opening angle of 25, a lateral inclination of 10 and frontal inclination of 10. The key columns were arched with QWERTY layout. For comparison a standard keyboard was constructed. Both keyboards had the limited coding key set of 26 letters, 10 numbers, and Pass, Shortcode and Cancel keys.

An independent subjects experimental design was employed, with three financially-rewarded volunteers using each keyboard. None had prior typing experience. A limited version of a Post Office training simulation was employed, with the addresses and 6 or 7 character codes to be entered scrolling right to left across a VDU. Each subject was put through a mixture of lessons (for training) and tests over 5 weeks with 2 half hour sessions per week. Fourteen tests were taken in all. The number of items to be coded in each test period varied between 20 and 95.

Results and conclusions
Group means for useful productivity and uncorrected error rates are shown in Figure 1. Although the standard keyboard appears to have consistently higher speeds (around 50 items/hr) the differences at any test are not statistically significant. Error rates show no consistent trend. On the evidence presented here and from other subjective results obtained, there seems no reason to suppose that after full training the split keyboard would eventually show worse performance than the standard. Indeed, the simulation display used in the lessons interspersed amongst the tests was based upon a standard keyboard and thus perhaps allowed better learning for the 'standard' group. Thus we are left with a reasonable alternative keyboard, which will probably meet performance criteria and which, on the face of it, would appear to hold less risk of musculo-skeletal disorders for the coders, since it was designed on biomechanical principles. Thus a

second experiment was proposed, originally intended to compare the physical workload between split and standard keyboards.

Figure 1. Useful productivity (in items per hour) and uncorrected error rates (%) for standard (△) and split (▲) keyboards.

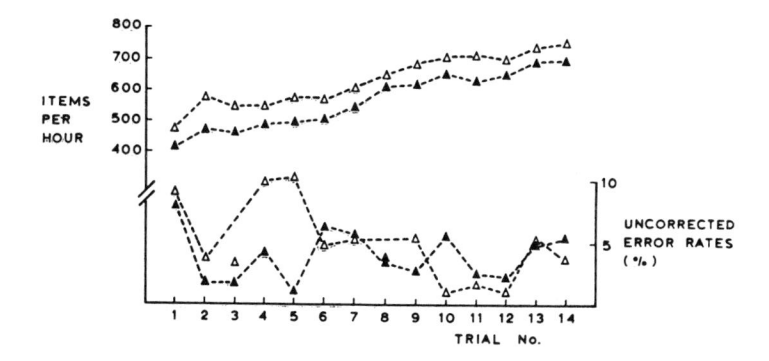

EXPERIMENT II - PHYSICAL WORKLOAD PREDICTION
Measurement methods

The use of electromyography was rejected on the grounds of the type of task, muscles, and muscle activity involved, and because of results in interpretation difficulties. Therefore, an approach of alternative task performance interference was decided upon. Experienced typists reported tasks they found progressively more difficult after increasing periods of intensive typing; most of these were "fiddly" tasks involving fine actions, steady hands and small parts (e.g. threading a needle). Tests of this kind, which could be carried out simply and quickly several times during an experimental session, were sought in order to show the effects that different times spent keying would have on coders. Tests which were developed a number of years ago for the assessment of psychomotor skills, especially in manipulative tasks, were considered. They have previously been used to assess the effects of tremor after static and dynamic work (eg Davies and Ward, 1977) amongst other uses.

The Crawford small parts dexterity test and the Purdue Pegboard were not used because of the slight variations in travel distance and grasps required for each element. Instead the One-hole Test (Salvendy, 1968) and a Finger Tremor Test were used. The former consists of gripping a pin positioned in a dish and moving it to drop it into a hole. The effect then is to push a column of pins downwards within the body of the device, such that another pin is awaiting in the same position in the dish. The Finger Tremor Test consists of a plate containing holes of different diameters, within which a stylus must be held by the subject without touching the sides. Both tests were instrumented to give microcomputer analysis of the number and duration of relevant events. In addition to these physical tests a version of the Corlett and Bishop (1976) body part discomfort rating technique, modified to concentrate upon upper limb discomfort, was employed to obtain subjective judgements.

Keyboards and experimental design
Comparison of physical workload was to be made between standard and split QUERTY coding keyboards. However, there was no evidence to show how good estimates of workload were the different measures employed, for the experimental sessions never mind as predictors of actual performance effects. Therefore a third, "poor" keyboard was used; it was hypothesised that if the measures had any utility at all they would show this layout to be worse than both the other two. To limit keyboard ´halo´ effects the same experimental keyboard unit was used throughout Experiment II. This consisted of two half keypads, with extremely wide ranges of movements allowed by use of ball and socket joints, and split adjustment achieved by positioning on runners. A hand/wrist rest was provided (Fig.2). Three adjustments of the basic unit were tested as follows.

	Standard (ST)	Split (SP)	´Poor´ (PR)
opening angle	0	25	-15
´G´-´H´	30mm	115mm	70mm
lateral inclination	0	10	0
frontal inclination	10	10	0

The same task simulation as for Experiment I was employed, within an ergonomically-acceptable work place layout. Four experienced female touch-typists were used as paid subjects with bonus payments for fast and accurate performance. The subjects where familiarised with the task, the tests, and the keyboards. Each then returned on three occasions, at the same time weekly, to use each of the keyboards for 90 minutes each session. Before the session and after each 30 minute period the subject was asked to complete the finger tremor and one-hole tests for one minute, and to complete the body part discomfort diagram. A stress and arousal checklist was filled in at the beginning and end only. The order of presentation of keyboards, and of tests between each session, were varied as appropriate.

Figure 2. Experimental keyboard.

Results
The results from Experiment II were largely inconclusive. No significant performance differences were found between the keyboards. Neither of the related task tests showed any keyboard differences, and only a visual analysis of the subjective discomfort ratings reveals a possible difference; overall scores at the end of 90 minutes were best for the split and worst for the poor keyboard. (Discomfort and stress were considerably higher at the end of experimental sessions than at the start for all keyboards).

DISCUSSION AND CONCLUSIONS
A design of keyboard has been proposed which could have performance advantages over the traditional form for the task of letter coding. Certainly, on the limited trials reported here it appears no worse in terms of speed and accuracy of coding in the simulated task. Indeed, with provisos about systematic subject variance error, it could be that improved performance would be shown after sufficient training. In either case, a powerful argument in favour of future implementation of split keyboards would be if they could be shown to reduce the chance of finding occupationally-related musculo-skeletal disorders amongst coders in the future. If such a phenomenen were to exist, in general or particular to this task, as a long term effect, then it is no use waiting several years for it to show itself in medical, absence, legal or other statistics. We need a predictor test now.

Assumptions underpinning the present work can be represented as below:

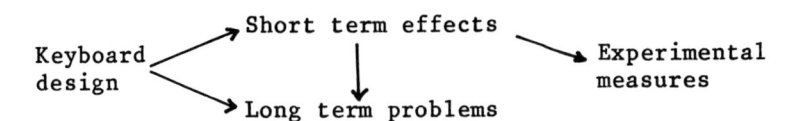

```
                    → Short term effects
    Keyboard                                    → Experimental
    design              ↓                          measures
                    → Long term problems
```

In other words, it is assumed that poorer keyboard designs will give rise to a greater chance of having long-term health problems. Thus if the experimental measures are selected so as to assess short-term discomfort and fatigue, and if such short-term effects can be assumed to be precursors of long term effects, then we have a means of predicting future health problems and also of evaluating keyboards. The second experiment reported here was to try to provide some evidence for some of these assumptions.

On the basis of our results the related task performance measures do not represent a good method of keyboard comparison let alone assessment. The subjective discomfort ratings allowed some comparisons but tells us little about the long-term.

Several cautions must be entered about the results and any conclusions drawn. The task simulation must be treated cautiously as regards real-world validity. Halo effects may have occurred due to the strange appearance of all the keyboards. Sample sizes were small and a particular difficulty is presented when just one subject's circumstances change across experimental conditions; twenty five percent of the results become unreliable. Furthermore, restrictions on resources limited the experimental period. More subjects, working longer periods and on more occasions may have produced different results. The more objective measures should be used again under such circumstances. More importantly, a start must be made on relating any appropriate methods of short term comparison to the potential long-term effects.

ACKNOWLEDGEMENT

The Engineer in Chief of the Post Office (who funded the work) is thanked for permission to publish this paper. However all views and conclusions are the authors' own.

REFERENCES

Corlett, E.N. and Bishop, R.P., 1976, A technique for assessing postural discomfort, *Ergonomics*, 19, 175-182.
Davies, B.T. and Ward, H., The effect of physical work on a subsequent fine manipulation task, *Ergonomics*, 21, 939-944.
Nakeseko, M., et al, 1985, Studies on ergonomically des. alphanumeric keyboards, Human Factors, 27, 175-187.
Salvendy, G., 1968, A comparitive study of selection tests for industrial operators. Unpublished PhD Thesis, University of Birmingham.

HIERARCHICAL TASK ANALYSIS: TWENTY YEARS ON

R.B. STAMMERS and J.A. ASTLEY

Applied Psychology Division
Aston University
Birmingham B4 7ET

Hierarchical Task Analysis (HTA) is a technique originally developed for use in a training applications by Annett and Duncan and first reported in 1967. Several adaptions have been made to the method but none has proven suitable for operator information needs analysis. The paper proposes a modification to HTA for such analysis in the context of an industrial process control environment.

INTRODUCTION

As part of a larger project on the design of improved VDU displays for coal preparation plants, financed by the ECSC, this work focusses on the development of task analysis techniques for determining operator information needs. Following an initial assessment of the range of task analysis techniques available, Hierarchical Task Analysis (HTA) was chosen as an appropriate vehicle for further development. In this paper a background to the technique is given and then a description of the adapted technique is provided.

The technique that has become known as HTA was initially promulgated in a paper by Annett and Duncan in 1967. The technique was originally developed with a view to determining the content of training programmes. In their paper the background to task analysis was reviewed and the basic idea of HTA as a progressive re-description of the task in terms of operations and sub-operations was put forward. Theoretical and practical justification for the approach was advanced and a rule for re-description, subsequently known as the 'P x C' rule was introduced.

This particular aspect of HTA has remained a problem area, but is basically a rule to enable re-description of a sub-task to proceed only when there is a need. It is based on the probability of operator error, given the current level of description (p), and the cost consequences to the system if an error is made (c). The rule is multiplicative to take into account high risk operations with a low probability of failure, or that an adequate level of description may be reached and further detail will not yield any cost savings in performance terms. The context in which the technique was developed included a research brief to develop a method that was in tune with current thinking in psychology, was economical in terms of description and led directly to training recommendations.

The hierarchical nature of the breakdown was influenced by the ideas of Miller, Galanter and Pribram for describing human performance. Economy in description was achieved through the iterative use of the P x C rule as the task analysis proceeded. The provision of training recommendations was attempted by the analyst noting training solutions at the time of analysis. These were included in a table to serve as an adjunct to the hierarchical diagram produced as the basic task representation.

Many of these latter details were set out in a more practical guide published by Annett et al. (1971). Guidelines on how to carry out the analysis were put forward in that paper. The technique was also described by Duncan (1974). This included details of a analysis of a complex chemical process control task. In this analysis archetypal task categories were identified, eg. the fault management skills of search within a structured system, and diagnosis from displayed symptoms. In addition, the Duncan paper articulates the various training alternatives that emerged for these types of sub-tasks.

Important revisions of the basic scheme were suggested by Shepherd (1976,1985). Based on his own use of the technique, and on the experiences of others in the chemical industry he suggested a number of revisions. Shepherd's work illustrated some of the problems of using HTA in practice. For example, there were difficulties over when to stop analysis and ambiguities in the P x C rule. Problems existed in isolating not just the structure of the task, but also in the rules or plans that were used to select and sequence sub-operations. In addition, Shepherd suggested a revision of the task analysis table of Annett et al. (1971), and put forward his own simplified version.

Since its inception there have been a number of reported applications of the technique both in training and in other human factors areas (A more more detailed bibliography on reported uses of HTA is availiable from the authors). A number of important theorectical issues remain to be explored, but the purpose of this paper is to report on a particular adaption of the technique for determining operator information needs in process control plants.

HTA FOR INTERFACE DESIGN

In its existing format HTA fails to provide an entirely suitable basis for determining operator information needs for human-machine interface design applications. Despite the adaptations that have been carried out to the original technique by Shepherd (1976,1985) and Piso (1981), the information made explicit is principally useful for training design purposes rather than directly applicable to interface design.

This section describes a proposed adaptation of HTA, that has been carried out in the context of an industrial process control environment. The initial development and testing of the method is in display design for coal preparation plants. The adaptation allows recommendations to be derived from the analysis that allow positive design decisions to be made.

The suggested modification to HTA does not aim to make design recommendations explicit, but allows them to be more easily derived from a task analysis. The objective is to produce an analysis which shows:-

(1) The operations (Task elements) the human has to perform.
(2) The conditions and temporal ordering under which they are performed.
(3) The information flows to and from the interface. (Needed to carry out operations).
(4) The information that is a pre-requisite to the operator being able to perform the task.
(5) The kind of task for which the information will be used. (i.e. the relationship between the operations and the information).

The tabular representation of HTA (Shepherd, 1985) is adapted to include this information, as in Figure 1. Details of information flows across the interface include, making explicit the information the operators need to perform each individual operation and detailing the information the analyst assumes the operators will already have at their disposal in relation to that operation. This category can include information from operating manuals, help pages on the displays and knowledge gained from training and other sources.

Super-ordinate	Plan	Operation	Information flow	Information assumed	Task classification	Notes

Figure 1. Modified HTA format for operator information needs analysis

The inherent hierarchical structure means that, as the level of task detail in each operation varies, so will the generality of the information related to it. This allows the analysis to be used as a complete analysis at any one level of the hierarchy. The hierarchial organization is a useful vehicle for display design as the relationships and groupings between items of information are explicit. The plans indicate how these relate to information used in other parts of the task.

The relationship between the operations, and the information needs that are specified in the analysis, are shown by classifying the information in terms of the type of task for which it will be used. Eight basic categories, describing the types of task encountered in a process control situation, have been identified. The categories do not aim to be mutually exclusive and exhaustive. What is proposed is an easily modified, yet workable scheme. The objective is to allow the information display to be designed in a way that is appropriate to the task type. If an item falls into more than one category the implications for display design can be drawn from the relationship between categories. The taxonomy of task types is as follows:-

Monitoring Information that is used in monitoring the process.
Prediction Information that is used to predict process trends.
Procedural Information used as a basis for carrying out part of, or a whole sequence of operations.
Fault diagnosis Information that is used to determine the reason for a fault or alarm.
Fault detection Determining where a fault has occured or is likely to occur.
Decision making Information allowing decisions to be made between possible courses of action.
Problem solving Using available information and knowledge to find a solution to a problem.
Operational Information used for practical plant operation.

HTA was selected as a suitable technique for modification for a variety of reasons. It has been used in a wide variety of applications and has been shown to have generality and flexibility. HTA has proved to be amenable to adaptation in the past (Shepherd, 1976,1985; Hodgkinson and Crawshaw,1985). In addition to this, the method can be applied to both existing and proposed systems and used for task synthesis purposes as well as analytically.

The suggested adaption of HTA provides a means of operator information needs analysis for human machine interface design and allows design recommendations to be derivable from the table. For full integration of the analysis into the design process it needs to be extended further. Current work is examining an intermediate form of analysis which allows direct mapping from the analysis to the design function.

RERERENCES

Annett, J. & Duncan, K.D., 1967, Task analysis and training design, **Occupational Psychology, 41**,211-221.

Annett, J., Duncan, K.D., Stammers, R.B. & Gray, M.J., 1971, **Task Analysis**. (London : HMSO) Training Information Paper No.6.

Duncan, K.D., 1974, Analytical techniques in training design. In, **The Human Operator in Process Control**, edited by E. Edwards & F.P. Lees, (London : Taylor & Francis) pp.283-319.

Hodgkinson, G.P. & Crawshaw, C.M., 1985, Hierarchical task analysis for ergonomics research, **Applied Ergonomics, 16**, 289-299.

Miller, G. A., Galanter, E. & Pribram, K.H., 1960, **Plans and the Structure of Behavior.** (New York : Holt, Rinehart & Winston).

Piso, E., 1981, Task Analysis for process control tasks : The method of Annett, et al. applied, **Occupational Psychology, 54**, 247-254.

Shepherd, A. , 1976, An improved tabular format for task analysis, **Occupational Psychology, 49**, 93-104.

Shepherd, A. , 1985, Hierarchical task analysis and training decisions, **Programmed Learning and Educational Technology, 22**, 162-176.

ERGONOMIC IMPLICATIONS IN THE DESIGN OF AN ENGINE ASSEMBLY LINE

W.L. CHAN[*], A.J. PETHICK[**] and R.J. GRAVES[**]

[*] National College of Art and Design
100 Thomas Street, Dublin 8, Ireland

[**] Ergonomics Branch, IOM
c/o British Coal, Stanhope Bretby
Burton-on-Trent, Staffordshire DE15 0QD

An ergonomic study, simulating four tasks on a new engine assembly line, indicated that their design would result in the target line cycle times not being achieved. Postural analysis also indicated a high potential risk of back pain problems. The proposed reduction in the number of repair loops on the line could increase repair times and had implications for the selection and training of repairmen. An ergonomic redesign of the workstations recommended changes in the workspace layout, the presentation of workpieces and component stock feeding systems which would improve the operator's health, as well as reducing the process cycle times for the tasks.

INTRODUCTION

Automation of the majority of operations in a new assembly line led to a proposed reduction in line cycle times. The Ergonomics Branch was asked to identify tasks containing elements with potentially adverse affects on the reliable achievement of target line cycle times (LCTs) and to evaluate the potential for redesigning workstation layouts to increase the probability of meeting the target LCTs.

Initial investigations identified four assembly tasks in particular which contained task elements potentially affecting the reliable achievement of the target line cycle time. Also, it was identified that some of the task elements could precipitate and aggravate musculoskeletal problems, especially back problems. It was therefore decided that detailed ergonomic assessment should be made of these four tasks.

THE SIMULATION

The new engine assembly line was still at the drawing stage. The outline layout of each of the workstations had been planned out. To evaluate the four tasks, it was necessary to construct mock-ups of the four workstations and to simulate the operations associated with each of these tasks.

A section of conveyor with an engine block, work benches and component containers were arranged so that the layout dimensions approximated each workstation on the new line.

<u>Subjects and Procedure</u>   Three subjects were used for the simulation. The subjects were selected from experienced operators, covering the 10th to 97th percentile stature of the workforce, and were among the "best" performers.

Each subject was carefully briefed as to the purpose of the study. They were asked to carry out the operations at their normal pace and in the sequence as specified by the company engineers. Each subject was given practice trials until he was satisfied and familiar with the operation. Each subject was asked to undertake at least two trials for each of the four tasks. The tools and components used were those intended to be used on the new assembly line.

On the new line, it was intended that some of the components would be packaged in bins (e.g. flywheels) or racks (e.g. manifolds). In order to study the interaction of the presentation of components, working posture and process time, subjects were required to pick up components from a range of positions within the boundary of their storage units. For example, subjects had to pick up from the bin flywheels which were nearest to and furthest from his body, or manifolds which were at the top and at the bottom of the storage racks.

The trials were video recorded. Working postures of the subjects were also photographed using 35 mm stills cameras, with particular attention paid to the postures of the trunk, forearms and shoulders.

ANALYSIS OF RESULTS AND DISCUSSIONS

Slow motion and frame-by-frame analysis was obtained from the video recordings. This provided the subjects' movement patterns, the methods used to pick up and grasp components, visual inspection patterns, postures of the limbs and the trunk and the time taken for each task element. Joint angles were estimated from the still photographs.

The Process Times

The process times for each of the tasks and the individual task elements were calculated for the subjects. A comparison between the mean cycle times for the simulated tasks and the target LCTs shows that the simulated cycle times of three of the four tasks exceeded the target LCTs; one of them by over 30% (manifold).

This could lead to the development of bottlenecks at these particular workstations or an increase in operational error owing to operators having to work at speeds substantially higher than their capability. The latter could lead to a drop in quality, and in turn resulting in an increase in repair rates and therefore an increased requirement for repair facilities and higher manning. All subjects on the flywheel task were able to consistently achieve the target LCTs.

Manual Lifting

For three of the tasks, operators had to lift and handle components in excess of 5 kg in a range of postures which exceeded ergonomic criteria. The combination of these factors together with the relatively high frequency of lifting prompted a more detailed investigation of potential manual handling problems.

Table 1 shows the weight of the components, the frequency of lifting and the angle of trunk flexion. The potential risks of back pain in three of the assembly tasks are indicated here and were assessed according to criteria derived by the Manual Handlng Research Unit (1980). These criteria took into account weight of the component, age of the operator, trunk posture and distance of the component from the operator's body.

Table 1. Risk of back pain in three assembly tasks.

TASK	COMPONENT WEIGHT (kg)	FREQUENCY OF LIFTS (per min)	TRUNK FLEXION (deg)	MAXIMUM LIMIT (kg)	BACK PAIN RISK
Flywheel	6.5	> 2	60-90	2.9	High
Clutch	5.2	> 2	60-90	2.9	High
Manifold	6.1	< 1	30-60	-	Low

The results suggest that the flywheel and clutch assembly tasks would exceed the maximum limits for safe handling. Therefore there is a high potential of back pain problems occurring. Although the manifold weighed 6.1 kg, that handling task did not exceed the prescribed limit owing to less severe trunk posture and lower frequency of lifts.

An assessment of the financial and manning implications of the potential back problem was carried out utilising estimates of incidence rates and the duration of absences from the existing line. The information, supplied by the company, showed that back pain was the largest single cause of absenteeism amongst assembly workers. It was predicted that absenteeism on the new line with 70 operators could cost approximately 50 man-weeks of wages per year.

On the new assembly line, the number of repair loops had been reduced by 75% and the number of spare-men reduced by 67%. Based on current absenteeism rate and taking account of the flexible leave days allowed for each worker, it can be shown that the proposed spare manning for the new line is underestimated by approximately 50%.

Increasing the number of spare-men is expensive. A more cost-effective alternative would be to reduce the overall rate of absenteeism by eliminating potential work-induced absenteeism. This could be achieved by removing the potential risk of back pain problems in tasks such as the flywheel and clutch assembly. Eliminating back pain related absenteeism could reduce the total absenteeism rate by about 20%.

The reduction in the number of repair loops also has implications for the skill requirement of the repairmen. With fewer repair loops on the new line, the number of operations between the repair loops is significantly increased. Therefore, there is a high probability that correcting faults will take substantially longer because of the larger number of dis-assembly and re-assembly tasks necessary. Furthermore, each repairman on the new line will require additional skills to deal with the increased number of assembly operations. This has obvious implications for the selection and training of repairmen and on the provision of skilled spare manning to cover for their absence. Any overloading of the repair loops could stop the line flow and therefore affect production rates.

Analysis of the Ergonomic Features of the Simulated Tasks

From the video recordings, the still photographs and dimensions of the workstation layout, it was possible to identify the features which contributed to increasing the simulated process cycle times. The features which caused the most common ergonomic problems were:

(a) Incorrect working heights of the workpiece and component stocks;
(b) Poor design of the component storage facility, e.g. Storage bins were too deep and wide forcing operator to adopt unsatisfactory postures;
(c) Component stocks and the engine block presented at sub-optimal angles causing visibility and alignment difficulties;
(d) Poor design of handtools resulting in poor arm and wrist postures which amplified inaccurate movements;
(e) For the manifold assembly task, shadows resulting from overhead lighting hindered visual location of the bolt holes.

Ergonomic improvement to workstation design. It was possible to specify alterations to the equipment design, the layout of the workstations, the presentation of component stocks and engine block and, in one case, altering the sequence of the assembly operations. Each of these improved the health, safety and effectiveness of operators.

For example, instead of stacking the flywheel and the clutch components in a deep-sided bin, a rack system was specified to store these components vertically. This eliminated a risk of potential back problems by removing the need to bend forward to pick up the flywheel from the bottom of the bin. Also, the rack system could incorporate a spring loaded device which would ensure that the component was delivered at the same location at the beginning of each operational cycle, reducing the time required to grasp and pick up the component. In addition, the time required to walk around the bin to reach components at the far corners would be saved. Tilting the engine block at $45°$ away from the operator would also reduce alignment difficulties during clutch and manifold assembly operations.

Using MTM1 (Currie, 1972) to calculate the new process cycle times resulting from the improved design of the workstations and task sequences showed that a reduction of between 6% to 16% in the cycle times for all four tasks could be achieved. If this reduction was achieved under operational conditions, it would mean that the clutch assembly tasks, which previously did not meet the target

LCTs, could be completed reliably within the target LCTs.

CONCLUSIONS

The ergonomic evaluation was able to demonstrate to the engineers and management of the company that:

(a) At least three of the assembly operations on the proposed new line would not be able to achieve the target line cycle times reliably.
(b) Two of the four tasks investigated contained elements which had a high potential risk for precipitating or aggravating back pain problems.
(c) Redesigning the individual workstations and the task elements could reduce the process cycle times as well as eliminating risk potentials for back pain problems leading to significant financial returns and improvements in system efficiency.
(d) The proposed spare manning provisions to cover for sickness and leave absences may be inadequate.
(e) The reduction of the number of repair loops by 75% coupled with the reduction in the number of repairmen could lead to overloading of the repair loop facilities. The reduction in the number of repair loops would also require additional training for existing repairmen to cover the full range of skills necessary to deal with the increased number of assembly operations between repair loops.

As a result of the study, the company design team recognised that the ergonomic findings would assist industrial, process and packaging engineers to make more cost-effective design decisions, especially if such ergonomic information was made available at the conceptual stage of the design process.

REFERENCES

Currie, R.M., 1972, Work study, (London, Pitman).
Materials Handling Research Unit, 1980, Force limits in manual work, (Luxembourg, The Secretariat of Community Ergonomics Action).

Equipment Design

DEFENCE STANDARD 00-25:
HUMAN FACTORS FOR DESIGNERS OF EQUIPMENT

R.S. HARVEY

Behavioural Science Division, ARE
Queens Road, Teddington
Middlesex TW11 0LN

A short history of Defence Standard 00-25 is presented together with the proposed programme to completion of the project.

Background

Historically, a large part of this Defence Standard (DEF STAN) has its origins in the two-volume handbook "Human Factors for Designers of Naval Equipment" published in 1971 by the Royal Naval Personnel Research Committee. This Handbook was intended to form a source of reference guide of human factors data for design staff. The success of this document in the ensuing years led to a proposal in 1977 from Senior Psychologist (Naval) for the production of a TriService Handbook, thus bringing together in a single source the individual human factors efforts of each Service Department. This proposal went ahead under the auspices of the Steering Committee on the Tri-Service Human Factors Handbook (SCOTSH) and subsequently it was arranged for the documents to be published under the umbrella of the Directorate of Standardization to appear as the constituent Parts of DEF STAN 00-25.

Advances in technology have inspired important advances in the design and development of equipment and facilities for use by the armed forces. In the light of such advances designers have tended to commit themselves to inexorable increases in equipment complexity, and more often than not it is equipment which at some point interfaces with or is controlled by one or more human operators. In these circumstances Human Factors has a crucial role to play in providing information on psychological and physiological aspects of human performance.

Experience has shown that this information must be effectively utilized in the _early_ design phases of equipment procurement. This optimisation of the man-machine combination will then feed through to performance figures which most closely match initial design performance predictions.

Aim of the DEF STAN 00-25 Series

The aim of DEF STAN 00-25 is to provide an up-to-date source-book of Human Factors data and guidance for designers of Navy, Army and Air Force equipment and facilities. Especial care has been taken to make the Standard applicable across all three Services.

The primary readership of the Standard is seen as designers representing a broad spectrum of technical background and knowledge. With this in mind the subject is correspondingly broad. The text has been structured to direct readers by means of bibliography lists to more detailed reference works as appropriate. However feed back from the use of the Parts published to date has shown that Human Factors specialists find them to be valuable data summaries.

Structure of the DEF STAN 00-25 Series

DEF STAN 00-25 has been planned in twelve parts, ranging broadly across the discipline of Human Factors as follows:

1. Introduction
2. Body Size
3. Body Strength and Stamina
4. Workplace Design
5. Stresses and Hazards
6. Vision and Lighting
7. Visual Displays
8. Auditory Information
9. Voice Communication
10. Controls
11. Design for Maintainability
12. Systems

The current publication status and projected completion dates of the constituent parts is shown at the end of the paper. Each part is published initially as an INTERIM Standard with revisions being undertaken at regular intervals.

Authors for the various parts have been commissioned by SCOTSH and selected for their specific specialist expertise from both Industry and the Universities. Subsequent editing is undertaken by the SCOTSH Working Group. This consists of MOD Human Factors specialists selected to represent the interests of all three Services. SCOTSH and the SCOTSH Working Group are served by representatives drawn from the following Service Establishments and Departments:

> Senior Psychologist (Naval)
> Admiralty Research Establishment
> Institute of Naval Medicine
> Army Personnel Research Establishment
> A Org & Sec (HF)
> HQ RAF Support Command
> Institute of Aviation Medicine
> Royal Aircraft Establishment
> Assistant Chief Scientist (Personnel) RAF
> Directorate of Standardization

Availability of DEF STAN 00-25 Series

As each Part is published it is distributed via an extensive mailing list to UK Defence Contractors, MOD Design Departments and Establishments, and also consultants, universities and polytechnics as appropriate. Distribution is undertaken by the MOD Directorate of Standardization, Stan 6b, Kentigern House, 65 Brown Street. Glasgow G2 8EX.

PUBLICATION STATUS OF CONSTITUENT PARTS

PART	TITLE	STATUS	DATE
1	Introduction	PUBLISHED	29 Nov 83
2	Body Size	PUBLISHED	30 Aug 85
3	Body Strength and Stamina	PUBLISHED	16 Apr 84
4	Workplace Design	To be Published	Apr 87*
5	Stresses and Hazards	To be Published	May 87*
6	Vision and Lighting	PUBLISHED	25 Aug 86
7	Visual Displays	PUBLISHED	27 Oct 86
8	Auditory Information	To be Published	Apr 87*
9	Voice Communication	To be Published	Apr 87*
10	Controls	To be Published	May 87*
11	Design for Maintainability	To be Published	Jul 87*
12	Systems	To be Published	Jul 87*

*Projected dates which should be considered as provisional and subject to revision in the light of editorial progress.

Copyright © HMSO, London, 1987

LABORATORY AND FIELD EVALUATION OF SELECTED
FIRE FIGHTING ASSEMBLIES

F.D. MAWBY[*], P.J. STREET[**] and C.J. NORMAN [**]

[*] CEGB, SW Region
Bridgwater Road, Bristol BS13 8AN

[**] CEGB, Technology Planning and Research Division
Marchwood, Southampton SO4 4ZB

A survey of commercially-available firefighting assemblies has been made followed by laboratory evaluations. All the garments were shown to possess serious deficiencies and this led to the development of two prototypes. These have been subjected to small scale testing, physiological assessment and field trials. An account is given of the techniques employed and of the results obtained. Implications for the future are discussed.

INTRODUCTION
Each major location within the Central Electricity Generating Board possesses its own firefighting team. The members of each team are volunteers and work in conjunction with professional and retained firemen when the need arises. Recently the question of providing suitable clothing for the volunteer force has arisen. All firefighters place particularly heavy demands on the clothing they wear and any would be successful designer of firefighting garments has to achieve a balance between an often conflicting set of requirements. The job of effecting a compromise is often made more difficult when volunteers are involved, since these workers are expected to don their suits over whatever clothes they might be wearing when the alarm sounds, and the latter might not be appropriate to the new task.

The present paper describes the approach made to developing suitable protective garments for a specific organisation, but their wider usage within the firefighting community is foreseen.

INITIAL SURVEY
Initially a survey was made of commercially - available

fighting assemblies and a representative selection of these was tested using a flammability cabinet and a shower cubicle.

Flammability

The technique has been described previously (Norman et al, 1985) but essentially consists of a fully instrumented manikin clothed in the test garment and exposed to a large gas burner flame (186kW: 90kW/m2). A standard exposure time of 20 seconds was used and the system was designed to simulate a prolonged flashover. Table 1 summarises the data obtained. Unless stated otherwise no under garments were worn.

Table 1. Manikin: Flammability and Simulated Burns Injuries.

Assembly	Results and Comments
1. 2-piece. Betaglass/ modacrylic with FR-cotton liner.	Ignition. Extensive 2° and 3° burns to legs, arms and top torso.
2. 2-piece. Betaglass/ modacrylic with modacrylic interline and liner.	Ignition. Very extensive 3° burns to legs, arms and torso.
3. 2-piece. Aramid with modacrylic liner: PVC overtrousers and cotton.	Ignition. Dense smoke. 3° burns to torso, arms and legs.
4. 2-piece. Neoprene/ aramid.	Ignition. Dense smoke. Extensive 3° burns to legs, arms and torso.
5. 2-piece. Melton wool (tunic) PVC overtrousers and wool trousers.	Upper torso and arms protected. 2° burns to lower torso. 3° burns to legs.
6. 2-piece. FR-wool with cotton liner (tunic). PVC overtrousers and FR-cotton trousers.	Upper torso and arms protected. 3° burns to lower torso and legs.

Abbreviations used: FR = flame retardant: 1° 2° and 3° = first, second and third degree: PVC = polyvinal chloride.

Water Penetration

Tests using a clothed manikin in a shower cubicle showed that all designs were vulnerable to water penetration around the neck; designs 1 to 5 allowed penetration up the sleeves; and in designs 1,2,3,5 and 6 water penetration occurred through the fabrics themselves. For the latter, design 6 gave the best performance. After 30 minutes exposure to the spray, the weight of the tunic had increased by 15% of its initial value. Design 3 was the worst: after 1 minute the weight of the tunic had increased by 80%.

NEW GARMENT

None of the designs tested were considered to be wholly suitable and the decision was made to develop a new design. Small scale methods were used to evaluate flame retardancy (BS5438: 1976) and heat transmission characteristics of a range of possible fabrics, and published data were employed to assess their likely wear properties. A two piece combination was finally chosen and possessed the following design and constructional features:

> Tunic, thigh length with (i) front zip fastener protected by two overlapping flaps, Velcro secured (ii) FR-cuffing at wrists (iii) one internal pocket (iv) madarin collar, Velcro secured trousers, bib and brace pattern with (i) adjustable braces (ii) side gussets (iii) ankle closures Velcro secured.

The construction of the tunic and trousers involved an outer layer of aramid/microporous membrane/modacrylic laminate and for the tunic alone, a single (for prototype I), or a double (for prototype II), FR-wool knit interliner and an FR-cotton liner. The total weights of the assemblies were 3.45 kg (prototype I) and 4.1 kg (prototype II).

NEW GARMENT ASSESSMENT

The following full-scale evaluations were then undertaken in parallel:

Shower Cubicle

After 1 minute both prototypes had increased in weight by 30%, predominantly by water absorption on the outer i.e. exterior side of the membrane.

Flammability

Both prototypes afforded complete protection to the

upper torso and arms. In the area of flame impingement, however, the outer fabric layer contracted and allowed flame to funnel up between the tunic and trousers. Moderate third degree burns occurred on the lower torso and legs and in a very restricted area above the buttocks.

Physiology

Assessments were made using a treadmill in the Laboratories of the Ergonomics Unit, the Polytechnic of Wales and have been reported previously (Tattersall and Thomas, 1985). Four fit subjects were employed and each tested in random order, the following four assemblies: prototypes I and II, design 3 and design 6 with the prototype trousers. Each was worn over a cotton boiler suit and cotton underwear and the total assembly included a fireman's helmet, heavy socks and rubber knee-length boots.

Among the more important conclusions arising from this work were that the physiological responses of the three fittest subjects to the four suits tested were not statistically different but that the response of the least fit and oldest subject suggested a risk of heat exhaustion when the same suits were worn.

Field Trials

It was recognised at an early stage that in practice, where fire sources are involved, additional physiological and psychological stresses might be imposed on the wearer. It was also anticipated that under field conditions design deficiencies might be revealed that had not been apparent in the laboratory experiments. Accordingly a test programme of work was undertaken which used the facilities of a fire school. A series of representative and carefully prescribed exercises were defined which included exposure to fires of realistic size and intensity. The tests consisted of three periods of work of approximately 15 minutes duration each, separated by two, twenty minute rest stages.

The same suits as were tested at the Polytechnic of Wales were worn by three subjects. These were a professional and two volunteer firemen, and their performances were recorded using a video camera and a small portable instrument package which had been specially developed for the task. The latter consisted of a data logger and means of measuring heart rate, aural temperature, skin temperature (on right chest) and a garment temperature (inner face outer layer, right chest).

In view of the wide range of experimental variables encountered, it was not possible in most cases to obtain comparison data of statistical significance for the suits. A number of useful results were produced, however, and these may be considered under the following headings:

Garment Insulation. Ambient air temperatures of between - 3° C and 17° C were recorded and near the fires could be considerably higher (but unmeasured). One subject wearing the lightest assembly (design 3) complained of shivering at sub-zero temperatures. Another wearing prototype I, lingered near a wall fire and was unaware that his garment was charring: the implication here is that the insulation was probably "too good".

Garment Damage. Subsequent investigation revealed that in the wall fire incident referred to above, the microporous membrane had become irrevocably damaged. The maximum garment temperature measured was 157° C.

Aramid assemblies became permanently marked on exposure to flames; reflective strips melted or tended to become sooty and ineffective.

Garment Acceptance. Overall subjective ratings (in which the comfort factor predominated) were:

Design 6 > Design 3 > Prototype I > Prototype II
(most preferred)   (least preferred)

These exactly tallied with the laboratory subjective assessments.

Psychological Factors. All subjects became apprehensive to some degree before firefighting and their heart rates increased: this was particularly the case with one volunteer. The professional was eager to demonstrate his prowess and his work rate was difficult to control.

Fitness. The subjects had been declared medically fit but the professional fireman, aged 44 years, showed a higher work capacity than the two volunteers (44 years and 30 years respectively. Heart rate is probably the most sensitive index of physiological cost and on occasion it approached the assumed maximum (220-age in years). (Andersen et al, 1971) for all three participants, but most frequently for the professional. The importance of fitness to the health and safety of firefighters of all ages is again emphasised.

THE FUTURE

The results of the tests cited above have led to a radical reappraisal of garment design and construction. A new, one-piece assembly is now being introduced for assessment. This is wool-based and offers improved all round protection and comfort. The membrane layer has been retained. It is waterproof, breathable, impervious to fine (>0.2 μm) dust and resistant to attack by most chemicals. However, it has been buried for protection (thermal, mechanical) beneath an outer layer of FR-wool. The single layer of FR-wool knit has been kept as a thermal insulator and the FR-cotton has been replaced by a thin liner of FR-viscose, which is slippery and facilitates donning. Other small design modifications have been made, such as the provision of leather elbow and knee patches and zipped closures from ankle to knee to allow boots to be more easily fitted. The total weight is 3.4kg. When tested (by a small scale technique), the arrangement of the constituent fabric layers has afforded an equivalent time-to-pain of 23 seconds and a time-to-blister of 30 seconds.

ACKNOWLEDGEMENTS

This paper is published by permission of the Central Electricity Generating Board.

REFERENCES

ANDERS, N.K.L., SHEPHARD, R.J., DENOLIN, N., VARNAUSKAS, E. & MASIRONI, R., 1971
  Fundamentals of Exercise Testing, Geneva: W.H.O.

NORMAN, C.J., STREET, P.J. & THOMPSON T., 1985
  Flame protective clothing for the workplace,
  Annals of Occupational Hygiene, 29, 2, 131-148

TATTERSALL, A.J. and THOMAS, N.T., 1985
  Flames and sweat - a physiological assessment of fire-fighting apparel. In Contemporary Ergonomics, Proceedings of the Ergonomics Society's Conference, Nottingham. D.J. Oborne (ed.), 171-180. Taylor and Francis.

ERGONOMICS FOR DESIGNING PRODUCTS

H. KANIS

Department of Product Ergonomics
University of Technology, Oude Delft 39a
Delft, The Netherlands

Design is generally seen as the relevant area of application for ergonomics. However, the relation between the two activities appears to be a problematic one. Approaches to deal with problems include the implementation of ergonomics in the curriculum of industrial designers, and the integration of future users in the design process. A theoretical perspective should be offered by the development of a conceptual framework which is explanatory of interactions between people and products in environments.

INTRODUCTION

Ergonomists and human factor specialists frequently wonder how their specialisms should be defined. Sometimes the study of human abilities, capabilities and limitations is considered to be the central issue. Other times descriptions focus on the design of objects, facilities or systems in order to meet human characteristics. Generally the environment referred to is man as an operator in working conditions; sometimes this environment is extended explicitly to the use of goods by the genral public. The result is almost as many descriptions as practitioners and researchers involved (Goldsmith 1985). What appears to be a central issue is the effectiveness for designers of knowledge gathered in the common field of interest: the interaction between man and product, machine, system. Here, the relevance of ergonomics (or human factors), conceived as 'what ergonomists do', is expressed by the applicability of ergonomic data and principles by designers of the material world.

This paper first focusses on the compatibility between ergonomics and design, secondly on the significance of ergonomics to the design of consumer products.

Are ergonomics and design incompatible?
The relationship between design and ergonomics seems generally to be considered as a problematic one. This is illustrated by the issues dealt with in the Forum on Design and Ergonomics as listed in Ergonomics (1986). Some of the topics discussed are: "the lack of a common language; the inadequacy of ergonomic methods for (...) problem solving; the inadequacy of ergonomic data sources (...) as easy-to-use devices".

The problems referred to are not new. Especially the applicability of ergonomic data by designers has been questioned repeatedly in the literature over a number of years. In a study, by means of interviews, Meister (1981) reports that practitioners involved consider the results of much human factors research as largely non-applicable, due to the theoretical orientation of that research. Here De Greene (1980) speaks of the lack of transferability of laboratory results to practical circumstances.

Actually, the problem of a certain mismatch between the needed design data and available research output has already been pointed out by Chapanis (1970), who elaborates on the lack of correspondence between experimental criteria in human factors/ergonomic research and functional aspects of systems and products. The absence of comparable publications that counterbalance these observations might easily suggest that ergonomics and design are practically incompatible. However, the evidence presented in the literature is much too unspecified to justify such a far-reaching conclusion.

The applicability of ergonomic data
In whatever philosophy a design process is framed, finally design activities always result in a number of practical solutions for a series of specific problems. Important data sources available for designers are current handbooks and corresponding operational computer design and assessment aids like SAMMY and ADAPS, see e.g. Case et al. (1980) and Hoekstra (1985).

Globally two categories of data can be distinguished. The first consists of real 'human factors' such as anthropometrics, sensorical thresholds and sensibilities, reaction times, admissible and attainable forces, endurance etc. These data generally touch only upon human aspects, without taking into account a concrete user-product situation. This means that designers have to interpret and translate these data for application in practical circumstances.

The other important category of ergonomic/human factors data consists of characteristics of products and physical use-conditions, e.g. (un)desirable handles, controls, displays, informational cues, illumination etc. Actually these data are geared to the human factors mentioned before, and sometimes also based upon human expectations (viz. the compatibility principle). However, the resulting guidelines are of a general nature compared to specific design problems. So translation and interpretation are needed again. This evaluation of existing data is all the more urgent as designers have to combine and compromise different - sometimes diverging - guidelines, not rarely within constraints that violate ergonomic principles a priori (viz. miniaturization).

This brief overview illustrates that easy-to-use devices generally are not of an algorithmic nature but include translation and interpretation of ergonomic data by the designer. This evaluation process will inevitably discourage the practical application of ergonomic data. Moreover, a disregarding of ergonomic principles by designers seems to cause serious problems only when human capabilities and limitations in physical, sensorical and cognitive respect are violated, resulting in obvious discomfort, faults, injuries etc. The fate of ergonomics is indeed that it strikes by non-appliance or, conversely, that the more it is put into effect the less it is felt. In this respect ergonomics appears to serve primarily as a restriction on possible design solutions. Within the often wide margins of human adaptability ergonomic achievements are hard to assess, as concepts usually associated with it - like comfort, safety, performance, efficiency - appear to be defined defectively. As long as this is so, a firm tool is lacking to identify possibilities for ergonomic upgrading.

Ergonomics attuned to design

Chapanis (1970) proposed to graft ergonomic/human factors research upon functional criteria of products and systems in order to attune ergonomics better to design. His plea has not left behind demonstrable traces in the literature. Undoubtedly an important reason for this is the difficulty to measure functional variables of products in a valid and reproducible way. This problem cannot be seen separately from the lack of a unifying conceptual framework to describe the functioning of products/systems, tailored to possibilities, limitations, preferences and experiences of human beings, within an environment that has physical, organizational, social and cultural features. Such a model, which must be explanatory of interactions between people and products in environments, as propagated by De Greene (1977), seems to be indispensible for delimiting and operationalizing central concepts like comfort and performance.

However, it should not be overlooked that such an approach, no matter how comprehensive it may be, will not completely cover the specificity and the variety of factors involved in a particular design problem. In most cases the gap designers face, in linking the 'general' with the 'specific', cannot be bypassed. There always remains a need for bridging this gap by what may be called 'ergonomic problem solving', which should be one of the inductive and reductive considerations of designers. A condition appears to be a certain commitment to and familiarity of designers with this approach.

This involvement is aimed at by the curriculum of the Faculty of Industrial Design Engineering of the Technical University Delft. This curriculum of four, in practice 5 1/2 years for a Masters degree including the undergraduate course, is based on the input from four equivalent areas: engineering, formgiving, innovation management and product ergonomics (Marinissen & Roozenburg 1982). Thus during their training industrial designers have continually been confronted with ergonomic principles, ergonomic design tools etc. Consequently, semantic problems as discussed in the Forum on Design and Ergonomics are excluded, allowing ergonomics to be applied preventively rather than curatively. However, the intended benefits of this curriculum might be presented more cogently if basic concepts like performance and comfort could be identified unambiguously. This can be seen as a main beacon to direct ergonomic research.

<u>Future research on behalf of designing consumer products</u>

Compared with products used in work conditions, the use of consumer products is characterized by e.g. an easy usability for large groups rather than job selection, job training, skills, and a relatively large freedom how, where and when to use a product rather than a specified task and environment (see also Mitchell 1981). How and where people use existing products is scarcely documented. Ideas about it are global and quite often mainly based on generalized individual experience. Variations are presumed to be considerable, but concrete insights are lacking.

For ergonomics a promising area can be identified here for research on behalf of designing products. For the way people manipulate products reflects the bodily human input; besides, both this input and environmental conditions, together with the product characteristics, determine the physical product output, such as product performance and (un)safety. So the extent to which functional requirements are met by newly designed products depends essentially on future usage. As far as suppositions about this usage are directive for the solutions chosen by designers it is important that these suppositions are checked and adjusted as early as possible in order to optimize the ongoing design.

Two approaches can be followed. The first consists of the integration of future users in the design process, as proposed by e.g. Harker & Eason (1984). This is practiced at the Technical University Delft, with the aim to develop methods that enable designers to anticipate possible/probable manipulations by users. The research is focussed on the handling of primitive models as a design cue for the way people use corresponding products. When this research is extended over a range of products it may reveal constants in the variables that explain differences between the man-model and the man-product interaction.

The second approach searches for invariant factors that determine the man-product interaction. Such invariances, if identifiable, can be of great help for designers. Analyses should be focussed on the linking of featural and functional aspects of products to environmental conditions, and to human manipulations, positions and postures, including explaining variables like antropometric, motoric, sensorical and psychological characteristics of users. A condition here is that data - mainly still to be gathered - are sufficiently precize. Framing those data into a model amounts to the theory building, plead for by De Greene (1977).

REFERENCES

Case, K., Porter, J.M. & Bonney, M.C., 1980, Design of mirror systems for commercial vehicles, Applied Ergonomics, 11.4, 199-206.

Chapanis, A., 1970, Relevance of physiological and psychological criteria to man-machine systems: The present state of the Art, Ergonomics, 13, 337-346.

Ergonomics, 1986, Forum on Design and Ergonomics, Ergonomics, 29, 1476-1477.

De Greene, K.B., 1977, Has human factors come of age? Proceedings of the Human Factors Society, 21st annual meeting, 457-461.

De Greene, K.B., 1980, Major Conceptual problems in the systems mangement of human factors/ergonomics research, Ergonomics, 23, 3-11.

Goldsmith, R., 1985, This month So what is ergonomics? Ergonomics, 28, 1407-1408.

Harker, S.D.P. & Eason, K.D., 1984, Representing the user in the design process, Design Studies, 5, 79-85.

Hoekstra, P.N., 1985, ADAPS Herkomst van de gegevens van het antropometrisch model, Faculty of Industrial Design, University of Technology, Delft.

Marinissen, A.H. & Roozenburg, N., 1982, Design projects in the curriculum of the Department of Industrial Design Engineering of the Delft University of Technology. Proceedings of the annual conference of ESEE, 44-50, Delft University Press.

Meister, D., 1982, The present and future of human factors, Applied Ergonomics, 13.4, 281-287.

Mitchell, J., 1981, User requirements and the development of products which are suitable for the broad spectrum of user capacities, Ergonomics, 24, 863-869.

GETTING TO GRIPS WITH HAND TOOL DESIGN

M.H. MABEY[*] and R.J. GRAVES[**]

[*] Department of Engineering Production
University of Birmingham, Birmingham B15 2TT

[**] Ergonomics Branch, IOM
c/o British Coal, Stanhope Bretby
Burton-on-Trent, Staffordshire, DE15 0QD

The results of a major study of the ergonomics of powered hand tools in the mining industry are discussed, as are complementary studies in other industries. Problems observed in the use of manual hand tools, particularly in food processing, suggest that insufficient guidance is available to manufacturers and users of hand tools concerning tool design in general. A coordinated programme of research is needed to validate the assumptions and confirm the results of previous research and to cover aspects as yet uninvestigated.

INTRODUCTION
   A search of the literature for ergonomic criteria against which mining industry powered hand tools could be assessed revealed that no comprehensive set of criteria was available for this purpose (Mabey et al, 1985). Tentative criteria had to be derived from general ergonomic data, criteria for analogous manual tools and tasks, and investigation of the demands of mining tasks involving powered hand tools. The tentative criteria so derived made it possible to make a rational assessment of the tools, subject to validation and confirmation in due course (Rushworth et al, 1985). It is surprising, surely, that such a procedure should be necessary after more than thirty years of ergonomics research and thousands of years of hand tool development.
   There is still no definitive and comprehensive set of ergonomic criteria for design of new tools or assessment of those already in use. Current tool designs rely heavily on tradition, manufacturing expedient and cosmetic styling, rather than on the needs and capabilities of users. There are exceptions, of course, but efforts to improve tool

design through research tend to be highly specific, to lack sound analysis of hand-arm anthropometry and function, and to fail adequately to examine the task in which the tool is used.

ERGONOMIC ASPECTS OF HAND TOOL DESIGN

Some criteria may be applicable to a wide range of tools (e.g. weight limits). Others may be specific to power tools, non-powered tools, general purpose tools or specialised tools. There are likely to be situations, as so often in ergonomics, where one aspect of tool design interacts with others in such a way that a compromise is necessary in order to achieve an optimal solution. To do so, the possible ill-effects of acceptance of a sub-optimal solution for any one aspect must be weighted, so that aspects which are most critical remain closest to their respective criteria.

The majority of powered hand tools used in mining were found to be too heavy, according to tentative criteria based on contours for maximum weights of lift (Materials Handling Research Unit, 1980). Though undesirable, this was not surprising, since they were heavy-duty tools for use in an environment where light alloy materials were not permitted. What is surprising, however, is that many power tools for surface industrial and domestic use are also heavy to lift and support, especially one-handed as is sometimes necessary, despite the availability of light-weight materials for their construction.

Drills and other pistol-type tools, besides being fairly heavy, are often poorly balanced, tending to tip or roll in the hand and placing additional static load on the wrist and arm. This is partly overcome in some types by having an optional support handle, usually to one side but reversible for left-handed use. Left-handers are not always well served in the domestic power tool field. For example, a power file recently introduced has part of the motor casing formed into a handle, so this tool can only be used right-handed. If left-handers are obliged to work right-handed, they may need to have lighter tools so that they can manipulate them more easily.

No experimental research appears to have been done so far on maximum weights of tools and comparable loads, taking into account not only lifting but all other aspects of tool use.

There is very little information available concerning the design of pistol grip control handles, and none on twist grips, even though these are fitted to large numbers of tools. For the mining industry study, tentative criteria were derived from recommendations for carrying handles given by Drury (1980), Mason et al (1980), Tichauer (1978) and Ayob and Lo Presti (1971), together with suggestions for grip angle and trigger displacement given by Fraser (1980). Grip angle in particular requires investigation since it is likely to vary considerably depending on hand anthropometry and handle cross section.

Although standard ergonomics texts such as Van Cott and Kinkade (1972) cover "knobs and dials" in considerable detail, the information mostly relates to control panels and such, and does not adequately cover controls used on power tools. Control location is another matter needing attention: again there is often a bias against the left-handed user, one example being the trigger lock on domestic power drills.

The handles of manual tools have received some research attention recently (Cochran and Riley, 1986a and 1986b; Konz, 1986; Mital and Sanghavi, 1986) but the fundamental unsuitability of many handle types for the tasks in which they are actually used has yet to be addressed. This is especially true of the typical knife handle, suitable only for a type of grip that is required for a specific operation and even then inappropriately orientated with respect to the blade.

This should be a matter of common observation, but is almost universally overlooked. This is because we all compensate for the deficiencies of the tools we use, however gross, often without realising it. The resulting compensatory postures and manoeuvres may lead to accidents and musculoskeletal strain injuries.

REQUIREMENTS FOR FUTURE RESEARCH

Research on hand tools needs to proceed on several fronts, which implies that research teams with differing interests and specialisations will be able to contribute in different ways. There could be some forum or coordinating body, however, so that results from the various approaches can be brought together for practical application. Lines of investigation that could be pursued include:

(a) handle and control design in relation to the surface and functional anthropometry of the hand and to the biomechanics of the hand-arm system;

(b) tool weights in relation to the lifting and supporting capabilities of the user population;

(c) tool guidance and feed force requirements in different tasks;

(d) design of controls and displays for setting up and operating power tools;

(e) presentation of information to the user concerning the limitation of exposure to noise and vibration from power tools;

(f) procedures for maintaining and servicing tools, particularly of the relatively inexperienced user;

(g) instruction or training requirements for typical tool applications.

The last point should be considered particularly in relation to the growing domestic market for powered hand tools, some of which could be almost as dangerous as firearms if used incorrectly or carelessly. Also, it should be kept in mind that handicapped people, the elderly or infirm might benefit greatly from being able to use manual and power tools designed to suit their capabilities. Tools could also be designed to be safely and comfortably used by children, say, or by adults with small hands.

A final, but fundamentally important point, is that tools should be equally-handed as far as possible. This means that they should fit either hand equally well, not equally badly. A case in point, observed by the authors, is the standard butchery knife used in poultry processing, which is equally unsuitable for use with either hand in this work, so that both right and left-handed workers have to make the same unconscious adaptation in gripping the knife handle, leading to greater risk of injury from laceration or musculoskeletal strain in both groups.

REFERENCES

Ayoub, M.M. & Lo Presti, P., 1971, The determination of an optimum size cylindrical handle by use of electromyograph, Ergonomics, 14, 509-518.

Cochran, D.J. & Riley, M.W., 1986a, The effects of handle shape and size on exerted forces, Human Factors, 28(3), 253-265.

Cochran, D.J. & Riley, M.W., 1986b, An evaluation of the knife handle guarding, Human Factors, 28(3), 295-301.

Drury, C.G., 1980, Handles for manual materials handling, Applied Ergonomics, 11, 35-42.

Fraser, T.M., 1980, Ergonomic Principles in the Design of Hand Tools, (International Labour Office, Geneva).

Konz, S., 1986, Bent hammer handles, Human Factors, 28(3), 317-323.

Mabey, M.H., Rushworth, A.M., Graves, R.J., & Collier, S.G., 1986, Development of ergonomic criteria for powered hand tools. In: Contemporary Ergonomics 1985, edited by D.J. Oborne (Proceedings of the Ergonomics Society's Conference, Nottingham, UK, 27-29 March 1985) (Taylor & Francis, London), pp. 117-123.

Mason, S., Simpson, G.C., Chan, W.L., Graves, R.J., Mabey, M.H., Rhodes, R.C. & Leamon, T.B., 1980, Investigation of Face End Equipment and Resultant Effects on Work Organisation, IOM Report TM/80/11 (Institute of Occupational Medicine, Edinburgh).

Materials Handling Research Unit, 1980, Force Limits in Manual Work, (IPC Science and Technology Press, Guildford).

Mital, A. & Sanghavi, N., 1986, Comparison of maximum volitional torque exertion capabilities of males and females using common hand tools, Human Factors, 28(3), 283-294.

Rushworth, A.M., Mabey, M.H., Graves, R.J., Collier, S.G., Nicholl, A.G.McK. & Simpson, G.C., 1985, A Study of Powered Hand Tools, IOM Report TM/85/9 (Institute of Occupational Medicine, Edinburgh).

Tichauer, E.R., 1978, The Biomechanical Basis of Ergonomics, (Wiley, New York).

SOCIAL AND ECONOMIC CONSEQUENCES OF AN INTERNATIONAL ERGONOMIC STANDARDIZATION

I. MATZDORFF

Anthropologisches Institut, Neue Universitat
Olshausenstrasse 40-60
23 Kiel, West Germany

ABSTRACT
Ergonomics is a field defined by technical and scientific spezialists; the results of their work sometimes lead to the establishment of new standards. The following study uses body measurements as an example in outlining the social and economic effects of such considerations, which are usually disregarded but must be seen as relevant factors if international standardization is to be attempted.

INTRODUCTION
It is a meanwhile generally accepted principle of ergonomics that working areas and manufactured items for use by humans need to be adapted to the body measurements and scope of movement of the user. This realization has resulted in many countries in prescribed standards suited to the requirements of the population. As body measurements and proportions of the people in different continents vary considerably, the individual national standards differ widely from each other.

Generally speaking, the body measurements taken to be ergonomically relevant have been differentiated in percentiles, and acceptable standards are generally agreed to lie between the 5th and 95th percentiles (except in the case of particularly dangerous issues, where the limits are set at the 1st and 99th percentiles).

Adapting products and working areas to the anthropometric variability of the population is possible in principle in three ways:

- A product may be manufactured in one size for all users, taking account of a given "critical value".

A typical example here is found in architecture: if a door or passageway is made high enough to accommodate the tallest possible user, it will also be high enough for everyone else.

- A product or a working area may be constructed in order to be adjustable, so that it can be altered to suit the body measurements and proportions of all possible users. This solution is found in numerous offices and factories and in particular in vehicles.

- The third method is to manufacture items in several sizes. This is the case, for example, in clothing and in school seating.

There are many intermediate stages between these three basic solutions, and some problems offer two possible answers: For example, in the case of beds, it is possible to build a fully functional bed on the basis of the critical value of the size of the largest user, or alternatively taking questions of cost and living space into account, in several different sizes.

CRITERIA DETERMINING ADAPTABILITY

The basic criteria determining the best method of accomodating the product to the user are of an economic nature. However, this principle is sometimes modified by tradition, habits and other factors in the field of social psychology. For example, it is highly economical to build living areas used almost exclusively by sitting and/or prostrate people so low that anyone entering or leaving the room has to walk with a pronounced stoop. The critical measurement determining the height of the room (the tallest person walking bent double) had its effect on architecture in some parts of Europe right up until the last century. Modern requirements with respect to comfort have now made this seem inacceptable to the Europeans of today.

Economic reasoning also lies behind the different standards in various countries, which are based on the anthropometric data of the individual populations. There would be no market in these countries for products too large or too small to be used by the population.

Problems always arise when the products manufactured to suit the needs of one population are introduced into other regions. A typical case is the export of cars and other vehicles. Such products are manufactured to suit social, economic and prestige tastes of a given population and to be adaptable to comply with the body measurements of the users. If this kind of vehicle is exported to a region where the population has a different kind of variability in body measurements and different economic and social standards and tastes, the range of adaptability is frequently found to be inadequate. It is sometimes more practical just to export the chassis and to adapt the coachwork separately to suit requirements in the destination country, as has been demonstrated in a number of cases in South East Asia or Africa ("mammy lorries").

There can be no question that this is only a makeshift solution, and that it would be far more practical, safer and in the case of a sufficiently large population also more economical to manufacture the products right from the start for a target regional population, rather than resorting to this kind of combination production method.

We are immediately faced with the question of cost effectivity, which can then only be answered on the basis of demographic and economic data: A large population with low purchasing power and little likelihood of increasing this cannot justify special manufacture of mass consumption articles any more than a small population with relatively high purchasing power.

But even the regions with relatively large populations and sufficient purchasing power to justify such adaptation to regional needs make high demands in terms of the possibility of adapting manufactured articles to suit their needs. If, for example, one wanted to construct a car that could be adapted to suit the needs of the whole world, this would involve a vast amount of adjustable parameters to cope with the body measurements of the populations of large and small people. On the other hand, this would give rise to high cost for the people in regions where body measurements have less tendency to extremes, because of the unnecessary range of adjustment possibilities. It is curious to note that, apart from some makeshift solutions, no-one has yet considered this issue from the point of view of making different sizes instead of a wider range of adjustment possibilities.

Especially in the case of products such as cars, which are built and sold in large quantities all over the world, it would be justifiable to consider differentiation into two size types, a "Northern car" and a "Southern car". This would of course necessitate adapting environmental parameters, such as drive-in facilities, to both sizes.

STANDARDIZATION OF PRODUCTS FOR THE POPULATION OF THE WORLD?

In recent years, rather than working on size differentiation (coupled with a restricted range of adjustments), attempts have been made to develop world-wide standards to account for all technical adjustment parameters for persons from the 5th percentile of the smallest-sized population to persons from the 95th percentile of large populations. Even without taking any of the economic or ergonomic considerations into account, this has serious implications. For example, the efforts made by the International Standards Organization (ISO/TC 23, 1986) to establish an international standard for agricultural tractors and machinery are related to the 5th and 95th percentile limits of the "world-wide operator population". As there are to date no regionally adapted tractors, this means that conditions of present-day operators are used to determine parameters. Products built for people in other regions have been designed according to the needs of Caucasoid men. This method of determining adjustment parameters would make the end product virtually useless for many non-European populations and for women all over the world.

As agricultural tractors are a product typically composed of parts made in mass manufacture (engine, gearbox, etc.), it should pose no very great technical problems to build articles in a range suited to specific different populations. It seems legitimate to ask why in such a case attempts are made to introduce an international standard, where regional solutions could be practical and reasonable from an economic point of view, especially if the possibility of differentiated combinations of the components was considered in the process of mass manufacture.

Another example of world-wide standardization is ISO 6682 (1986), which concerns the user-oriented adjustment of earth-moving machinery. These machines are considerably more complex than tractors, and are only built in relatively small numbers for use all over the world. This standard was not based on the body measurements of present users, but on anthropometric data from a number of different populations, which were combined to result in parameters for a large operator (95th percentile) and a small operator(5th percentile). Although the international body measurement data provided were compiled in a rather coincidental manner for use by the nations concernd in the relevant ISO Committee, comparison with data given by Garret (1971) and Pieper (1978) shows a fair approximation of the values, at least for populations whose size and market relevance justify their being considered. But here again we have to ask whether the solution is really sensible: adaptation of earth-moving machinery to the operator's requirements is performed by providing for sufficient adjustability. The high variance world-wide in body measurements necessitates a fore-aft adjustment for the seat of 150 mm, a vertical seat adjustment of 75 mm and additionally a 25 mm adjustment of the foot controls.

Earth-moving machinery built according to these specifications can be used just as well in South East Asia as in Northern Europe, but the operators in both regions will only make use of a very restricted range of the available adjustability. Once again it seems justifiable to ask whether it is really practical to aim for such international standardization, or whether it might not be more sensible to propose different sizes with less adjustment range.

CONCLUSION

Standards and international standardization should be considered in a very positive light if they help mankind and improve environmental and working conditions. However, the steps taken so far in an attempt to produce first international definitions of human body measurements in metrical terms with respect to the operators of agricultural tractors and earth-moving machinery raise some doubts and a new field of problems. These first examples in attempted internationalization have resulted in the metrical definition of a world-wide earth-moving machinery operator and a very different but just as internationally defined agricultural tractor operator. The car manufacturing industry has separately defined a partial user by developing other parameters according to SAE. If this trend continues, we will soon be surrounded by numerous patterns of human type contradicting each other in both definition and metrics, which depending on the social attitudes of their inventors will take certain sub-groups of the world population such as women or non-Caucasoids into account or will disregard them. Ergonomics is intended to help mankind; ergonomic standardization is an important instrument, but the real aim can only be reached by considering, in addition to purely technical parameters, the more far-reaching factors of the economic and social implications as well.

REFERENCES

Garret, J.W. et al, 1971, A collation of anthropometry, Aerospace Medical Res. Lab. Wright-Patterson Air Force Base, Ohio.

ISO/TC 23/SC 3, 1986, Draft spezification for reach volumes for the location of controls on agricultural tractors and machinery.

ISO 6682, 1986, Earth-moving machinery, zones of comfort and reach for controls.

Pieper, U., 1978, Rassenmerkmale in industrieanthropologischer Sicht. Anthropologischer Anzeiger 36, 177-182.

Vehicle Ergonomics

MOTORCYCLE ERGONOMICS: AN EXPLORATORY STUDY

S. ROBERTSON[*] and J.M. PORTER[**]

[*] Transport Studies Unit, Oxford University
11 Bevington Road, Oxford   OX2 6NB

[**] Department of Human Sciences
University of Technology, Loughborough
Leicestershire   LE11 3TU

This paper describes a questionnaire survey of the ergonomics problems of motorcycles, and a pilot experimental study of the preferred riding position for motorcyclists. A high incidence of discomfort was evident and there are strong indications that the seat height and the location of handlebars and footrests do not adequately cater for the preferences of the users.

INTRODUCTION
If the motorcycle is considered as a workstation, then a number of important points arise:
- The rider's posture is very constrained with very little flexibility
- the environment is very varied, especially in terms of thermal conditions and wind velocity
- the rider is very often clad in protective clothing which may well be bulky and restrict movement and vision
- the task of controlling the machine is a complex one, both in terms of maintaining balance/directional control and the tasks involved in driving.

However, the literature on the ergonomics of motor-cycling has concentrated on the road safety aspects, mainly conspicuity (Fulton, J., et al., 1980). The engineering journals tend to ignore the rider when considering motor-cycle characteristics (Rice, R., 1978) and there is a large gap in knowledge of the needs of the rider. This study was designed to identify the problem areas within the field of motorcycling, with particular emphasis upon comfort.

QUESTIONNAIRE SURVEY

A comprehensive questionnaire was designed to elicit information about many aspects of motorcycling. Respondents were obtained through an article in a motorcycle magazine (Robertson, 1986) and through local clubs and training schemes.

General findings

The sample consisted of 120 riders (110 males, 10 females), the mean ages were males 30.8 years, females 27.5 years. The mean statures were males 1772mm, females 1645mm. The stature of the male riders was significantly greater ($p<0.0001$) than that of the general population as described in Pheasant (1984). The riders were generally experienced, the mean number of years experience was 11.3 for males and 7.6 for females and 86% of the sample rode either every day or several times a week.

The types of motorcycle used by the respondents were classified into five groups, obtained through discussions with motorcycle users. Touring machines are those which tend to be used for long distance work and performance and handling are often of secondary importance. Commuter machines are usually of relatively low capacity (less than 250cc) and are economical to run. Sport bikes have high performance with good handling and are most influenced by current trends. Combinations are motorcycles with a sidecar attached. The trail bikes are machines which may be used on or off road and are often styled on the motocross/scramble machines. 44% of the sample owned sport bikes, 34.5% owned tourers, 15% commuter, 3% trail and 1.5% owned combinations. The sample was therefore biased towards experienced riders on large capacity machines when compared to the sample of Hobbs et al., 1983.

The respondents tended to use their machines for leisure purposes and as a form of cheap transport. 91% of male respondents quoted "being in the open" and 74% specified "freedom" as reasons for riding. Nine out of the ten females specified "freedom". Cheap transport was given as a reason for riding by 49% of males and 80% of females.

33% of the respondents were found to have modified the riding package, the main reason being to improve comfort. The more frequent changes were the addition of a fairing and a change of handlebars.

Discomfort of some nature and degree was reported by 78% of riders. The owners of touring machines experienced less discomfort than owners of commuter or sports machines (66%, 83% and 84%, respectively). Motorways were found to have the highest incidence of discomfort (60%) compared to major roads (36%) and minor roads (13%). The most frequently reported sources of discomfort were muscular aches and pains (60%), thermal problems (33%), noise (27%) and vibration (22%).

Figure 1. Areas of the body in which discomfort was reported by more than 20% of the sample (n=120).

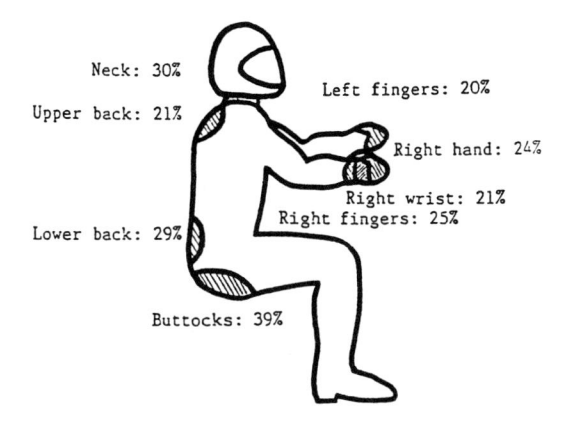

Neck: 30%
Upper back: 21%
Left fingers: 20%
Right hand: 24%
Right wrist: 21%
Right fingers: 25%
Lower back: 29%
Buttocks: 39%

Comfort

Riders were asked to rate their comfort separately for 20 body areas using a body area diagram and a 5 point rating scale (very comfortable, moderately comfortable, neutral, moderately uncomfortable, very uncomfortable). Figure 1 identifies the main areas of reported discomfort (i.e. 'moderately uncomfortable' or 'very uncomfortable' scores). No significant differences were identified between class of machine, engine capacity or the age of the rider. However, the taller riders were found to report discomfort on significantly more occasions than the shorter riders in the middle and lower back, knees and fingers. Whilst the discomfort in the knees might be taken to indicate a more cramped leg posture, it is not immediately clear why the taller riders should experience more discomfort in the other areas. A likely reason is that the taller riders suffer because of their increased surface area exposed to the wind. This would result in either discomfort of the fingers, which would effectively take the extra load, or back discomfort due to the rider bending lower to minimise his wind resistance.

Bike assessment

The riders were also asked to assess the locations of the primary components of the riding package (i.e. seat, handlebars and footrests). The most frequent changes suggested were to decrease seat height (30%), make the seat softer (36%) and wider (26%) and to move the footrest either further rearwards (23%) or forwards (16%) and to make them larger (18%). The changes required in seat height were fairly constant across the classes of bike but they were strongly related to stature with only the tallest riders requesting a higher seat. Additional adjustment for the handlebars was requested (18% higher, 11% lower, 10% forwards, 15% rearwards) and it was evident that riders of touring machines would prefer their handlebars to be higher and closer to them. 17% of all riders indicated a preference for larger diameter handlebar grips.

The riders considered the majority of the controls and instrumentation to be satisfactory. The main exceptions to this were the mirrors (21% dissatisfied with their poor location), the choke control (18% dissatisfied with the choke control when mounted on the carburettor as this location made adjustment difficult whilst travelling) and the fuel gauge (15% dissatisfied with its inaccuracy and/or inappropriately set fuel level warning lights). There was considerable variation in preferences for the different types of switchgear.

Summary

The questionnaire survey indicated that the motorcycle is not a comfortable mode of transport, although it is often selected for leisure purposes because of its other attractions. There is evidence that smaller people are excluded from riding a motorcycle, possibly because the seat is too high. The design of the riding position clearly requires a more detailed examination and a preliminary experimental study was undertaken for this reason.

EXPERIMENTAL STUDY

Aims

This study was designed to investigate the preferred riding posture of motorcylists using a static test rig based upon the rolling chassis of a production machine with a large range of adjustment provided for seat height and the locations of the handlebars and footbars.

Method

Nine experienced male motorcyclists, covering the stature range from 3rd - 95th percentile (Pheasant, 1984), were used as subjects. Each subject was asked to select their optimium seat height and then predict their optimum posture for a touring machine. A touring machine was specified because comfort considerations are important for this class of bike and the chosen posture is not so compromised by aerodynamic factors due to typically lower speeds and the fitting of protective fairings. The measurements of the location of handlebars and footrests (centre of contact on upper surface) were made relative to the seat reference point (SRP), a location defined by the intersection of a horizontal plane through the lowest point of the seat (uncompressed) and a vertical plane through the front edge of the seat.

Recommendations

From the results of this pilot study the following locations and ranges of adjustment would cater for 90% (mean $\pm$ 1.65 SD) of male riders for a touring machine. These recommendations must be considered as tentative due to the small size of the sample and, of course, female data are also required for the complete picture. The data were based upon a static set-up and there may well be differences in the perceived optimum riding package in the dynamic situation, even for a touring machine.

The range of handlebar positions indicates that the handlebars are preferred closer to the rider as the height increases (correlation coefficient -0.81). This finding is in agreement with the comments from riders of touring machines in the questionnaire survey. The traditional method of handlebar adjustment (an arc centred on the handlebar mounting point, giving a forwards movement when they are raised higher) does not cater for this preferred range of locations.

Seat height:	SRP at 750 mm ± 55 mm above ground with the suspension compressed.
Footrest location:	480 mm ± 40 mm below SRP X 0 mm ± 180 mm forward/rearwards of SRP.
Handlebar location:	300 mm ± 110 mm above SRP 450 mm ± 120 mm forward of SRP Adjustments should be independent or arranged so that the handlebars adjust closer to the rider as they are raised.

At present it would appear that many people, particularly women, are excluded from enjoying motorcycling because the seats are too high. Those who become riders often complain of muscular discomfort as a consequence of an unsatisfactory riding package. It would do the manufacturers well to consider how many potential sales are lost through the machines not being tailored to fit the whole population of potential customers.

REFERENCES

Fulton, J., Kirkby, C. & Stroud, P., 1980, Daytime motorcycle conspicuity. TRRL Supplementary Report 625.

Hobbs, C., Galer, I. & Stroud, P., 1983, Attitudes, opinions and knowledge of motorcyclists. Institute of Consumer Ergonomics Report No G354.

Pheasant, S.T., 1984, Anthropometrics: an introduction for schools and colleges. BSI Publications.

Rice, R., 1978, Rider skill influences on motorcycle manoeuvering. SAE Paper 780312, SAE Transactions 87, No. 2, p 1419.

Robertson, S.A., 1986, An assessment of the ergonomic problems of motorcyclists with special reference to the riding position and seat height. MSc. dissertation, Loughborough University of Technology.

Robertson, S.A., 1986, Motorcycle Ergonomics. *Motorcycle Rider*, Vol. 2 No. 23 p 30.

THE ERGONOMIC LONDON BUS

G.N. DAVIS[*] and T.J. LOWE[**]

[*] Davis Associates, 40 Hale Close, Melbourn
Royston, Hertfordshire SG8 6ET

[**] London Buses Ltd, 566 Chiswick High Road
Chiswick, London W4 5RR

This paper describes a major human factors research and development programme commissioned by London Buses Ltd., and executed by Ogle Design Ltd.. Comprehensive field research, extensive user trials, and the use of the SAMMIE II CAD system, lead to a human factors specification for the next generation London Bus. The paper lists the key conclusions and recommendations of the study, many of which have already been incorporated into new vehicles.

INTRODUCTION
In the latter half of 1983, London Buses Ltd. (LBL) commissioned one of the most comprehensive human factors research and development programmes ever undertaken for the passenger service vehicle industry.
  The main aim of the programme was to provide a better service to the travelling public and thereby maintain or increase patronage. To meet this aim, the following objectives were established:

- To improve the attractiveness of using one-person-operated (OPO) buses in London.

- To improve the accessibility to the system for a wider range of the population.

- To improve the driveability of the vehicle.

The scale of the programme dictated its subdivision into six project areas, and the adoption of three main avenues of research.

RESEARCH
- Field Research
- User Trials
- Computer Aided Design

PROJECT AREAS
- Front Entrance
- Driver's Cab
- Interior Interface
- Passenger Information
- Ticketing System
- Exterior Design

RESEARCH METHODS
Field Research
From the outset it was realised that comprehensive field research was necessary before any meaningful input could be made. The information to be gained by thoroughly analysing the existing system would identify any current problems together with the constraints that would affect their solution.

The field research included the following:

- Familiarisation with the bus system, vehicles, routes, bus stops, operating procedures, and passengers.

- Literature research, including LBL's own reports and previous studies.

- Observation and video analysis of passenger behaviour during queueing, boarding, payment, within the bus, and during alighting.

- Discussions with drivers, conductors, and staff unions.

- Discussions with operation, engineering, and maintainance departments within LBL.

- Liaison with London Regional Transport's Medical Service.

- Video analysis of driver activity in the cabs of existing vehicles.

- Measurement and assessment of existing vehicles.

- Discussions with manufacturers of vehicles and ticketing equipment.

User Trials
A series of user trials was devised to evaluate and develop new design solutions to the critical problems identified during the field research. A full-scale, double-deck test rig was constructed with the ability to be configured in any arrangement of entrance, exit, and staircase. Features such as handrails, steps, and ticketing equipment, could be installed, evaluated, and modified as required. Evaluation of specific design features was by video analysis and timing of the complex behavioural interactions of passengers.

In all, 32 user trials were conducted, each with at least 40 subjects. Two trials were conducted with over 80 subjects to investigate the effects of peak loads. Each trial lasted two hours (including suitable rest and refreshment periods), and consisted of up to 30 boarding and alighting cycles. All 'passengers' were sourced and pre-selected through the local job centre, and were representative of London bus users in terms of age, sex, and agility. Suitable numbers of shopping trolleys, pushchairs, and other luggage were carried. Severly disabled passengers were not represented in the main user trials but separate trials and interviews were held with local authority groups of disabled persons.

Computer Aided Design

The SAMMIE II CAD system proved to be a useful tool for the rapid evaluation and development of design features which could not easily be evaluated in the field or through user trials. The Man Model could represent the entire range of the current and potential driver populations, and was particularly useful for evaluating vision related features.

The system was used to:

- develop a windscreen profile that would cause the minimum internal reflections at night.

- minimise obscuration of the driver's view from the vehicle.

- eliminate obscuration of the displays and controls within the cab.

- specify the size, radius of curvature, and position of the rear view mirrors.

- investigate the important eye-to-eye relationship of driver and boarding passenger.

RESULTS
Front Entrance

The front entrance of the bus is the most critical area in terms of boarding time, accessibility, and the overall attractiveness of using the system. Emphasis was placed on the development of this area and the key results were:

- 2½ stream boarding can be achieved (in which, two streams of boarding passengers pass automated ticket validation devices, and the ½ stream is created by a small number of cash payers paying the driver).

- An elaborate plug door arrangement on the front corner of the vehicle proved to have none of the expected advantages during boarding trials and was replaced with a more conventional door.

- A split step arrangement provides a step height 100mm lower than conventional steps. The 250mm step allows unaided access by 95% of the general population, and by 93% of the ambulant disabled. This compares with 74% and 65% respectively for the conventional step height of 350mm (extrapolated from Brooks et al 1974).

- A unique arrangement of steps extends further into the bus thus creating larger tread areas which are safer for all passengers and encourage more rapid boarding.

- Handrails are positioned to provide maximum support at steps and to provide horizontal guidance into the bus.

- A green handrail colour has been recommended as the best compromise to meet the conflicting requirements for clearly visible handrails, and for a colour which will not cause excessive windscreen reflections at night.

- The location and arrangement of driver ticket machine and automated validators was optimised for ease of use and speed of boarding.

- Lighting is to be closely controlled in the front entrance area to aid night driving. General lighting here will be extinguished when the doors are closed.

Driver's Cab

The driver's workstation is of vital importance and must allow a range of drivers to safely drive the vehicle, and aid his ticket selling and passenger supervision tasks.

The key results in this area were:

- The driver package accomodates drivers from 95th percentile male stature to 25th percentile female.

- Maximum pedal forces for this range of drivers were recommended at 8 lbf for the accelerator, and 30 lbf for the brake (no clutch is required).

- Pedals are to be angled at 30° depressing to 15°.

- All hand controls were positioned within easy reach of all drivers, and many were mounted in finger tip pods around the steering wheel (a first for buses!).

- Video analysis lead to the optimum arrangement of the frequently used handbrake, door buttons, and indicator switch.

- Displays were grouped in a single binnicle which was positioned using SAMMIE to ensure that all drivers could see all displays.

The Ergonomic London Bus

- Mirrors were also specified using SAMMIE and consideration was given to both the legal requirements for rear view vision, and to the practical requirements of supervising passenger alighting and road trafic.

Interior Interface

The interior of the bus has been designed around an optimum arrangement of entrance, staircase, and exit.
The key results were:

- The forward position of the staircase and the relative position of the exit door create the most efficient passenger flow during simultaneous boarding and alighting.

- The staircase design avoids the use of angled treads and contains consistant riser heights. The handrails are designed to closely follow the natural path of hand movement.

- The exit design features a third step which reduces the height of the lower step to 260mm from 350mm. Handrails are improved versions of those described by Brooks and retain the elliptical cross section to aid grip.

- Seating layout was partially determined by the overall layout of the lower deck. Other considerations were: the avoidance of inward facing seats (known to be unpopular), and seat pitch was to be 680mm or greater.

- Provision for luggage storage has been more closely matched to the needs of passengers. Several smaller bays have been provided adjacent to certain seats.

- Handrails have been strategically placed throughout the bus and are to be between 30mm and 40mm in diameter.

- Bell push (stop request) buttons are to be in contrasting colours and are mounted on staunchions at 1500mm.

- Lighting in the interior ranges from 60lux to 250lux depending on the proximity to the driver and to the doors. Diffusers minimise direct glare and distribute light more evenly.

Passenger Information

Requirements for passenger information were considered in co-operation with a specialist Graphic Design company.
The key results were:

- Passengers require information concerning route and destination, fares, timetable, vehicle location, and instructions in the use of the system (i.e. ticketing).

- Several high technology solutions are available, such as Dot Matrix or Light Emiting Diode based systems, but these were dismissed on cost grounds.

- Improvements to standard LBL Johnston typeface enhanced ledgibility. The route number was mounted nearer to the kerb side of the bus front so as to be visible even when a row of buses are present.

- Public address systems are to be re-considered with the possibility of taped messages or links to the experimental BUSCO (BUS COntrol) system.

- A 'Bus Stopping' sign is recommended for the interior which indicates to all passengers when a stop request has been made, thus reducing anxiety for passengers.

Ticketing Systems

Development of a new ticketing system was not covered by the brief, but comparison trials were performed using real and simulated equipment.

The key results were:

- Boarding times per passenger of 2½ seconds were obtained with swipe-read type validators (this compares with around 4 seconds on existing OPO buses).

- Insertion type validators were not significantly slower but passenger flow characteristics were less smooth.

Exterior Design

The exterior styling of the bus was also influenced by human factors considerations, including:

- Filler caps are to be mounted between 950mm and 1400mm for can filling, or up to 1750mm if hoses are used.

- Access panels should be easily secured open or closed, they should be very obvious when open, and should be completely removable for major maintainance.

- The specified location of doors and staircase determined the window geometry and therefore the appearance.

SELECTED REFERENCE

Brooks,B.M., Ruffle-Smith,M.P., & Ward,J.S., 1974,
An investigation of Factors Affecting the Use of Buses by Both the Elderly and Ambulant Disabled Persons, TRRL contract report by Leyland Vehicles Human Factors Group.

ERGONOMICS OF THE STANDARD CITY BUS CABIN

M. KOMPIER, F. VAN NOORD, H. MULDERS, T. MEIJMAN

Institute for Experimental and Occupational
Psychology, University of Groningen
Groningen, The Netherlands

ABSTRACT
Both the ergonomics of the Dutch standard city bus and working postures of city bus drivers are studied. Deficiencies in the design and construction are demonstrated, in particular concerning lack of uniformity, freedom of movement, the driving seat, steering wheel and pedals. As a result of such shortcomings drivers are not able to obtain an adequate adjustment of the seat, steering wheel and pedals, fitting individual body dimensions. Several measures affecting the design and construction of cabin components are recommended.

INTRODUCTION
Musculoskeletal disorders are a major work-related health problem of city bus drivers (Kompier et al. 1987, also Backman 1983, Gardell et al. 1982). In order to explain this phenomenon some authors have tried to investigate physical working conditions by means of survey studies (e.g. Feickert & Forrester, 1983). There is, however, a surprising lack of research on the ergonomics of the bus cabin itself. Furthermore, it has been reported that clear (inter)national standards regarding the design and construction of cabin components do not exist (Kompier et al., 1986).
It was decided to study the ergonomics of the standard bus cab in Dutch urban transport.

METHODS
1. A representative sample - stratified towards year of construction - of 14 standard city buses (DAF Company, chassis type SB 201, coachwork Hainje Company of the municipal bus company in Groningen is studied. This most common type of bus in urban transport is used in six of nine municipal companies in the Netherlands namely the cities of Groningen, Amsterdam, Rotterdam, Utrecht, Den

Haag and Nijmegen. Each cab is drawn to scale 1:5. Cabin components are measured and form, colour, function and position of knobs, handles and metres are recorded. These specifications will be compared with relevant directives advocated by UITP (International Union of Public Transport).
2. The use of the cabin and working postures of six city bus drivers, differing in sex, height and weight, are systematically photographed during their daily work routine.

Table 1. Global measures illustrating variation in body dimensions (six drivers).

		weight (kilo)			height (cm.)			leg (cm.)			lower leg (cm.)		
male	n=3	70	96	85	171	189	195	80	87.5	96.5	57	60	65
female	n=3	58	78	65	164	169	177	77.5	-	-	50	54	58

RESULTS
- Working space

Figure 1. Measurements of mean cabin area in centimeters (scale 1:20).

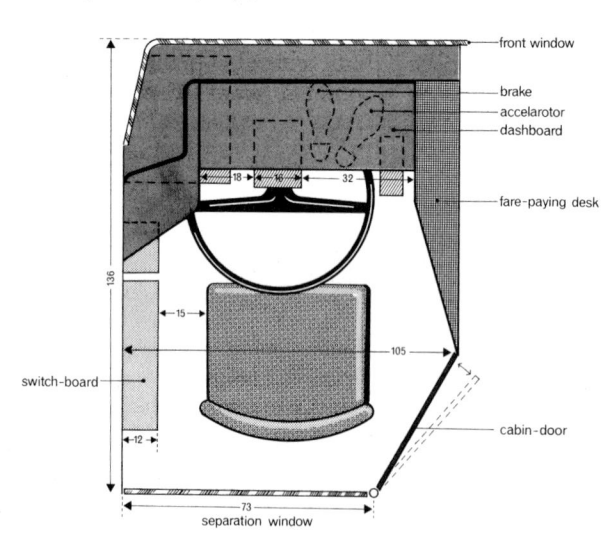

The bus cabin is a small open working place (136x100 cm.) When seated, the driver's latitudinal freedom of movement is limited by the steering column and by dashboard, fare-paying desk and switchboard (a.o. knobs for opening and shutting doors). In case of the left knee it is approximately 18 cm.; for the right knee ± 32 cm.
Vertical freedom is limited by the position of the steering wheel, as shown in Photo 1.

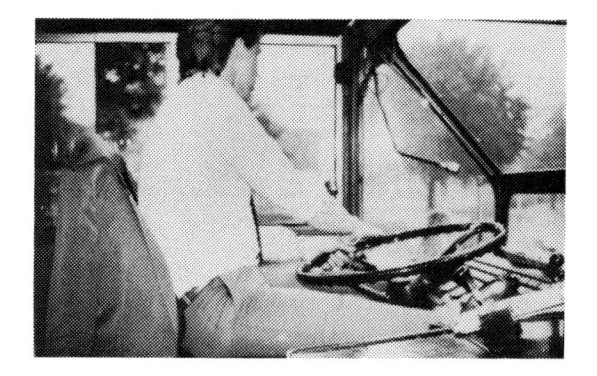

Photo 1

As a result, it is difficult to change the position of upper or lower legs in order not to charge continually the same muscles. Limited freedom of movement is illustrated by wear and tear marks on the cabin walls.
- Dashboard and switchboard.
There is a lot of variation in the design of the dashboard and switchboard. In most cases this has merely nuisance value. However, the variation in placing of the horn, emergency signal and regulator of the intercom system - controls that should be under automatic control of the driver - is a source of potential error in dangerous situations.
- Driving seat.
The seated posture is mainly determined by the seat. It is also affected by the required sight height and by steering wheel and foot pedals. Eleven of the buses in the sample are equipped with a Widney-seat (Type Brilithe SW); three with a Bremshey-seat (Type 404-2). Neither of these possesses an adjustable lumbar support, nor adjustable springs to prevent damage due to whole body vibrations. Vibration levels on Dutch standard buses are a major occupational hazard (Kompier et al., 1987). In a recent study (Oortman Gerlings et al, 1985) it has been demonstrated that vibration levels on Dutch standard city bus cabins exceed ISO-2631 eight hours fatigue decreased profiency boundaries and, in most conditions, even exceed eight hours exposure limits.

Horizontal and vertical adjustability are measured on the basis of the seat reference point (intersection of back rest - maximum vertical adjustment - and bottom), the steer-o-point (lowest edge of steering wheel) and foot reference point (turning point of foot bottom). Vertical adjustability falls below UITP directives. Minimum heights vary from 430 to 455 mm. (mean 441 mm.), whereas maximum heights go from 510 to 550 mm. (mean 525 mm.) The vertical range of adjustment (maximum - minimum height) is determined for each seat, its mean being 84 mm. (min. 70 - max. 95 mm.). Since UITP advocates a vertical range of 100 mm. it is clear that no seat does meet this criterion. The adjustability backward and forward is, according to UITP directives (⩾ 150 mm.), too limited on two of the buses in the sample. The seats are positioned inaccurately, since neither their bottoms nor their back rests are positioned exactly behind the steering wheel. Nor is the back rest positioned exactly behind the bottom.

In our study of working postures it is demonstrated that all six drivers, irrespectively of their anthropometric dimensions, adjust the seat to its maximal forward and upward position (photo 1).
The reason for this choice is found in the physical characteristics of the steering wheel.
- Steering wheel. The vertical distance between the steer-o-point and the floor is 730 mm. With a diameter of 550 mm. the fixed steering wheel is too large, according to the UITP directive of 500 mm. The angle of the wheel is 12°, sloping upward. UITP recommends an inclination, adjustable from 15° to 32°. Neither this minimum angle, nor the criterion of adjustability is met. UITP also advocates vertical adjustability. Since the standard steering wheel is fixed it deviates from this directive. A large fixed steering wheel has many disadvantages. Even if a maximum upward andforward seat position is chosen (photo 1) drivers are not able to turn the wheel around without bending the trunk forward, as is illustrated in Figure 2.
Turning the wheel is a frequent task operation in city bus driving. Therefore all drivers often move their position forward away from the back rest, thus putting an extra demand on muscles of the back, neck, legs, arms and shoulders (photo 1).
- Pedals.
Since the standard bus has an automatic gear it has only two pedals: the accelerator and the foot brake, both operated by the right foot. The inclinations to the horizontal are recorded in two conditions: non-operated and maximally pushed-downward. Angles differ from bus to bus (non-operated: mean accelerator 42.7°, range 39° - 46°; mean brake 44.4°, range 37° - 46°; pushed-downward: mean accelerator 20.8°, range 15° - 25.5°; mean brake 17.5°, range 14.5° - 21°). Also within the same type of bus differences between the angles of inclination of the pedals exist, the largest difference being 11°. As to the accelerator the mean non-operated versus pushed-down difference on the same bus is

Figure 2. Diameter of steering wheel related to reach dimensions.

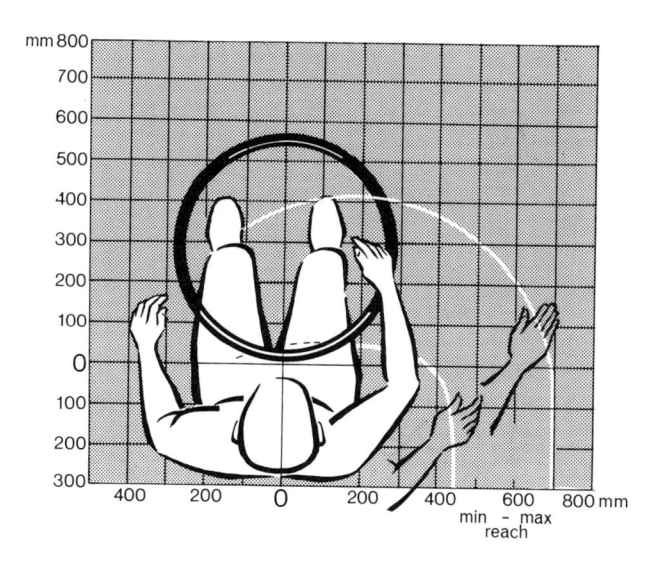

22° (range 17° - 26°). In the case of the brake this difference is 26.4° (range 18° - 30.5°). According to Kellerman's standard (Kellerman et al., 1982) an angle exceeding 25° is too large to prevent the ankle joint from overstretching. Moving the seat forward and upward to compensate for a large, fixed steering wheel results in uncomfortable seat-to-pedals and wheel-to-pedals relationship (photo 2).

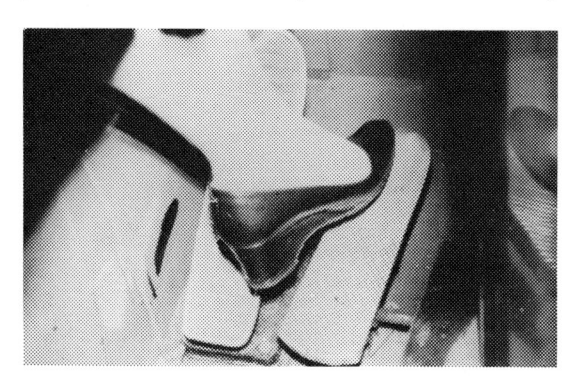

Photo 2

DISCUSSION AND RECOMMENDATIONS
It is demonstrated that there is no uniform design nor uniform construction of the various cabin components. Shortcomings in design and construction are demonstrated, in

particular those concerning freedom of movement, the seat, the steering wheel and the pedals. It turns out that all drivers experience difficulties in working positions and movements. As a result of shortcomings in the construction of the cabin, the drivers are not able to obtain an adequate adjustment of the chair, steering wheel and the pedals, fitting individual body dimensions. As a result of our study several measures affecting the design and construction of cabin components are recommended. Consequently, cabin components will be more adapted to individual anthropometric characteristics.

1. Driving seat
   - range of adjustments needs to be enlarged: vertical 100 mm.; fore-aft ≥ 150 mm.
   - adjustable springs
   - lumbar support, adjustable in height and thickness
2. Steering wheel
   - diameter ≤ 500 mm.
   - adjustable to the vertical, and fore-aft
   - independent adjustment of angle of inclination: 15°-32°
3. Pedals: equal angles; range of angle ≤ 25°
4. Dashboard: uniform design
5. Working space
   - larger cabin; since it is an open work place, more effective heating and cooling systems are needed
6. With respect to these recommendations training and evaluation are important issues.

Backman, A.L., 1983, Health survey of professional drivers. Scand. J. of Work Environment and Health, 9, 30-35.

Gardell, B., Aronsson, G., & Barklöf, K, 1982. The working environment for local public transport personnel (Report from the Swedish Work Environment Fund).

Feickert, D. & Forrester, D., 1983. Stress factors in urban public transport (University of Bradford).

Kellerman, T., Klinkhamer, H., Wely, P. van & Willems, P., 1982. Vademecum Ergonomie (NIVE/Kluwer, Deventer).

Kompier, M., Noord, F. van, Mulders, H. & Meijman, T., 1986, Ergonomische evaluatie van de cabine van de stadsbus. Tijdschrift voor Ergonomie, 11, 3, 2-6.

Kompier, M., De Vries, M., Van Noord, F., Mulders, H., Meijman, T., & Broersen, J., 1987. Physical work environment and musculoskeletal disorders in the busdriver's profession. In: P. Buckle (Ed.), Musculoskeletal disorders at work. Proceedings. Taylor and Francis, London.

Oortman Gerlings, P., Drimmelen, D. van, & Musson, Y., 1985. Trillen en schokken tijdens het werk (TH-Delft).

THE EFFECTS OF POSTURE AND SEAT DESIGN ON LUMBAR LORDOSIS

J.M. PORTER and B.J.NORRIS

Department of Human Sciences
University of Technology, Loughborough
Leicestershire LE11 3TU

A standardised method was developed to record the external profile of the spine whilst standing and sitting. The provision of a lumbar support, adjusted for individual comfort, was found to help somewhat to preserve lordosis although the maximum displacement of the lumbar spine from the backrest was only half that when standing. This finding conflicts with previous work by other researchers who have suggested that the lumbar support should be designed to minimise any loss of lordosis from standing.

INTRODUCTION
The Vehicle Ergonomics Group has evaluated over 70 car seats for manufacturers over the last six years. The major complaint with many of these seats has been the high levels of discomfort reported in the lower back. The reason that sitting is considered a factor in the cause of low back pain is that the lumbar curve flattens on sitting (Akerblom 1948, Andersson et al 1979). The pelvis rotates backwards, due to bodyweight and the pull of the posterior trunk-thigh muscles, forcing the lumbar spine to straighten and thereby considerably increase intradiscal pressure (Nachemson 1963). This pressure is not evenly spread over the discs (a wedging pressure) and can lead to the disc bulging posteriorly and thus stretching surrounding ligaments causing back pain. This is most often found in the discs closest to the sacrum (i.e. 4th and 5th lumbar discs). Epidemiological studies have shown that those people who spend more than half their working life driving are three times more likely to develop an acute herniated disc (Kelsey & Hardy 1975).

A backrest reduces muscle activity by supporting the torso but muscular effort is required to resist the loss in lordosis. Keegan (1953) recommended a trunk-thigh angle of 135° for the most 'natural' lordosis (i.e. muscles in balance and lower intradiscal pressure than when standing). However, this large angle makes any work or activity difficult whilst seated. To compromise and reduce the backrest angle, a lumbar support is important in reclined sitting to maintain lordosis by preventing pelvic tilt.

Several manufacturers now offer seats with adjustable 'in/out' lumbar supports (i.e. variable support at a fixed height on the backrest). We have found that many people would like additional adjustment to the support, particularly an 'up/down' adjustment (i.e. support at a variable height on the backrest), in order to be comfortable.

Consulting the literature it is evident that most studies use objective measures such as electromyography and disc pressure to produce their recommendations for seat design. For example, Andersson et al. (1974) concluded that, for minimum muscle activity and disc pressure, vehicle seat design should aim for a seat pan 14° to the horizontal, a backrest reclined at 30° to the vertical (i.e. 120°) and a lumbar support protruding 50 mm from the backrest. Further work (Andersson et al, 1979) using an experimental chair and X-rays to identify the orientations of the sacral end plate and the superior surface of the 1st lumbar vertebra (L1), the angle thus formed being the total lumbar angle, showed that a lumbar support protruding 40 mm from a vertical (90°) backrest changed the total lumbar angle from a mean of 28.9° (no support) to 46.8° (with support). The total lumbar angle whilst standing being a mean of 59.8° (using a separate group of subjects). The authors therefore recommended a 40 mm support to maintain a similar extent of lordosis during sitting and standing. Furthermore, they found that the height of this 40 mm support within the range of L1 to L5 had little influence upon the total lumbar angle.

The findings from the above studies show considerable agreement but people, whether sitting for relaxation or to perform work, rarely assume or maintain an anatomically 'correct' posture. Seat design should encourage a good posture which allows the occupant to achieve a high standard of task performance whilst remaining comfortable.

AIMS OF THE STUDY

This study was designed to replicate the work by Andersson et al. using subjective comfort instead of objective measures. More specifically, they were to:-
* collect anthropometric data on the lumbar spine
* determine the preferred position of a lumbar support (both 'in/out' and 'up/down').
* investigate the effects of posture (i.e. standing vs sitting upright vs sitting reclined) upon the lumbar curve.
* investigate the effect of the chosen position of a lumbar support upon the extent of lordosis.

DESIGN OF EQUIPMENT

An experimental chair, standing frame and 2 profile recorders were designed which enabled standardised vertical profiles of the spine to be recorded.

The chair was modelled on the chair used by Andersson and colleagues (see figure 1). The profile recorders were similar in design to that used by Branton (1984) and consisted of wooden blocks with holes drilled along their lengths at intervals of 10 mm. These blocks can be fitted to the chair and standing frame so that a subject's spine rests against them from the sacrum upwards. Plastic probes (6 mm diameter, tapering to 1 mm at the end) can then be inserted through the holes until they come into contact with the subject's spine. The spinal profiles can be quantified by removing the recorders, laying them flat on paper and tracing along the curves and then measuring the displacement of each probe from the backrest.

Figure 1. Experimental chair with spinal profile recorder.

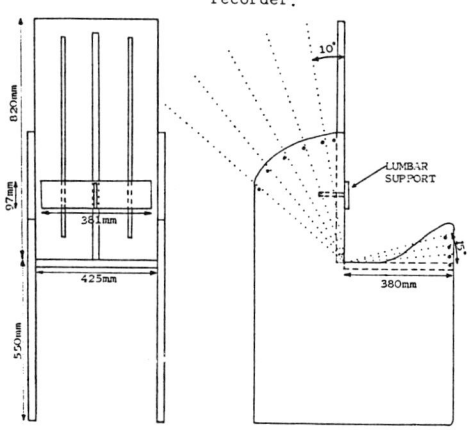

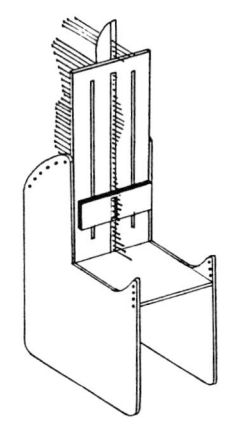

METHOD

Spinal profile data were recorded for 20 young healthy undergraduates (10 males, 10 females) whilst standing upright and in three sitting conditions. The first ('90°') was a backrest angle of 90° and a horizontal seat pan. The second ('120°') was a backrest angle of 120° and a 15° seat pan angle with the lower legs positioned vertically (as recommended by Andersson et al (1974) for minimum disc pressure and myoelectric activity in a vehicle seat). The third ('car') being identical to the above but with the legs extended and with the feet on a raised floor to stimulate a typical car seat height. In this inclined posture, the subjects were asked to keep their heads vertical. Profiles were taken both with and without the provision of a lumbar support. When a lumbar support was used, it was carefully positioned to be in the preferred location as decided by each subject. A second study using a further 42 subjects (27 males, 15 females) investigated preferred lumbar support position only for the above sitting conditions.

The spinal profiles were transferred to paper and then digitised with respect to the seat back at 10 mm intervals up the vertical length of the spine.

RESULTS AND DISCUSSION

The main findings of this study are shown in tables 1 and 2 and figure 2.

Table 1. Anthropometric details and preferred location of the lumbar support.

	Males n = 37 $\bar{X}$ (SD)	Females n = 25 $\bar{X}$ (SD)	All n = 62 $\bar{X}$ (SD)
Stature (mm)	1806 (61)	1663 (80)	1748 (99)
Weight (kg)	75 (9)	62 (7)	70 (11)
Sitting height (mm)	942 (29)	874 (42)	914 (49)
[1] Height of L1 (mm)	298 (20)	293 (18)	296 (18)
[1] Height of L5 (mm)	210 (18)	211 (16)	210 (17)
[2] Preferred support height (mm)			
'90°'	245 (16)	234 (20)	240 (18)
'120°'	223 (16)	210 (19)	218 (18)
'car'	219 (19)	209 (21)	215 (20)
[3] Preferred support distance (mm)			
'90°'	20 (4)	20 (4)	20 (4)
'120°'	20 (4)	20 (4)	20 (4)
'car'	20 (4)	20 (4)	20 (4)

Notes 1. Males n = 10, Females n = 10.
2. Lumbar support height measured from seat pan to middle of support.
3. Lumbar support distance measured from surface of lumbar support to surface of backrest.

The heights of L1 and L5 were found to be nearly identical between the 10 males and 10 females measured. However, the females consistently preferred the lumbar support approximately 10mm lower than the males in all 3 seating conditions. The chosen height was lower in the reclined backrest conditions because of the differing rotational axes of the backrest and the body (ishial tuberosities).

The preferred 'in/out' location of the support was 20mm forwards of the backrest for both sexes and all 3 seating conditions. This chosen value is exactly half that recommended by Andersson and his colleagues. We recorded the lumbar curve with a 40mm lumbar support with 4 of our subjects and found that it nearly restored lordosis to its value when standing but it was considered unacceptable in terms of comfort.

Sitting upright was found to cause a large decrease in lordosis compared to standing erect. For example, the maximum displacement of the lumbar spine from the backrest changed from a mean of 60.5mm when standing to 20.3mm when sitting upright. This was reduced to 15.9mm when reclined and a further slight reduction to 11.6mm with the legs extended. Male and female data were nearly identical. The preferred lumbar support positions were found to produce mean maximum displacements of 33.2, 28.7 and 27.3mm for the three sitting conditions (upright; reclined; reclined with legs extended). These are approximately half the value observed when standing erect.

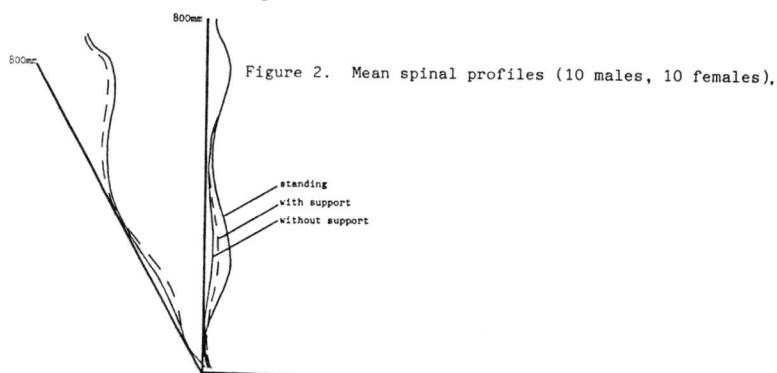

Figure 2. Mean spinal profiles (10 males, 10 females).

standing
with support
without support

CONCLUSIONS

Recommendations for lumbar supports based upon objective criteria were not found to be in complete agreement with our subjective data. We have found that the height of the support affects comfort, even though the lumbar angle may not change, and we would recommend a range of height adjustment from 195-260mm from the compressed seatpan to the centre of the support. The requirement for an adjustable 'in/out' support is secondary and would cover the range 13-27mm forward of the backrest, even though lordosis is only maintained at approximately half that observed when standing.

REFERENCES

Akerblom, B., 1948, Standing and sitting posture with special reference to the construction of chairs. Thesis, Stockholm Nordska, Bockhandelm.

Andersson, B., Murphy, R., Ortengren, R., Nachemson, A., 1979, The influence of backrest inclination and a lumbar support on lumbar lordosis. Spine, 4, 1, 52-58.

Andersson, B., Ortengren, R., Nachemson, A., Elfstrom, G., 1974, Lumbar disc pressure and myoelectric activity during sitting - studies on a car seat. Scand. J. Rehab. Med., 6, 128-133.

Branton, P., 1984, Backshapes of seated persons - how close can the interface be designed? Applied Ergonomics, 15, 2, 105-107.

Keegan, J.J., 1953, Alterations of the lumbar curve related to posture and seating. Journal of Bone & Joint Surgery, 35, 3, 367-389.

Nachemson, A., 1963, Influence of spinal movements on lumbar intradiscal pressure and on the tensile stresses in the annulus fibrosus. Acta Orthopaedica Scandinavia, 33, 1, 183.

Kelsey, J., Hardy, W., 1975, Driving of a motor vehicle as a risk factor for acute herniated lumbar disc. American Journal of Epidemiology, 120, 1, 63-73.

Table 2. Measurement of lordosis: mean displacements of spinal profiles at L1, L5 and the maximum displacement of ther lumbar curve for the various postures without (A) and with (B) a lumbar support (10 males, 20 females).

		Standing	'90°' A	'90°' B	'120°' A	'120°' B	'car' A	'car' B
Displacement at L1 (mm)	male	55.2	19.9	32.6	10.1	20.5	7.3	21.7
	female	55.0	15.5	27.7	7.2	14.6	5.6	15.1
	all	55.1	17.7	30.2	8.7	17.6	6.5	18.4
Displacement at L5 (mm)	male	51.7	14.7	28.6	15.0	28.1	8.8	25.9
	female	50.4	15.4	27.2	12.8	25.8	10.2	25.0
	all	51.1	15.1	27.9	13.9	26.9	9.5	25.9
Maximum displacement (mm)	male	60.1	21.8	34.5	16.6	30.2	10.8	27.4
	female	60.9	18.7	31.9	15.1	27.2	12.4	27.1
	all	60.5	20.3	33.2	15.9	28.7	11.6	27.3
Height of max. displacement (above seat pan)	male	263.5	307.0	276.0	237.5	228.0	243.5	215.0
	female	253.5	251.9	240.5	193.0	195.6	196.5	199.5
	all	257.0	279.5	258.3	215.3	211.8	220.0	207.3

Working Posture

INDUSTRIAL MAINTENANCE TASKS INVOLVING OVERHEAD WORKING

C.M. HASLEGRAVE, M. TRACY and E.N. CORLETT

Department of Psychology
University of Nottingham
Nottingham NG7 2RD

Many heavy manual jobs still exist in repair and maintenance work, even in high technology industries. These aspects are often forgotten in the design and layout of equipment, so that problems occur because workers have difficulty in access to components and may have to adopt awkward body postures while applying high forces. Some of these tasks involve overhead working while either standing or lying supine. The strength and reach capabilities of workers are being studied in the laboratory to provide better techniques for assessing such tasks and to develop guidelines for the design of fasteners and the use of handtools in overhead working.

INTRODUCTION

Industry still has many tasks which require heavy manual work and many examples of these occur in repair and maintenance. These tasks are necessary even in high technology industries, but unfortunately they are often forgotten or ignored during the design and layout of equipment. Maintenance workers often have to work in difficult or confined spaces and problems with these have been reported by various industries, including mining (Ferguson et al 1985) and nuclear reactors (Seminara & Parsons 1982). Maintenance tasks of course occur in all industries and frequently in agriculture and construction work at remote sites, where there may well be less access to powered assistance in the tasks and so involve more manual work and greater use of handtools.

The problems arise mainly because workers carrying out maintenance have great difficulty in obtaining access to the components they have to remove or adjust, and may have to adopt awkward body postures while applying high forces.

Postures which involve twisting or bending combined with force exertions are particularly likely to cause injuries. Another cause of injuries is a sudden overload or release of force. One recent report from the United States (Mital & Aghazadeh 1985) indicated that overexertion was the second most frequent cause of injuries which occur during the use of handtools.

In order to provide designers and planners with better information on the body loads and stresses which occur during maintenance work, these types of tasks are being studied in the laboratory. The objectives of the study are to investigate the strength and reach capabilities of workers and to develop better techniques for assessing such tasks when workers are liable to be required to adopt awkward working postures while operating controls or using tools. A set of recommendations will then be developed for safe levels of force application as well as guidelines for the design of fasteners and the use of handtools.

Two of the tasks which have been studied involve working overhead, either standing or lying supine. Such tasks occur for instance in vehicle maintenance, where fixings or access points may be difficult to reach, and frequently involve working at arm's length. Reach distance and direction of operation of controls both have a large influence on strength capability, and the relationships have been studied in the experiments.

EXPERIMENTAL MEASUREMENTS

Strength and reach capabilities were measured for groups of male subjects covering the normal range of stature and weight, in working postures which are typical of overhead maintenance tasks, measuring the forces which people could exert on controls in various directions of operation. Single-handed isometric forces were measured on a strain-gauged handle, which was mounted on a rigid framework and could be placed at various locations and orientations relative to the subject being tested. The maximum force was measured because it can be used to determine the upper limit to the forces which can be demanded and also because this measure can be used further to evaluate endurance times for particular tasks, using the research findings which relate these two parameters.

The postures of the subjects were recorded at the same time using a CODA-3 optical scanner, which gives accurate co-ordinates of the various body reference points in space. Both force/time history and body co-ordinates were recorded on-line, and further biomechanical analysis of the loads on the spine and joints could then be carried out to evaluate the internal loadings on the body. The biomechanical model which is being used is based on the lower back model developed by Chaffin (Chaffin & Andersson 1984), but the use of the other models is also being studied in the project and new information on the location of muscles is being collected and incorporated in the model.

The resultant loadings on the body, calculated by biomechanical analysis, will be compared for different postures and control locations, in order to identify postures which are harmful in terms of the stresses on the spine and other joints and control locations where forces can be applied most advantageously.

STRENGTH CAPABILITY WHEN WORKING OVERHEAD

The strength of subjects working overhead was measured both while they were standing and while they were lying supine. The position and orientation of the control was defined (for right handed subjects) relative to the position of the right foot for standing subjects and right shoulder for subjects lying on the floor. Otherwise, subjects were allowed to choose their posture quite freely, so that they were in the optimum position for applying maximum force. It may be seen from the stick diagrams in Figure 1 (which give side views representing the positions of the subject's head and limbs relative to the control location in the right hand) that the subject changes his posture to gain maximum advantage when applying forces in different directions. Such movements are necessary both to bring the strongest muscle groups into play and to brace the body to exert higher forces.

The results of the various tests of overhead force applications are shown in Tables 1 and 2, which give the mean maximum force for the group of six subjects tested in each posture. The position of the control is defined relative to the subject's shoulder and at a distance given as a proportion of arm reach. Thus, when standing, the control was placed forward, rearward and to each side of the subject so that arm reach was at 15° to the vertical. It was placed at a maximum arm reach, and for a sub-set of the subjects at 75% and 50% of maximum arm reach. Similarly, for subjects lying supine, the control was placed directly above the shoulder and at 15° to each side. The force capability was taken as a percentage of the force which the standing subject could exert when pushing at shoulder height.

It may be seen that both when standing and when lying supine, subjects could exert very much higher forces when lifting upwards or pressing downwards than they were able to exert in the horizontal plane, although the difference was considerably greater when standing. The force applied vertically was greatest at maximum reach distance (as shown in Table 2), so that when space is restricted the worker may not be able to exert such high forces in this direction. This was less marked for standing subjects since they were able to increase the effective reach distance by adopting a crouching posture (as may be seen in Fig. 1). By contrast, for both postures, maximum force capability in the horizontal directions increased as the reach distance was decreased, so that a worker may find it easier to work closer when applying forces in a horizontal direction. The effect of direction of force application and interaction between direction and reach distance were statistically significant (at $p < 0.001$).

Subjects were also asked to rate how awkward they found each posture for working, on a scale from 0 - not awkward at all to 10 - extremely awkward. The mean ratings are shown in Table 1 for the maximum reach distance. This shows that lifting and pressing controls were perceived to be easier than force applications in horizontal directions, when the subjects were less able to generate high forces. A similar pattern was found for 75% reach distances (although the results are not shown here), but was less apparent at the 50% reach distance.

CONCLUSIONS

The aim of this study is to investigate the strength and reach capabilities of workers and to develop techniques for assessing such tasks when workers are liable to be required to adopt awkward working postures. The results presented here are preliminary findings in this investigation, dealing with tasks involving working overhead, and illustrate some of the factors which will need to be considered. It is important for designers to be aware of how workplace constraints may reduce a worker's strength capability below the levels he can achieve in normal working postures.

ACKNOWLEDGEMENTS

This project is being funded by the ACME Directorate of the Science and Engineering Research Council.

REFERENCES

Chaffin, D.B., & Andersson, G., 1984, Occupational Biomechanics, (John Wiley and Sons), pp206-209.

Ferguson, C.A., Mason, S., Collier, S.G., Golding, D., Graveling, R.A., Morris, L.A., Pethick, A.J., & Simpson, G.C., 1985, The ergonomics of the maintenance of mining equipment (including ergonomic principles in designing for maintainability), IOM Report TM/85/12, (Edinburgh: Institute of Occupational Medicine).

Mital, A., & Aghazadeh, F., 1985, A review of handtool injuries, In Ergonomics International 85, edited by I.D. Brown, R. Goldsmith, K. Coombes & M.A. Sinclair, (London: Taylor and Francis), pp85-87.

Seminara, J.L., & Parsons, S.O., 1982, Nuclear power plant maintainability, Applied Ergonomics, 13, 3, pp177-189.

Table 1. Variation in force capability at maximum reach (as % push force capability in standard posture).

Posture	Push	Pull	Across Body	To Side	Lift	Press
FORCE CAPABILITY (%)						
Lying Supine, working overhead						
0°	41	44	38	33	165	145
15° Left	42	42	34	35	142	134
15° Right	45	39	51	34	136	139
Standing, working overhead						
15° Forward	37	43	37	45	244	177
15° Rearward	36	28	36	45	196	187
15° Left	35	42	41	48	253	172
15° Right	39	43	41	49	211	193
SUBJECTIVE RATING OF AWKWARDNESS						
Lying Supine, working overhead						
0°	7.0	7.2	7.2	7.1	3.3	3.3
15° Left	6.8	7.0	7.2	7.2	3.9	3.3
15° Right	7.5	7.3	7.4	7.6	3.2	2.8
Standing, working overhead						
15° Forward	8.2	7.8	8.8	8.6	5.1	5.1
15° Rearward	6.8	7.5	6.4	7.6	3.0	3.0
15° Left	6.6	7.5	7.5	7.1	4.1	4.2
15° Right	6.4	6.0	6.8	6.3	4.2	3.8

Figure 1. Variation in posture with direction of force exertion (control positioned at 50% maximum overhead reach).

PUSH   PULL   ACROSS BODY

TO SIDE   LIFT   PRESS

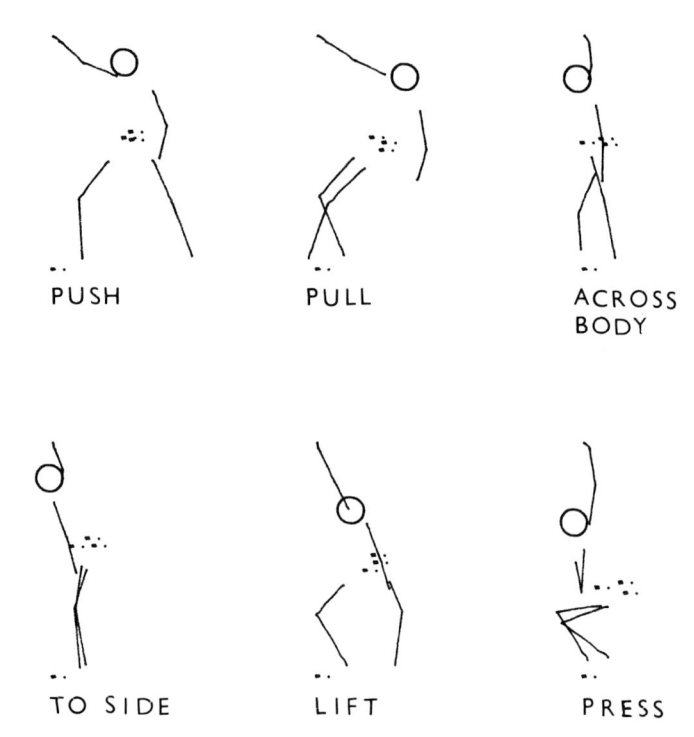

Table 2. Variation in force capability with reach distance and direction of force application (as % push force capability in standard posture).

Posture	Direction of Force Application					
	Push	Pull	Across Body	To Side	Lift	Press
FORCE CAPABILITY (%)						
100% REACH						
Lying Supine, 0°	44	47	35	34	149	132
Standing, 15° Forward	37	43	37	45	244	177
75% REACH						
Lying Supine, 0°	53	48	35	48	108	115
Standing, 15° Forward	62	57	46	63	266	195
50% REACH						
Lying Supine, 0°	80	58	42	54	68	81
Standing, 15° Forward	75	66	51	68	160	143

AN EVALUATION OF OFFICE SEATING

I.G. KLEBERG and J.E. RIDD

Ergonomics Research Unit, Robens Institute
University of Surrey, Guildford
Surrey GU2 5XH

Postural aspects appear to be crucial in the onset of musculoskeletal disorders amongst sedentary workers. Ten different chairs were examined in an attempt to identify beneficial design aspects that might help to ameliorate the problem. This study indicated that the design aspects of the operating mechanisms were critical if the chairs were to be adjusted frequently and correctly. To assist in this a number of suggestions are made for future chair design.

INTRODUCTION
   The number of people working in a seated posture has increased during the twentieth century and particularly in recent years with the growing use of new technology. Similarly, there has been an increase in complaints of musculoskeletal disorders of the low back and neck associated with the seated posture and the task being performed (Magora 1972). Clearly the design of the furniture being used is critical in this context. Whilst many authors have recommended preferred heights and angles for seats and seat backs (Grandjean 1980, Bendix 1986) no in depth studies could be found relating to the design and location of the adjustment mechanisms; these are considered to be important since the degree of difficulty associated with their operation will govern the frequency of their use and the accuracy of adjustment achievable. The study reported here considers this point and also examines the problems of selecting an appropriate chair for office work. Two aspects were considered in relation to chair selection, firstly the question of which type of chair was preferred by a defined user group, and secondly the question of whether or not this group's perceived design requirements for office seating equated with those of ergonomists.

METHODS

Ten different chairs were allocated to ten test subjects according to a random (Latin square) pattern. Each subject used each chair for one day; each trial day being separated from the next by a single day on which the subjects returned to their own standard issue chair. At the end of each day's trial the subjects were asked to fill in a questionnaire concerned with a number of points relating to the comfort and design of the chair and to the day's work. The data were then coded for subsequent analysis on the University's PRIME computer system using the SPSSX package.

As a control these subjects were also required to complete questionnaires relating to their own normal chair.

A second subject group was asked to evaluate the same chairs from an ergonomic viewpoint only. Each of the test chairs was examined in random order for approximately 10 minutes, by each of ten ergonomists. After each assessment one of the questionnaires described above was completed but with two sections omitted - those dealing with questions of the comfort attainable and with the type of work undertaken. Hence the specific aspect under consideration was chair design particularly in relation to the shape, location and method of operation of the various adjustment mechanisms.

Where cam-operated levers or hand wheels provided the means of release for adjustment then, the forces required to move these from the locked position were also measured.

SUBJECTS

Ten female office staff involved in secretarial and clerical duties took part in the study (group 1). Ten ergonomists, nine male and one female, constituted group 2.

The age, height and weight of each of the subjects were recorded.

RESULTS

Table 1. Anthropometry.

	GROUP 1 Mean	SD	50th %ile WOMEN *	GROUP 2 Mean	SD
AGE (Yrs.)	35.1	10.9	-----	32.7	6.6
HEIGHT (cm.)	164.9	5.9	161.0	174.6	9.8
WEIGHT (kg.)	56.6	7.6	62.5	68.9	7.6

* From 'Bodyspace' (Pheasant, 1986; Table 4.1.)

Table 2. Analysis 1: Mean rating for the overall assessment of each chair by Group 1.
Analysis 2: Comparison of the mean ratings for each chair from the two subject groups.

	ANALYSIS 1. GROUP 1			ANALYSIS 2. GROUP 1			GROUP 2		
Chair	M	SD	Order	M	SD	Order	M	SD	Order
G	**47.4	15.4	11	48.3	18.3	11	60.4	10.8	11
H	37.3	22.6	6	35.4	22.5	5	41.4	8.4	7
I	37.4	17.3	7	37.5	16.3	7	36.0	9.5	6
J	25.9	13.7	3	28.2	14.6	2	34.8	10.2	4
K	*45.3	11.5	10	39.5	14.9	8	31.1	10.3	3
L	31.0	13.7	4	28.4	14.4	3	35.2	20.1	5
M	*21.9	9.4	1	24.5	16.0	1	27.5	13.2	1
N	38.4	15.7	8	41.2	17.6	9	42.3	15.0	8
O	40.5	17.2	9	42.3	15.9	10	46.9	12.1	10
P	24.6	12.3	2	37.3	21.7	6	46.2	17.0	9
Q	32.8	16.4	5	29.8	12.7	4	29.4	10.7	2

(Chair G was the normal work chair for Group 1.)
For Analysis 1:-
* Assessment is significantly different from mean, p=<0.05.
** Assessment is significantly different from mean, p=<0.02.

DISCUSSION

The chairs may be grouped in a number of ways, (eg) those with and those without arms, high back or low back, single lever/multiple function and multi lever/single and multiple function (see figure 1); however, no significant correlation could be found between any of these groupings and the overall assessment by the subjects.

Table 2, Analysis 1, shows the mean overall assessment of the ten chairs by the user group. The results to specific questions are not reported here for reasons of brevity, however it is interesting to note that the ranking for overall assessment is closely reflected by that for only one of the specific assessment points - namely chair appearance. There was no evidence for any other criterion to have been commonly used to judge the chairs.

For questions relating to design, location and ease of use of the chair and its operating system the comparison between subjects' and ergonomists' assessments did not reveal large differences of opinion for most chairs (Table 2, Analysis 2). However, there were two examples (for chairs K and P) where the difference in the relative assessment was greater than with the others, although

these were not significant. Chair K was rated more highly
(and chair P was rated lower) by ergonomists than by the
user group (Group 1). Chair P, rated second by Group 1 in
the overall assessment, was considered to be comfortable and
aesthetically pleasing - the problems in adjustment (the
cause of its bad rating by Group 2) was secondary. On the
other hand Chair K, rated tenth by Group 1 for the overall
assessment was considered uncomfortable and "ugly";
however, when considering design factors alone Group 2
marked this chair third. It is hypothesised that
considerations beyond the specifically asked question
influenced the responses in respect of these two chairs in
particular, as supported by the ´appearance´ relationship
mentioned earlier. If so the results appear to indicate
that ergonomic design is given a relatively low level of
importance in the perception of the user population.

Figure 1. The ten test chairs used in the study:

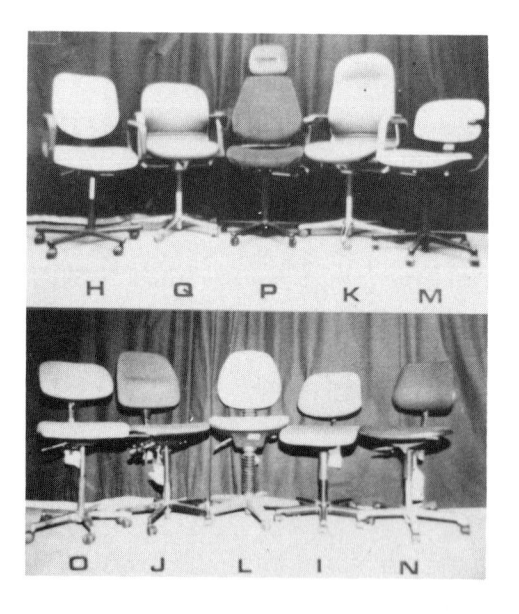

The test chairs exhibited between them ten different
operating mechanisms for effecting any single adjustment.
This variety of operating systems alone would lead to
confusion for any user moving between chairs, but this is
further complicated by the non-compatability with population
stereotypes in terms of the relative movement of levers and
chair - some systems required the lever to be moved in a

forwards direction and others in a backwards direction to effect a similar chair movement. However, the most common mechanism was one where an upward movement on a lever would unlock the position enabling the chair or chair back to be moved either against or with a spring to the required position. Some activities were gas or spring assisted, some simply used a lever and cam to lock a movement, others employed a hand wheel and screw for this same purpose; generally the last two systems were to be found on the cheaper chairs, however, it is interesting to record that the rank orders given in no way correlate with the purchase prices (the range being from £70-£300).

The gas and spring assisted mechanisms were preferred; indeed, of particular concern was the force required to operate some of the cam levers and hand wheels. These were found to be of a similar order to the values quoted by Pheasant and Scriven (1983) for the female maximum capacity when operating cylinder screws. However whilst the subjects in that study could adopt the most appropriate posture to exert the necessary force our subjects could not, and were often forced into disadvantaged postures in order to adjust the chair to the appropriate position, necessarily whilst seated. Clearly, by requiring sections of the user population to perform actions in excess of their capacity some manufacturers are consigning their products to a lifetime of misuse.

CONCLUSIONS

Whilst this study did not set out to look specifically at the area of operational design of these chairs the questionnaire analysis (including a consideration of the many appended comments) did reveal this important problem which is not properly covered by the present literature and which has apparently been given insufficient thought by many of the manufacturers.

In an attempt to produce a framework for design considerations the following recommendations based on this study's findings are made:-

The forces required to operate any mechanisms should be at a level for a precision, rather than a power grip, since these exertions are invariably made in awkward postures.

Levers should be within easy reach, should operate with uniformity and in accordance with population stereotypes.

The lever should be designed to fit the user's grasp in preference to designing for appearance.

Where one lever controls more than one chair movement these must be independently adjustable; where there are a number of levers for different operations these should be easy to differentiate and to identify.

Where the tension of the springs (that assist a chair movement) can be altered by the user to suit their own requirements proper account must be taken of the likely range of body weight of the projected user group; (ie) sufficient range of adjustment should be incorporated to ensure both that heavy subjects can retain posture support and that light subjects do not need to exert high static forces to maintain their position.

Similar account should be taken where body weight and the position of it's centre of gravity are used to assist chair adjustment.

It was clear that the more simple a system was to operate, the more often it would be used and the better the chances that the user's posture would be correct. Where these points are ignored there is a tendency not to use such facilities; this in turn means that the chairs are used incorrectly - the required postures are not supported, muscles become tense and musculoskeletal discomfort is experienced.

This study serves to emphasise the philosophy that all aspects of a task and its operation must be considered if the finished article is to be well designed, effective, efficient in use, and pleasing to the user - unfortunately one does not automatically follow from the other.

ACKNOWLEDGEMENTS

The authors wish to thank the subjects who participated in the study and also the companies which supplied the chairs, in particular - NKR, RH Form, Ergonom, Kinnarps, Grammer and Nordpatent.

REFERENCES

Bendix T., 1986, Chair and table adjustments for seated work. The ergonomics of working postures, (Taylor & Francis).

Grandjean E., 1980, Fitting the task to the man, (Taylor & Francis).

Magora A., 1972, Investigation of the relation between low-back pain and occupation. 3. Physical requirements: sitting, standing and weight lifting. Industrial Medicine, 41,5-9.

Pheasant S., 1986, Bodyspace, anthropometry, ergonomics and design, (Taylor & Francis).

Pheasant S. & Scriven J., 1983, Sex differences in strength. Some implications for the design of handtools. Proceedings of the Ergonomics Society's Conference, edited by K. Coombes, pp. 9-13.

**THE EFFECT ON NURSES' BACKS OF A SWITCH FROM
BEDS TO CHAIRS FOR USE BY KIDNEY DIALYSIS PATIENTS**

M. PORTER[*] and A. FARROW[**]

[*] School of Occupational Studies
Newcastle Polytechnic
Newcastle-Upon-Tyne NE1 8ST

[**] School of Applied Consumer Science
Newcastle Polytechnic

There is currently a move within Dialysis Units to replace some of the traditional hospital beds with reclining chairs. This study investigated the effects of this change by using standard pain/discomfort diagrams and observations of the patients undergoing treatment. It is concluded that the nurses bend and twist more often when attending to patients in the chairs. The design of the chairs used could be improved.

Introduction
In 1983 the local Renal Dialysis Unit concerned was treating ten patients per session. To overcome an increased demand for renal dialysis, some of the beds were replaced in 1985 by Parker Knoll "NORTON RECLINER" chairs. They were selected, by the hospital, as the only chairs easily available with acceptable fire-resistant and hygiene properties. The chairs have major dimensions which are smaller than the beds (Table 1), so allowing five extra patients to be treated each session.

A patient's vein is connected to the dialysis machine via a hollow needle and plastic piping, in a manner similar to that of a blood donor, except that the purified blood is returned, usually via a second needle. The dialysis needles must be put into the arm at a shallow angle to, and along the line of, the blood vessel. As the arm level of a patient on a chair (580 mm) is some 280 mm lower than on a bed and the area around the chairs congested with equipment and other patients, the nurse is forced to work in an awkward posture. The nurse will usually bend down forward and laterally creating a twisted bent posture. The nurses thought that this bending, despite the low loads that were involved, created additional stress on their backs which might lead to back pain. There is evidence in the literature that this could be so. (eg van Wely 1970)

Table 1. Chair and Bed Dimensions.

Dimension	Chair (mm)	Bed (mm)
Height (arm rest)	580*	860
Width	890	900
Length	990	2200
Length (Reclined)	1660	2200

*Including a 50 mm allowance for the pillow used to rest The patient's arm on.

The major problems of backpain within the nursing profession have been discussed elsewhere; for example by the Ergonomics Research Unit (1986). These studies are mainly concerned with the heavy manual work of patient handling while the nurses working in the Renal Unit are mainly involved in light, fine motor tasks.

The patients mostly undergo dialysis 3 times a week and for 3-5 hours at a time. This treatment will usually continue until either a kidney suitable for transplantation becomes available or the patient dies. During dialysis both waste products and fluid volume are reduced, the latter possibly leading to a fall in blood pressure so large that the patient may "black out". The fluid loss that has occurred is determined by calculating the patient's weight loss during the session.

Patients who, as their blood volume is reduced, are not expected to "black out" are selected to use the chairs. Thus it can be expected that the patients in the chairs are the more stable and should require less attention than those in the beds. It was originally expected that these patients would usually sit upright but most prefer to sit back and rest their feet on "ordinary" chairs as no purpose designed footstools are available. These add further to the congestion within the unit.

Unlike the beds used in the unit, the chairs have no built-in weighing mechanism which displays the patient's weight. A portable set of scales were therefore introduced in to the unit and the patients then must get up to stand on them. They cannot, of course, move far from their chairs because the blood lines are kept connected. The scales weigh 7 Kg and are awkward for the nurses to move around the unit. This would appear to create stress on the nurses' backs additional to that found in a bed only dialysis unit.

Methodology

After initial visits to the unit, four areas were selected for investigation.
1. Do the patients find the chairs comfortable during the treatment?
2. Did the nurses have back problems?
3. Was the back problem (if present) one of discomfort or pain and did it increase in intensity with time?
4. Did the nurses bend more frequently at the beds or at the chairs?

The chairs were found uncomfortable by staff who sat on them to simulate a patient undergoing dialysis during a hospital open day, but this was not supported by the results of a small study of four patients over two sessions. Patient posture changes were observed and interviews conducted but these did not lead to the expected conclusions. It is hypothesised (Porter 1985) that because the chair is essential to the patient's life preserving treatment; they did not consider the concept of comfort important. This, of course, was not true for the nurses who sat on the chairs during the demonstration. This work is ongoing.

To determine the extent that the nurses would report problems, a "body regions diagram" (Corlett & Bishop 1976) was adopted and all the nursing staff asked to report areas of discomfort/pain on it. The postures and the length of time for which they were held by the nurses were recorded. Bending levels at $30°$ intervals were found to be the most easily recognised and were recorded together with details of other postures used; eg kneeling, crouching, sitting and bent knee bending.

Results

The unit had been visited by both researchers on several occasions prior to the data collection session. In all sixteen patients, eight dialysing on chairs and eight on beds, were observed being put on, during and being taken off dialysis.

There were 21 nurses questioned and given the pain/discomfort diagram; 15 (71%) identified areas of pain and 20 (95%) reported areas of discomfort.

Table 2. Discomfort/pain reported in the sites specifically associated with bending and lifting

	Nurses Affected		
	Renal Unit		Stubbs & Buckle (1985)
Area	Pain	Discomfort	Pain (% only)
Upper back & neck	2(22.2%)	2(14.3%)	(4.4%)
Mid back	0(0%)	2(14.3%)	(4.5%)
Low back	6(66.7%)	8(57.1%)	(53.7%)
Buttocks	1(11.1%)	2(14.3%)	(9.9%)
Multiple back sites	0(0%)	0(0%)	(27.5%)
TOTAL	9(100%)	14(100%)	(100%)

Table 3. Reported discomfort/pain related to the length of time on task

Increase in pain/discomfort reported by the nurses to be related to the length of time on shift.

	Discomfort	Pain
Related	17(80.9%)	13(61.9%)
Unrelated	3(14.3%)	2(9.5%)

Table 4. Observed bending postures.

Posture	Bends at Chairs	Bends at Beds
Bend at 150°	4 (1.0%)	0 (0%)
Bend at 120°	3 (0.75%)	3 (0.75%)
Bend at 90°	77(19.3%)	8 (2.0%)
Bend at 60°	91(22.8%)	25 (6.3%)
Bend at 30°	73(18.3%)	80(20.0%)
Twisted upright	3 (0.75%)	10 (2.5%)
Sitting	1 (0.25%)	0 (0%)
Crouching	12 (3.0%)	6 (1.5%)
Kneeling	3 (0.75%)	0 (0%)
TOTAL	267(66.9%)	132(33.1%)

Table 5. Observed frequency of Twisting.

	Chair	Bed
Number of occasions twisted	17 (4.5%)	23 (6.1%)
Number of occasions bent but not twisted	234 (62.1%)	103 (27.3%)
TOTAL	251 (66.6%)	126 (33.4%)

Discussion and conclusions

Although the nurses were not generally involved in heavy lifting they did report a large number of sites that gave discomfort or were painful. (Table 1.). The levels of pain/discomfort were thought, by the nurses, to increase during the shift. The data in Table 3 was found to be significant (Chi^2=8.45, 1 d.f., $p<0.01$ and Chi^2=5.0, 1 d.f., $p<0.05$ for "discomfort" and "pain" respectively).

The difference between the frequency of bending at the chairs and beds (Table 4) was found to be significant (Chi^2=17.46, 8 d.f., $p<0.05$). The degree of twisting and the duration for which the posture was held was not found to be significantly different between the chairs and beds.

The location of the chair/bed in the unit was found to have no effect on the nurses' postures although some were clearly more difficult to work at than others. The nurses, for example, generally disliked working at the corner chair and one even described it as "agony and lethal to work at".

It can be concluded that the nurses within the unit have back problems despite the low loads involved in their work (Table 1); 16.8% of the total sites of discomfort and 11.1% of the total areas of pain reported were those related to back problems. The other frequently reported site of pain/discomfort was the feet but this was expected as they spent most of their shift on their feet and often walked over 3 Km around the unit during the day.

The nurses were found to bend and twist more often when dealing with patients on the chairs and this must give cause for concern as the trend is to replace beds with chairs.

It is hoped that it will soon be possible to produce a prototype chair for renal dialysis patients. However, the methodological problems to validate the design and the prescriptive fire and hygiene regulations make this a difficult task.

References

Corlett, E.N. and Bishop, R.P., 1976, A Technique for Assessing Postural Discomfort, Ergonomics, 19, 175-183.

Ergonomics Research Unit. 1986, Back pain in nurses – summary and Recommendations, Robens Institute of Industrial and Environmental Health and Safety, University of Surrey.

Porter, M., 1985, Preliminary Report to the medical staff working in the Renal Unit on the suitability of available chairs for use in Dialysis. (unpublished).

Stubbs, D. and Buckle, P., 1984, The Epidemiology of Back Pain in Nurses, Nursing, 2, 935-936.

van Wely, P., 1970, Design and Disease, Applied Ergonomics, 1, 262-269.

Human-Computer Interface Design

**SOFTWARE ERGONOMICS AT THE SHARP END:
DEVELOPMENT OF A NEW INTERFACE FOR A CAD PACKAGE**

S.E. POWRIE

PAFEC Ltd
Strelley Hall, Strelley
Nottinghamshire NG8 6PE

Aspects of the development of a new front-end for a computer-aided design package are described. It is concluded that the usefulness of design guidelines, where they exist at all in this context, is limited without knowledge of the underlying data. An evaluation strategy for the new interface is outlined.

THE PRESENT SYSTEM
PAFEC's DOGS (Design Office Graphics System) is a well established package used in its standard form mainly for engineering design and draughting. We are concerned here with the standard 2D interface which has remained substantially unchanged since its launch some 7 years ago. The work described has been carried out on the screen display component of the interface: for the most part the content and structure of the underlying dialogue is unaltered. A brief account of the current mode of interaction with DOGS follows.

Once the user has entered the main part of the program - i.e. creating or editing a drawing - he is able to choose from 20 menus of drawing and management options. Each menu may contain up to 40 options. All menus are immediately accessible. Menu names are listed to the left of a large graphics area with a vertical display of the numbers from 1 to 40. Options are selected by hitting the chosen menu name and option number with the screen cursors, by typing an abbreviated version of the name and the option number, or by selecting from a card menu on a tablet. Many users will not have a tablet available but will use the menu card as an aide-memoire. The screen displays typed input and system responses in a text area located below the graphics area. Also shown are a help option, an accept option used in various drawing operations and a protractor which can be used to input angles.

The impetus for changing these arrangements came from the marketing side of the company. It was suggested that DOGS should make use of features such as pop-up menus and icons to provide a more attractive display and improve user friendliness. Ideally the system should be capable of providing a display tailored to individual customers' and users' requirements.

SPECIFICATION

After some consideration it was decided that the new front-end for DOGS should be implemented by an independent module, which became the focus of development effort. The module, later known as MICE (Menu Input Control Environment), would include stacking pop-up menus, icons, dynamic status display items, and multiple text and graphics areas. MICE would be capable of handling its own internal actions relating to screen display and of sending other actions back to the applications program. The applications program would also be able to trigger selected MICE actions. A major criterion was that MICE should continue the house tradition of terminal or workstation independent software and to this end it should work with new in-house packages for graphics generation and general purpose functions. Both these were still in the course of development.

A preliminary specification, subsequently somewhat modified, for the DOGS interface was drawn up. (The new interface covers the displays on entering and exiting the program as well as the main drawing screen: only the last is discussed here). This replaced the menu names with icons and introduced stacking pop-up menus; four separate text areas for prompts, error messages, help messages and typed input; a clock; cursor co-ordinate display; screen keyboard and protractor; and a dog which would adopt different poses depending on the state of the system.

At this stage the author joined the project team and became responsible for the detailed specification of the interface. The essential elements of the design problem now were a preferred set of screen display techniques, a well established existing interface, and a heterogenous user population. As well as users whose first experience of the package would be through the new interface, there would be many (but not all - MICE would be optional) existing users transferring from the old interface to the new. There were users in colleges and universities whose main aim was to learn about the potential of CAD: this group were likely to use most of the options available. By contrast many industrial users performed specialised tasks, operating within a small sub-set of the menus. Finally, as with other systems, there were novice, expert and partially expert users. In some areas the interests of these groups conflicted: radical reorganisation to produce a more coherent interface would, for example, in the short term at least, make transfer harder for existing users. It was therefore doubly important to ensure that any changes made were genuine improvements. A further complication arose from the need to produce similar interface features on terminals or workstations of widely differing capabilities.

Progress to date (December 1986) on two facets of the problem is now examined in more detail.

ICONS

Aside from their space saving properties, the usefulness of icons lies in their potential for rapid interpretation and hence reduction in cognitive load (Foley et al, 1984). To achieve this it is desirable that the item depicted in the icon can be identified, and that there is some link between the image used and the function represented (Hemenway, 1982). (Note that it is possible to use icons of completely arbitrary significance, but in this case presumably the advantage in terms of cognitive load is lost.) There are also suggestions that ambiguous icons can be improved by the incorporation of textual clues (Bewley et al, 1983), and that icons preferred by the user are more memorable (Funt, 1984). Since the initial work on the DOGS icons was undertaken, Rogers (1986) has shown that icons depicting operations on concrete items are more effective than images relying on analogy, and Gittins (1986) has summarised existing data and provided guidelines for design.

Prototypes of icons to represent menus had already been displayed on internal noticeboards with an invitation to identify them. It was apparent from this that not all were identifiable, and a number of alternatives were generated. These attempted to take into account available data and considerations such as the need to conform with emerging industry stereotypes, to be internationally comprehensible, and to be as simple as possible. This last feature was important: on some terminals the space for icons would be small, and they would be displayed using vector graphics techniques in a small space under less than perfect resolution. Furthermore, complicated icons take longer to produce on screen. In some cases these considerations conflicted. For example the original choice of a bin icon for the 'delete' menu was potentially misleading since this image is often used for deletion of entire files, whereas the items to be deleted here were individual lines etc. on drawings, but it proved very difficult to produce a simple but recognisable eraser, the proposed alternative. It was possible to generate images for a number of the menus which were both concrete and fitted a general drawing office metaphor (e.g. a filing cabinet for the archive menu) but more abstract functions were much more problematic. The worst case, still unresolved, was the derivation of an image for the 'symbol' menu which does not rely on a possibly untranslatable pun. The extent of this difficulty raised the issue of whether icons were appropriate at all in this context.

It was agreed that some user evaluation of the icon set would be useful. Accordingly a questionnaire asking subjects to identify the menus represented by proposed icons was distributed to a number of university DOGS users. The methodological limitations of this technique and this subject population are acknowledged, twenty sets of usable data resulted, however, for a mimimal investment of resources. Even on a generous interpretation of results, many of the icons needed revision, particularly where they had been positively identified as belonging to the wrong menu.

It was decided to include the abbreviation for the menu name as part of each icon. This solves the problem of enigmatic icons, but may have costs in speed of recognition: conceivably linguistic and graphic processing may interfere with one another. Bewley et al suggest that labels are helpful, but the basic icons used in that study already contained text; it would be helpful to have clarification on this point. Other changes were the grouping of icons by broad function (shape coding produced too cluttered an appearance), the development of a set of icons for drawing functions based on operations on a standard shape, and the use of the skills of a graphic artist in producing the designs. The icon set has yet to be finalised and the further question of how far it is feasible to use icons for individual menu options needs determination.

POP-UP MENUS

The experienced user has little to gain from pop-up menus and will be able to set a 'menus-off' option which displays option numbers alongside the menu icons instead of popping a full menu. He can then operate much as before, reverting to 'menus-on' mode if moving to an unfamiliar area. For the novice, their potential advantage is the elimination of the need to consult off-screen documentation. The menu therefore has to convey enough information to make this possible, in this case without overwhelming the area available for the primary task of producing a drawing. In the current environment the amount of text which can be displayed on screen in a reasonable space is dictated by the use (for speed) of hardware text in a very limited range of sizes and thus it has been necessary to describe sophisticated features in one or two words only. The abolition of menu numbers in the display would facilitate the design task, but would make it harder for the user to graduate to the use of abbreviations as a command language, and cause further difficulties in displaying options meaningfully when menus are partially overlapped in a stack. Consideration is being given to the use of icons where possible, although many menu options do not facilitate iconic imagery. A better long-term solution would be a combination of terse menu items and an on-line manual: this will be considered for future releases.

Consideration of menu content revealed some underlying problems. DOGS menus had grown unwieldy and less than logically organised as options had been added over the years; this became very obvious in the context of a screen menu. It was eventually decided that the introduction of a new interface should be taken as an opportunity for an improved menu structure and this revision is now under way. Various principles for menu organisation were possible: a choice has been made to arrange options within menus by mode of action (i.e. separating switches and operators) and then by function, but other arrangements, e.g. by actions necessary for a particular task or by frequency of use could be equally valid. Guidelines were of little help in assessing relative merits and ideally a specific investigation of menu organisation in this environment should have been undertaken, perhaps taking user models as a starting point.

EVALUATION

There has so far been little opportunity for investigation of user needs or evaluation of interface features. The first task when time allows is to check that the proposed design is workable now MICE is operational: if the menus do not appear virtually instantaneously, for example, a radical rethink will be required. It is also hoped to conduct some research into problems encountered with the old DOGS interface: relevant changes can be easily made to the new interface using MICE. Benchmark measures for ease of use and learning under the existing system will be collected for comparison purposes, and some prototyping of the new interface carried out, but it is unlikely that MICE and therefore the new interface will be linked into DOGS much before the release date, so assessment of the fully operational system may have to take place post-release.

Some passive monitoring of users using an automatic user log can be set up, and it is also hoped to use more intrusive techniques such as observation of user sessions, ideally with standard tasks in some cases, and structured interviews or questionnaires, but there is clearly a limit to the amount of cooperation that can be asked of paying customers. It should be possible to set up some small scale evaluation of non-task related features on in-house users.

Most decisions have been taken on an ad hoc basis relying on intuition and, where available, research data and guidelines. Usable data has been lacking in many areas, some of which are mentioned above. Others include the relative merits of positive and negative displays for graphics applications and for 3D applications, the form of status indicators for indicating view orientation. Guidelines such as those produced by Smith and Mosier (1984) have been helpful as checklists of matters for consideration, but over-general or too context-dependent to be useful as exact models. This type of information might better be organised as a database of detailed research reports accessed by subject areas or linked to the guidelines themselves. In this way it would be possible to judge more accurately the application of any particular finding, but there will inevitably be many circumstances where prototyping of the system in context is unavoidable.

REFERENCES

Bewley, W., Roberts, T., Schroit, D. & Verplank, W., 1983, Human Factors Testing in the Design of Xerox's 8010 'Star' Office Workstation, Proc. CHI 83 Human Factors in Computing Systems, (ACM), 72-77.

Foley, J.D., Wallace, V.L. & Chan, P., 1984, The Human Factors of Computer Graphics Interaction Techniques, IEEE Computer Graphics & Applications, November 1984, 13-48.

Funt, K., 1984, Human Factors in Icon Design, unpublished MSc thesis, Loughborough University of Technology.

Gittins, D., Icon-based human-computer interaction, Int. J. Man-Machine Studies, 24, 519-543.

Hemenway, K., 1982, Psychological Issues in the Use of Icons in Command Menus, Proc. Human Factors in Computer Systems Conf., Washington, D.C., (ACM), 20-24.

Rogers, Y., 1986, Evaluating the meaningfulness of icon sets to represent command operations, Proc. 2nd Conf British Computer Soc. Human Computer Interaction Specialist Group, (Cambridge University Press), 586-603.

Smith, S.L. & Mosier, J.N., 1984, Design guidelines for user-system interface software, Technical Report ESD-TR-84-190, U.S.A.F. Electronic Systems Division, Hanscom Air Force Base, Mass., U.S.A.

WHAT IS A GOOD CAD DIALOGUE?

M.A. SINCLAIR

Department of Human Sciences
University of Technology, Loughborough
Leicestershire LE11 3TU

The paper outlines some of the high-level desirable features of CAD dialogues of the future for mechanical engineering design. These have been derived with respect to the context in which design normally occurs.

Before we can discuss the requirements that must be met for a CAD dialogue to be considered 'good', we ought to make sure that the starting assumptions are clear. What follows is predicated on design within a mechanical engineering environment, and that the organisation within which the design activity takes place is medium-sized or large; say, more than 100 employees.

With these assumptions, Fig. 1 becomes a reasonable representation of a manufacturing organisation; on the left hand side is a raw materials world, while on the right is a finished goods world. In between is the organisation. It manufactures goods by moving raw materials from left to right at a cost, which it recoups by a cash transfer in the opposite direction, hopefully making a profit at the same

Figure 1. Representation of a manufacturing company.

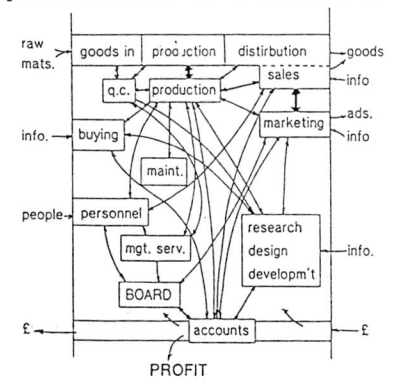

time. All the other functions in the diagram indicate what is necessary for the organisation at least to maintain homeostasis, and hopefully to grow within a changing, somewhat hostile environment.

It is within such a context that we must consider design. If there is one criticism that can be levelled at current research into human-computer interaction in CAD, it is that most of it seems to assume that design occurs in a vacuum; that there is only one seat in front of the screen, and it is permanently occupied by a normative designer. In fact, it is a multilevel process, that must be considered holistically.

All organisations must manage their design. This requires that there be some form of formalised design process at the organisational level, to allow timely and efficient allocation, and subsequent control of resources. At this level, timetabling is required for problem analysis and requirements specification, for conceptual design, for conceptual choice, detail design, prototyping, etc. This is an iterative process, with frequent formalised reviews. CAD systems of the future must support these activities.

At a lower level, there is the design activity carried out by individual designers. While it is true that designers are given responsibility for their part of the design, and that they tend to work on it alone, it is also true that other designers (their superiors, for instance) will also have occasional inputs, and at a later time may take over the design. Each of these designers will have an iodiosyncratic way of working; this may be at the level of strategy (top-down, middle-out, breadth-first, etc.) or it may be tactical (messy with lots of notes, etc.). Future CAD systems must accommodate the very wide range of design styles that are apparent within a single organisation (despite the existence of 'house style'), let alone across organisations.

It should also be remembered that the essential purpose of the exercise is to capture information for downstream activities. From this perspective, the production of elegant images on a screen would be an irrelevance, were it not for the fact that design is an exploratory activity, necessitated by the constraints of human cognition. What is required from design is information that allows manufacture, assembly, sales, and field support to occur in the most efficient manner. While part geometry is an essential requirement, so too are surface finishes, materials, tolerances, centres of mass, etc. This information has to allow the production of process plans, schedules, bills of materials, and so on to occur, within the constraints of manufacturing resource planners and business planning systems, and of factory control systems.

A dialogue for CAD, then, must be more than an interactive language for drawing tidy pictures. It must support the various phases of the design activity as each occurs, and it must enable information to be captured at each stage, so that it may be used in subsequent stages, even if they do not occur in a recognised order. It must enable designers to carry out 'back of the envelope' designs, as well as detailed drawing activities, and it must enable the administrative activities to occur as well. That CAD systems must encompass these things is because of the advent of artificial intelligence concepts into CAD systems. At the time of writing, this author has personal knowledge of systems being developed by three different manufacturers that will include some degree of AI within them, that will be on the market by the early 1990's. There is an interesting circularity here; AI is being introduced in order that CAD systems might extend further into the organisation's mainstream activities, but the AI applications require such an extension before they can make a useful contribution. It should also be noted that once CAD systems contain more than vestigial reasoning capabilities, dialogues will have to accomplish more than the transfer of commands and data; meta-level information regarding purpose, meaning and planning will also have to be transferred.

With one bound, let us now move to considering some of the criteria that a dialogue for CAD systems of the future must meet. In what follows, the word 'dialogue' is used fairly loosely to refer to transactions between the user and the CAD system, as well as some of the management functions of the system.

1. It must enable all relevant information for downstream activities to be accreted to basic shape information. Downstream information includes functions, materials, spatial relationships, tolerances and surfaces, tools, jigs and fixtures, designer's notes, etc. It should also be remembered that because a product may well undergo redesign during its life cycle, the information captured should also include the design history of the part and extensive library facilities. The provision of this information will require a much richer interaction between the designer and the CAD system.

2. It must be easily tailored to meet the needs of the client organisation. While it is quite likely that the vendor will carry out the initial parametrising of the system, if the dialogue is to have a long-term future it must be capable of change to meet the evolving needs of the client organisation. It is quite likely that many of these changes will be accomplished by users in the client organisation, not by the supplier, and the dialogue must support these people.

3. It should allow easy access by different classes of users, such as production engineers, technical writers, etc. This need is less clearcut; most of these transactions will occur after the designer has completed the design, and therefore a dialogue based on database query languages may be sufficient. However, as was said earlier, designers do not design in a vacuum, and have many discussions with engineers and others as the design progresses. Under these circumstances it is likely that the design dialogue must accommodate other users as well.

4. The dialogue must be fast-reacting. For a CAD system to provide the kind of downstream information discussed above, some sort of solid modelling is essential. Current solid modellers have some restrictions in what they can represent, and some of them are unacceptably slow in that the response time can be in hours rather than seconds. While parallel processing in one form or another offers great improvements in this, it still seems likely that anticipatory processing in some form as a background procedure will be necessary. This will not be restricted to solid modelling; several analytical techniques are equally timeconsuming, and would present an equal demand for background anticipatory processing.

5. The dialogue must be individually user-friendly. Designers are there to design, and this frequently requires intense cognitive effort. The last thing required of a dialogue under such circumstances is that it should be a distraction and/or a hindrance. We may define the following requirements:

5.1 The dialogue must allow the designer to work in an idiosyncratic, personal manner, while still retaining sufficient control of the process to ensure that appropriate information is captured for the system. This indicates a need for a user interface management system with the dialogue operating at two levels, those of design and of management.

5.2 At the design level, it appears that designers operate very much in graphic mode, using shapes as a visual code for functions, properties, relationships, etc., and using sketching as an extension to working memory. It should be noted that typically designers sketch in 2D while they think in 3D, and that what they sketch is not necessarily a 1:1 mapping of what is in their minds. This indicates that dialogues that mimic current 2D methods, while being reassuring to designers, may not be appropriate for the kinds of knowledge capture required, whereas ones that are, may not be easy for designers to use.

5.3 In many situations designers require quantitative information about their designs. Often, an approximate answer is all that they require, rather than a detailed one, and often the designers do not know the values for all their variables and parameters. Equally, the answer required is not always a precise number, but whether the number is comfortably within some region or interval. The dialogue must allow such 'ballpark' transactions to occur, typically when the designer wishes to pose the query in mixed graphics and text modes.

5.4 Most engineering design places heavy reliance on previous design ideas, often embodied in old designs. In future CAD systems these will be held in systems databases both as the design idea and its developmental history, to which designers will require access. The dialogue must support this sort of database query and browsing activity as a fairly frequent need on the part of the designer.

5.5 At various times designers deliberately transgress the established rules when designing. Thus, they must allow two components to occupy the same space, or they may conceive of a virtual component (which will be distributed across two or more real components), while they explore a particular idea. Systems incorporating AI or other deductive algorithms should be prevented from crashing under these circumstances, while such transgressions must be trapped by the management system to ensure they are subsequently eliminated.

5.6 Because of the clear and unequivocal need for designers to understand their design, and because this will only happen effectively if there is mental sweat upon the designer's brow, most of the design decisions will be allocated to humans. This reasoning will apply to much of the detail design as well, since in mechanical engineering design there is a very close link between form and function. Therefore, it is unlikely that there will be many design synthesis functions built into the CAD system though there may well be many analysis techniques available. These thoughts lead to the concept of the CAD system being both passive and submissive in its behavioural characteristics; passive because it must leave design decisions to the designer, and submissive because it should not interfere with the designer's train of thought, even though the designer may be transgressing some known design rules at a particular time. Only at a later stage, when the designer wishes to evaluate the work so far should the system assume some level of equality in its response. It is not immediately apparent how such characteristics should be build into the dialogue, though there is a need for them.

5.7 The dialogue must be able to answer the standard user's questions:

(i)   What's going on?
(ii)   What isn't going on?
(iii)   Remind me, how did I get here?
(iv)   What did I just do to get this?
(v)   How do I change this?
(vi)   What do I do next?
(vii)   What can I do next?
(viii) Where is ...?
(xi)   What if ...?
(x)   When will ...?
(xi)   Why are you telling me this?
(xii)   What does it all mean?

These criteria are not comprehensive, nor are they explicit in what they say. However, what should be apparent is that they go far beyond the capabilities of dialogues that are currently available, and if implemented would represent a significant advance in the usability of CAD systems. Perhaps this is one of those occasions where Ergonomists could genuinely be ahead of other technologists, right at the beginning of a systems design venture.

PRESCRIPTION, DESCRIPTION AND EVOLUTION: DESIGN OF USER-COMPUTER INTERFACES FOR CHANGING SYSTEMS

H. DAVID

EUROCONTROL, Apt 160
15 Avenue Gabrielle d'Estrees
91830 - Le Coudray-Montceaux, France

The design of ergonomic user-computer interfaces requires a definition of the information needed by the user and the means of controlling the system that may be required. These may be obtained from a job description or task definition or from systematic observation. The advantages and disadvantages of these methods are discussed, with particular reference to systems where the nature and quantity of work may change during the life-time of the system.

INTRODUCTION

It is generally accepted that the design of a user-computer interface cannot be properly undertaken without knowing how the system will work, what it will work on and what it is required to do.

This paper addresses the problem of how this information can be obtained in a useable form within a reasonable time at a reasonable cost, bearing in mind the rate at which user-computer systems are currently evolving.

There are many possible approaches to this problem, which can be placed in a range from formal job description to informal observation. In most practical situations, some combination of techniques will probably be necessary, but they can be treated separately for the purposes of this paper.

PRESCRIPTIVE METHODS

Prescriptive methods begin by defining the overall task, then consider how it may be achieved.

Job Description

A job description, in the most general sense, states what is required to be done by the person doing the job. It is usually phrased in general rather than specific terms (for example, "the safe, orderly and expeditious conduct of air traffic"). A job description specifies the results required, rather than the way in which they are achieved, or the tools to be employed.

Task Definition

A task definition is a definition of the tasks that are to be carried out at a specific working position. It is usually more detailed than a job description (for example, the FAA 'Operations Concept for the Advanced Automation System Man-Machine Interface' (Phillips et al 1984) covers 260 tasks in seven major categories in the course of approximately 400 pages).

DESCRIPTIVE METHODS

Descriptive methods begin by examining the way in which a task is carried out, from which a description of the system can be extracted.

Simulation

Simulation is essentially a means of observing a system without interference with the actual task, free from safety constraints, with reduced costs and greater freedom to explore extreme cases. Simulations may be extremely detailed, extending over weeks and involving dozens of participants, or restricted to very specific skills. Lafon-Millet (1981) is a good example of the latter approach.

Observation

Observation, in general, covers not only the direct visual observation of a task, but all methods used to find out how the task is actually carried out in practice. Some of these methods are well established - such as the collection of statistics on actual system operations, recording the number and duration of communications, informed observation of specific processes, questionnaires and interviews (David 1983). Others, such as eye-movement recording, (LeGuillou et al 1981) are less obvious.

COMPARISON

Figure 1 summarises, in very general terms, the characteristics of the prescriptive and descriptive approaches.

Table 1. Prescription vs Description.

	Prescriptive	Descriptive
Scope	General	Specific
Source	A Priori	Evolutionary
Complexity	Simple	Complex
'Attitude'	'Mechanical'	'Organic'
'Objectivity'	'Objective'	'Subjective'
Cost	Low	High
Blunders	Occasional	Rare
Extreme cases	Overemphasised	Neglected

Prescriptive methods tend to begin with relatively simple, general principles, which are often difficult to apply in practice, while descriptive methods usually produce a mass of information from which general concepts can be extracted only with difficulty. There is a tendency for prescriptive methods to appear rigid, while descriptive methods adapt to reality - sometimes to the extent that they inhibit future development. Because most prescriptive techniques rest on general rules, they may neglect vital components not included in their 'model', while, on the other hand, descriptive techniques may misinterpret their observations, and may miss rare but important activities.

Whatever method is used, it is never practically possible to be absolutely certain that all the activities of a complete job have been identified and accurately described. Even where a task exists, experienced and articulate performers of the task do not always know exactly how they do it. Unable to express their own 'internal models', they may accept an incorrect model presented by 'scientific experts' and struggle to provide values for parameters that do not correspond to reality.

TASK ALLOCATION

In the design of a user- computer interface, one of the first steps is to decide how tasks should be allocated between the user and the computer. The choice is not entirely free, since there may be legal constraints (although bank note numbers may be read by an OCR device it cannot at present appear in court to swear to their identity!) or mechanical constraints (listening to messages and translating freely spoken language are beyond our current technology - and seeing infra-red is beyond our biology).

Within these constraints, allocation criteria have passed from a stage of 'give the machine what it can do - the user can do the rest' through a stage of 'give the machine what it is best at, and the user what he is best at' to (ideally) 'give the user a satisfying job he can carry out efficiently - the computer can do the rest'.

TASK VARIABILITY

Unfortunately, work, tasks or responsibilities rarely remain the same in either the long or short terms. The conventional methods of task allocation tend to neglect the requirements for flexibility and adaptability. There are several sources of change in workload that can be anticipated (to say nothing of unanticipated changes). Many jobs have inherent rhythms (daily, weekly or yearly) to which the system must adapt. In jobs where task arrivals are independent, the operator is often able to vary his mode of operation to allow for faster operation, less cost-efficient operation to increase capacity. (Sperandio 1972).

Many users acquire familiarity with a system by using it, so that while they initially need detailed input menus, they graduate to faster, more compact 'command' modes with experience. They may not be equally familiar with all modes of operation, and they tend to forget unused system commands, so that a well-designed system should revert to more prolix modes when necessary.

In modern civil aircraft design (Airbus Industrie 1983), the pilot is seen as similar to a stage-coach driver, "the driver holds the reins, but the horses follow the road". The military approach is slightly different - the electronic control and guidance systems are seen as a 'Pilot's Associate' - although the description, involving a 'navigation manager', a 'communications manager' and several other managers, suggests that the pilot may well be outranked by his aircraft. (He is not yet required to salute it.) It is proposed that the aircraft should be able to take over those tasks that the pilot is unable to carry out in the rush of action and even to navigate itself out of danger when the pilot is incapacitated.

All these cases illustrate the trend towards increased automation, and the need for increased flexibility that accompanies it. The 'user-system' interface is not a fixed line on a diagram, but a threshold over which some aspects of the task are passed from one partner in a symbiosis to the other. Two potential problems have already been identified: - who decides which tasks are passed across the threshold, and can the human operator accept them adequately?

Obviously, the answers to these problems depend on the context, but some general principles may be suggested. The responsibility for for task allocation may remain with the operator, although the system must be able to continue operating when the operator is absent or disabled. The problem of 'unsafe operation' remains, and will often depend on context. (In the aftermath of Chernobyl, one might expect that nuclear power station systems might be designed to refuse unsafe orders, but a aircraft may, in emergency, have to perform unsafe manoeuvres lest worse befall.)

The second problem, although less dramatic, is probably more directly relevant to interface design. As the operator becomes more remote from routine operations, he may find it harder to remain aware of the current situation. (Routine verbal reports from aircraft draw the attention of the controller to aircraft in his sector. 'Omit Report' procedures require deliberate effort to maintain awareness.) When his intervention is required, it may well be an emergency, so that he must be able to acquire very rapidly the information necessary to resolve the emergency. This may require very different displays from those currently in use.

TASK CHANGE

These forms of variability are already being assimilated by system designers. There is however a further step that poses more difficult problems.

This is that tasks may change qualitatively in the life-time of the system. In aviation, for example, radical changes in navigation methods lead to pressure for direct routeings of aircraft, and make the traditional airways structures obsolete.

The continued increase in traffic makes the original routines laid down for coordination too time-consuming. As a result there is an increasing tendency for operators to communicate via the system, rather than by the traditional methods of nudges, gestures, shouts and telephone calls. At least one instance has been observed in which operators used spare fields in a display as a way of storing extra information not foreseen by the system designers.

CONCLUSION

In systems which will remain in use for several years, provision should be made, not only for different modes of interaction to suit more or less skilled users, and for flexible allocation of tasks to accomodate greater or lesser workload, but for changes and adaptations to suit variations in the nature of the task with time.

ACKNOWLEDGEMENT

I am indebted to the Director General of the European Organisation for the Safety of Air Navigation for permission to publish this paper, which reflects only the opinions of the author and in no way represents EUROCONTROL policy.

REFERENCES

Airbus Industrie, 1983, The New World Airbus Industrie BP 33 F-31700 Blagnac AI/VF:AI/CS-P50

David, H., 1983, Potential Measures of Strain on Controllers Eurocontrol Experimental Centre Report No. 164.

Huddlestone J H F, 1985 Colour Coding for Computer-Driven Displays Ferranti Computer Systems Report 5370

LeGuillou M., Halliez B. and Nobel J., 1981 Etude du Travail du Controleur Organique au CRNA/N par Analyse de la saisie Visuelle Centre d'Etudes de la Navigation Aerienne, Orly-Sud INRIA CO 810 R66 - CENA R.80-28

Lafon-Millet M-T., 1981 Representation Mentale de la Separation Verticale au cours du diagnostic dans le controle aerien Institut National de Recherche en Informatique et Automatique CO 8107 R66

Phillips M D., Tischer K., Ammerman H. A., Jones G. W., Kloster G.V., 1984 Operations Concept for the Advanced Automation System Man-Machine Interface Federal Aviation Administration, Washington, USA DOT/FAA/AAP-84-16

Sperandio J-C, 1972 Charge de Travail et Variations des Modes Operatoires These pour le Doctorat d'Etat es-Lettres et Sciences Humaines Universite Rene Descartes - Paris V

USING VIDEOTEX TO ORDER GOODS FROM HOME

S. FENN[*] and P. BUCKLEY[**]

[*] Department of Management Science
Imperial College of Science and Technology
Exhibition Road, London   SW7 2BX

[**] Department of Computer Science and Statistics
Queen Mary College, Mile End Road, London   E1 4NS

'Teleshopping' has become available using videotex technology. Users can order goods by entering details on a 'response frame'. An experiment was carried out to assess the usability of four types of response frame. Fastest performance times were associated with 'tailored' and 'menu' response frames when ordering single items, and with 'tailored' and 'generalized' frames when ordering several items. Conclusions are expressed as suggestions for videotex dialogue designers.

INTRODUCTION

Shopping from home via a videotex system is now possible. Such a facility has clear advantages in being convenient and time saving, particularly for people who do not have easy access to shops. Users may acquire details of goods, evaluate the available items, and place orders using a home terminal or an adapted television set. However, the task of teleshopping imposes demands on the user which are not experienced in normal shopping. (For a model of general transaction tasks, see Long & Buckley 1984). Whereas in normal shopping, the shopper's resources (cash, cheque etc), and the shopper's transfer conditions (item required, quantity, colour etc.) are displayed by actions or are spoken (eg handing over cash, asking for 'so many' of an item), in teleshopping, shoppers have to enter correctly the details of their transaction into a particular type of videotex screen format called a 'response frame' (RF), and then effect system operations which transmit the completed frame to the retailer offering the relevant goods (information provider or IP).

Gilligan & Long (1984) identified three types of RF, distinguished in terms of the relationship between the RF and the information pages (which provide details of the items available for purchase). The 'tailored' RF is associated with a single information page and is unique to the item on that page. The 'menu' RF is associated with up to nine information pages. It presents a short list of items, and the user is required to indicate which item is desired. The 'generalized' RF is associated with potentially many information pages, and the user is required to fill in all the relevant details. It provides separate labelled fields for each piece of information required. A fourth type of RF referred to as an 'open' RF has also been found in use on the British Prestel system. It can be regarded as a special type of generalized RF in that it is linked to many information pages. It is however, less structured than the generalized RF, providing only a single blank field into which all relevant details must be entered.

It is suggested by Gilligan & Long (1984) that the effort involved in completing the RF may differ according to its type. For example, the number of keypresses involved in entering the data and the memory demands, would be greater for generalized and open RFs than for tailored and menu RFs. This is because with the latter the user need only indicate an affirmative decision and enter details of payment in order to complete the ordering task, whereas with the former it is generally necessary to enter such information as the item's reference number, colour, size etc. as well as the credit card number. In addition the user might need to remember details of the item from a separately displayed information page (eg the reference number R313/2).

We might expect these differences to be influenced by the number of items ordered at any one time. If several different items are ordered the completion of a separate RF for each item is necessary in the case of the tailored type. The same is also true of the menu type, unless it contains more than one of the desired items. In contrast, generalized and open designs typically allow several different items to be ordered on a single RF. In such cases, overall effort may be equal to or possibly less than the effort involved in completing several tailored or menu RFs for the same items.

An experiment investigated and attempted to quantify the effect of RF type on task performance. Four different types of RF were compared for single and multiple ordering.

METHOD

Twenty four subjects (all naive videotex users), each performed 24 shopping tasks on a simulation of a Prestel-type teleshopping system. The shopping tasks comprised three single item and three 3 item trials on each of the four RF types. Transaction details were matched in all RF conditions (all involved quantity, price, item description or reference number, and credit card number) and were either contained on the RF or had to be entered by the subject, depending on RF type. In order to familiarize subjects with the operation of the equipment and how to perform the task of RF completion, a training session preceded the experimental trials. Following the set of shopping tasks, subjects completed a questionnaire.

RESULTS & DISCUSSION

RF completion time was the time taken to enter all details requested on the RF, and to remove the RF from the screen by pressing the appropriate key. The dependent variable was the time taken to order a single item ('time per item'). This was calculated by dividing RF completion time by the number of items ordered per frame (1 or 3).

Subjects (as instructed) corrected any errors made during the experimental trials. An error in RF completion is defined as any entry or omission that would prevent the IP from evaluating correctly the intended order. To quantify error free performance, time spent making and correcting errors was subtracted from total time.

Time per item ('T') and time per item less error ('T-E') mean scores for each condition over the two sets of scores.

Figure 1. Mean performance times ('T' and 'T-E') for single and multiple ordering as a functon of RF type.

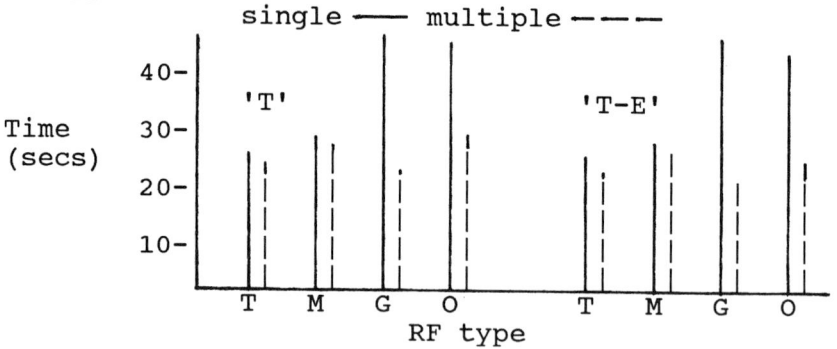

[Figure 1: Bar chart showing single (solid) and multiple (dashed) performance times. Y-axis: Time (secs), 10-40. X-axis: RF type (T, M, G, O) shown for both 'T' and 'T-E' metrics.]

The effect of 'RF type' was significant on both metrics, T ($F(3,69)=31.3, p<0.001$) and T-E ($F(3,69)=24.6, p<0.001$). There was a significant interaction on both metrics between 'RF type' and 'number of items ordered', T ($F(3,69)=51.1, p<0.001$) and T-E ($F(3,69)=95.2, p<0.001$).

A Tukey test was used to compare the condition means of the four RF types for single and multiple ordering. For single item ordering, on both metrics there was a significant difference between all paired comparisons except between the tailored and menu, and between the generalized and open RFs. For multiple orders, on the 'T' metric, performance on the tailored and generalized RFs was significantly faster than on open RFs, and performance on generalized was significantly faster than on menu RFs. All other paired comparisons were not significantly different. On the 'T-E' metric, performance time for multiple orders on tailored and generalized RFs was significantly faster than on menu RFs. All other paired comparisons were not significantly different.

Thus, RF type affected performance for single item ordering. Performance time was faster when using tailored or menu RFs than when using generalized or open. There was a trade-off between 'time per item' and 'number of items ordered' when generalized or open RFs were used. Best performance times for multiple ordering occurred when tailored or generalized were used.

Best performance times for multiple ordering occurred when tailored or generalized were used.

When error time was included in the analysis (the 'T' metric), performance for multiple orders on open RFs was significantly slower than on tailored or generalized RFs. The advantage of the trade-off between 'time per item' and 'number of items ordered' was diminished by error time.

The number of errors made on generalized and open RFs was approximately the same (30 and 28 respectively). The significant difference in performance time between generalized and open RFs on the 'T' metric can be partly explained by the rather different effects of the 'cancel error' command on the two RF types. Its general effect is to delete the contents of the current field. Since goods and payment details are entered in a single field in the open RF, and in a set of fields in the generalized RF, the effect of the command is to delete more or less material.

In the experiment, multiple item ordering on generalized and open RFs was found to reduce time per item, as compared with single item ordering. If a greater number of items were to be ordered, it would be expected that time per item on generalized and open RFs would be further reduced. However, this advantage would be limited by restrictions on the text capacity of single frames and the maximum number of fields.

Subject opinions of RFs based on their experiences during the experiment seemed to favour the tailored RF for single ordering, and the generalized for multiple ordering.

SUMMARY & CONCLUSIONS

Based on subjects' performance and opinions it appears that the suitability of RFs depends on the number of items being ordered at any one time. This finding has implications for the IP when selecting RF type. The choice may depend upon the type of goods on offer, as some goods are characteristically purchased as single items, whereas others are generally not purchased on their own. In selecting RF type, IPs would be advised to choose tailored (or menu) RFs for items likely to be purchased singly (eg a washing machine), but for items that are likely to be purchased a few at a time (eg some groceries) generalized (or tailored) are advised.

An RF type which is suitable for the user may not be the most suitable for the IP. Gilligan & Long (1984) suggest that the effort exerted by users and IPs is inversely related. For example, the tailored RF requires least effort from the user, but maximum effort from the IP who has to create a unique RF for each item. The generalized RF presents an opposite distribution, requiring less effort from the IP who has to create only a single RF for all items, but more effort from the user, who has to enter all the transaction details. Our work here suggests that this only holds for single items.

If the choice of the teleshopping ordering dialogue is considered from the IPs point of view, a form based on a generalized or open RF is probably preferable. Using generalized RFs for goods that are likely to be purchased a few at a time would suit IPs and (as is indicated by our data) users. For items that are likely to be purchased one at a time, there may be a conflict between the IP's and user's needs. Our results indicate that for single orders tailored RFs appear to be most suitable for users. For an IP who supplies a large number of items, the effort of producing a large number of RFs (one per item) could be reduced by the use of software designed for providing RFs.

ACKNOWLEDGEMENTS

This report forms part of a research project GR/C/23032 funded by the SERC and British Telecom.

REFERENCES

Gilligan,P.,& Long,J.B.,1984, Behaviour and Information Technology, 3, 1, p.41-71.
Long,J.B.,& Buckley,P.K.,1984, Transaction Processing Using Videotex: or 'Shopping on PRESTEL' in B. Shackel (ed) Interact '84, North Holland, p.251-256.

**COMMUNICATION FAILURE IN DIALOGUE:
NON-VERBAL BEHAVIOUR AT THE USER INTERFACE**

N.P. SHEEHY, A.J. CHAPMAN and M.A. FORREST

Department of Psychology
University of Leeds
Leeds   LS2 9JT

This paper presents a selective review of some literature examining nonverbal behaviour at the user interface. The potential for enhancing the user interface by accommodating non-textual input is considered. A study is reported in which an imaging analysis of users' behaviours was conducted. The results are encouraging for future developments.

INTRODUCTION
There is remarkably little known about what users do when seated at terminals. Although there is a vast amount of literature on the ergonomics of terminal use, (cf. Meyer, Crespy and Rey, 1980), this relates principally to anthropometric measurement and the effect of computer use on other personal and organisational variables. The literature does not consider the relationship between the user's task and the user's behaviour. Reviewing the scant literature in this area, Delvolve and Quiennec (1983) found no studies which examined the spatio-temporal organisation of task constrained gesture and movement. They conducted an observational study of six experienced terminal keyboard operators over a complete working day. The observations were not video recorded so the results may not be as reliable as Delvolve and Quiennec suggest. Their results indicated:
- o increased 'fidgeting' as the period progressed
- o increased frequency of self-oriented gesturing as the work period progressed
- o relative short-term constancy of verbal and nonverbal behaviour, (e.g. self-gesturing increased slowly rather than rapidly)

o significantly more grimaces and sighs, etc, when the
     operators were working on long documents; length may
     have been confounded with difficulty

   These data suggest that the verbal and nonverbal
behaviour that accompanies operators' task directed movements remain relatively constant over a working day, but
long or problematic documents are characterized by statistically significant increases in nonverbal non-task-related
movements. These may reflect cues to stress and distress and
and the fact that they have a stochastic quality suggests
that they may be easily discriminated from other task-related
behaviour. If this is the case then image analysis of users'
movements may be able to discriminate between small,
unimportant changes in posture, and larger, more significant
movements which potentially may convey information about
difficulties the user is encountering. Of course, the
detection of differences in face position and posture do
not reveal the content of the difficulty. They provide a
cue, the value of which depends on the way in which the cues
are communicated to other system components and failure
avoidance or remedial action taken.

METHODS
     In order to obtain recordings of user behaviour across
similar kinds of tasks, requiring different interactions
with the computer, a study was conducted which investigated
different ways of imparting knowledge to new users of a
micro computer. Thus, the material for the image processing
and analysis was recorded in the context of a well-defined
experimental framework.
     Two 15-20 minute audio-visual recordings were made of
how to perform some simple tasks on a BBC micro computer.
One recording consisted of a demonstration of how to use
some basic commands in command mode and the other recording
consisted of a demonstration of how to achieve identical
effects with the use of a package utilizing windows and icons,
controlled by step keys. In the other two conditions these
subjects received just the documentation accompanying the
video demonstrations that is, a transcription of an audio
tape. Immediately afterwards, subjects were asked to
perform some of the tasks they had either seen demonstrated
or which they had read about in the documentation. Subjects'
performance was recorded in a dimly lit room using a low
light level camera, placed at eye-level and to the right of
the micro computer. The experimenter was seated behind and
away from the subject.

There were two subjects per condition (8 subjects). They were told that the study was about learning some basic functions on the BBC micro computer with the aid of either a video demonstration or documentation.

RESULTS

The recordings taken from the low light level camera were analysed using the following categories: Gross body movements; screen looking; keyboard looking; note looking; rapid up-down movement head movement; frown; pursed mouth; any other expression with mouth; smile; face/head touch.

Video recordings were made of six subjects and these were image analysed. Table 1 contains a summary of recorded movements for each of the subjects.

Table 1. Descriptive analysis of users' movements.

	S1	S2	S3	S4	S5	S6
Gross body movement	0	0	3	0	1	0
Screen looking	85	49	53	30	46	63
Keyboard looking	70	44	34	22	51	53
Note looking	24	35	14	14	35	42
Rapid up/down movement	57	3	56	28	7	21
Frown	14	3	6	4	3	3
Pursed mouth	13	2	7	2	4	5
Other facial expression	0	5	3	0	8	0
Smile	4	2	4	0	3	0
Face/head touch	3	2	6	12	3	16

Image Analysis

The purpose of the image analysis was to examine instances of user movement and determine whether it is possible to correctly assign a behaviour to its appropriate category on the basis of its statistical features.

In order to reduce the quantity of data contained in each digitized image a histogram of the pixel intensity values was computed. Histogramming entire frames in real time is only possible at hardware speeds. On the system used in this study, (DEC LSI-11/23 with pipelined analog processor), it is not possible to exceed speeds of about 2.0 seconds per frame. Higher speeds can be achieved by reducing the number of data points sampled. Thus, by sampling every 20th pixel on the X and Y coordinates of the frame buffer, it was possible to keep processing speeds low, (0.2 seconds per frame), and also reduce the quantity of data sampled. An inter-pixel sampling rate of 20 was selected arbitrarily. In addition to sampling data points within frames it is possible to sample between frames. An arbitrary sampling rate of one frame per second, or one frame in every 30, was taken over ten minutes of videotape. This yielded 600 frames at 512^2 which was reduced to 600 frames at 20^2.

Rather than presenting extensive tables of histograms and correlations, the statistical analyses and outcomes will be reported with reference to the behaviour of individual subjects and then general trends will be described. Table 2 presents summary statistics on 30 frames collected for subject three. Similar statistics, for all 600 frames, are not reported here. Some statistics are more sensitive to change than others. For example, the mean is not a particularly sensitive measure: it varies between 26.38 (Frame 30) and 25.27, (Frame 22). The median, on the other hand, varies rather more and is more sensitive to change between frames. Other analyses, not reported here, have examined the inter-correlations between pairs of adjacent images, and the differences between pixel intensity histograms using a Wilcoxon Matched Pairs Signed Rank test. Neither test was particularly sensitive to observable differences between images. Thus, movements which are known, from visual inspection, to suggest informative or communicative significance are not detectable on a criterion of statistical significance which uses conventional confidence limits.

Table 2. Summary statistics on thirty frames.

FRAME	MEAN	MEDIAN	STANDARD DEVIATION
1	26.00	15.00	35.20
2	25.92	15.00	33.70
3	25.96	14.50	30.49
4	26.00	16.50	27.02
5	25.96	21.00	26.15
6	26.00	15.50	27.85
7	25.81	14.50	27.96
8	26.00	15.50	25.65
9	26.00	14.50	27.80
10	25.96	13.00	34.54
11	25.77	19.00	25.38
12	26.00	15.50	28.18
13	26.04	14.50	30.85
14	26.00	13.50	31.95
15	26.00	13.00	31.59
16	26.00	13.00	31.59

Table 2. Summary statistics on thirty frames (continued)

FRAME	MEAN	MEDIAN	STANDARD DEVIATION
17	26.00	17.50	29.56
18	26.04	16.00	28.82
19	26.00	12.50	30.56
20	25.69	16.00	25.89
21	25.92	13.00	31.19
22	25.27	16.00	29.73
23	26.00	17.00	31.28
24	25.88	14.50	29.70
25	25.92	17.50	29.32
26	26.00	9.50	30.10
27	25.69	16.50	28.55
28	26.00	15.50	30.35
29	26.00	16.50	28.17
30	26.38	13.00	33.50

Table 2 suggests that the median may be sensitive to changes due to the user's movements. In fact, one can easily construct a simple heuristic to decide whether or not a user is looking at, and probably attending to, the terminal screen. Essentially the rules for obtaining a classification are these:
(1) Take a sample of frames and measure the median of the distribution of pixel intensities.
(2) Assume that the lowest medians suggest that the user is looking away from the screen and the highest suggest that the user is looking at the screen.
(3) For medians near the lowest median classify as 'looking away from the screen' and for medians near the largest median classify as 'looking towards the screen'.
(4) Medians in the mid-range indicate that the user is probably looking at the screen, though probably the head is turned slightly away from the central line of regard.

In order to evaluate the accuracy of these simple rules each of the 600 digitized frames was viewed by a judge who recorded what the user was doing, and in porticular whether s/he was looking towards or away from the screen. Independently another judge classified each of the six hundred frames (looking at the screen/not looking at the screen), on the basis of the sample of medians alone. The percentage agreement was 60 per cent. That is relatively low and is due to the fact that the first judge provided detailed behavioural descriptions of the users' movements, not simple a dichotomous classification. Thus one is comparing judgements elicited against non-equivalent scales.

If one allows the second judge four categories: 'definitely not looking', 'probably not looking', 'probably looking' and 'definitely looking', then percentage agreement rises to 79 per cent.

If one were to automate this classification process, and allow the decision to be taken as part of the image analysis, then one would need to build in a training component. That is regular practice when installing image inspection systems in industrial processes and effectively calibrates the machine to the environmental conditions.

CONCLUSIONS

The evidence suggests that it is possible to accurately determine whether the user is attending or not to the terminal screen. Further work with the system has produced accurate classifications at the 80 per cent level. The accuracy of the system might be retained with still further reductions and a narrower grey-scale, but that would dramatically restrict the flexibility of the system. In this study a variety of lighting conditions was used, ranging from a well illuminated forground to one in which the only light was emitted from the terminal itself. Even low light cameras have difficulty functioning in very dim environments and one needs as wide a dispersion of pixels and as broad a grey-scale as possible in order to accurately detect users' movements. Further refinements in the system should be pursued with a view to determining what kinds of simple, but useful feature extraction might be accomplished in real time. This is essential if the large volume of 'mid-range' data, indicating moderate but important movements, is to be used to enhance the image identification and classification process.

REFERENCES

Devolve, N. and Queinnec, Y., 1983, Operators activities at Terminals: A behavioural approach. Ergonomics, 26, 329-346.
Meyer, J.J., Crespy, J. and Rey, P., 1980. L'Analyse Ergonomique des Postes de Travail avec Ecran de Visualisation. Geneva: Ecotra.

Acknowledgement: The work reported in this paper was undertaken as part of ESPRIT Project 527: Communication Failure in Dialogue.

A FRAMEWORK FOR USER MODELS

J. LONG

Ergonomics Unit
University College London, 26 Bedford Way
London   WC1H 0AP

A framework is presented whose aim is to help eliminate the confusion associated with the term user model. First, users´ interaction models and agents´ models of users are distinguished. Next, the term model is defined. Last, a format is presented which allows for the inclusion of the genesis and purpose of models. The application of the framework is briefly illustrated. The framework serves to classify individual user models and so to facilitate comparisons between them.

1. Background and Problem

User models figure prominently in current human-computer interaction (HCI) and intelligent knowledge-based systems research and development. The models range from researchers´ and developers´ models of the user, to users´ models of the computer, and computers´ models of the user. User models exhibit potential as an aid to developing more usable computer systems. They help make design issues explicit and form a basis for design choices. They encourage the designer to focus on aspects of the system affecting usability and facilitate communication between members of a design team.

One obstacle to realising this potential is the confusion associated with the term user model. First, it is not always clear what constitutes such a model. The "designer´s user model" of Hammond (1985) is almost certainly equivalent to the "designer´s conceptual model" of Norman (1986) - which is not even considered to be a user model. Second, equivalent models may be assigned different descriptions. The "user´s model of the system" of Whitefield (1986) is equivalent to the "conceptualised user´s mental model of the system" of Norman (but not to his "user's mental model of the system"). Third, even the same model may attract different descriptions. The Keystroke-level Model of Card et al (1983) is a model of the user for Young (1985), and a model of the system for Whitefield. The aim of this paper is to help eliminate this confusion by proposing a framework to describe all user models. The framework will make it possible to classify and so to operationalise models in a standard way.

2. The Framework

A framework is an analytic structure which characterises the objects of its scope. By providing a common description for objects, a framework makes possible the establishment of identity, similarity and difference relations between them. User models in general constitute the scope of this framework (including non-cognitive e.g. anthropometric models and non-computer associated e.g. control models) to ensure the widest application within ergonomics. HCI user models, however, remain the primary concern. The framework follows. First, a distinction is made between user models. Second, the term model is defined. Third, a format for characterising user models is proposed.

2.1. A Distinction between User Models

One of the major sources of confusion concerning user models derives from the use of the same term to refer both to users' models (which users have) and users models (which others have of users). Whitefield (1986) makes clear the difference in his general classification of HCI models by dividing them according to who or what is doing the modelling and who or what is being modelled. Thus, in the case of users' models, users do the modelling, but in the case of users models, users are modelled. This distinction is emphasised here by describing the two types of model differently and by motivating them separately. When users do the modelling, their models are of the interaction entities. These can be separately enumerated e.g. user, computer, task and domain (Long, in press). These models might be termed: "users' models of interaction entities" or "users' interaction models" for short. If known, the users might be identified e.g. word processing operators' interaction models. When users are modelled, however, others (or "agents", following Whitefield) do the modelling and these agents might be separately enumerated e.g. in terms of the paradigms associated with HCI: science (human and computer scientists), engineering (human factors and software research engineers) and system development (human factors and software engineering practitioners) (Long, 1986). Models developed by "paradigm agents" might be termed: "agents´ models of interactive users" or "agents models of users" for short. If known, the agents might be identified e.g. human factors research engineers´ models of users.

The framework, then, recognises the confusion between users´ models and users models, but accepts the distinction and attempts to clarify the difference by providing separate terms each with its own motivation.

2.2 A Definition of the Term Model

Given the distinction between users´ interaction models and agents´ models of users, a definition of "model" is needed which applies to both. Inadequate definitions also constitute sources of confusion. The framework here extends previous definitions by specifying the attributes and features of a model and its relation with the entity modelled. The definition is consistent with usage outside ergonomics. The additional details help characterise models more completely.

The definition tries to capture the following aspects of a model: (i) a model is not the "real thing" - it is different from the entity modelled; (ii) the difference always involves a reduction of some sort; (iii) the reduction is systematic; (iv) the systemisation is expressed as a set of relations; (v) the relations are determined by the purpose of the model; and (vi) the relations are explicit. The definition is as follows:

(i) a model $\underline{M}$ is a representation $\underline{R}$ of an entity $\underline{E}$: i.e. $\underline{M} \longrightarrow \underline{R}(\underline{E})$ (where ---> means "is rewritten as").

(ii) an entity $\underline{E}$ has attributes $\underline{A}$ and a representation $\underline{R}$ has attributes $\underline{a}$: i.e. $\underline{E} \longrightarrow Ai....An$ and $\underline{R} \longrightarrow ai....an$.

(iii) representation attributes $R(ai....an)$ are a reduced set of, but systematically related for some purpose to, entity attributes $E(Ai....An)$: i.e. $R(ai....an) < E(Ai....An)$.

A representation can be physical (a manikin), or abstract (algebraic). Entities can likewise be physical (a terminal) or abstract (a program). Hence, a model can itself be modelled (an agent´s model of a user´s interaction model of the task). Likewise, attributes can also be physical (display colour), or abstract (database structure). Attributes have values (red colour; hierarchic database structure). In addition, models can also be distinguished according to global dimensions. These might include coherence, completeness, explicitness, formality, etc. Thus, models with identical representations, attributes, features and relations might nevertheless differ in their completeness or formality.

To illustrate the application of this definition, Card et al´s Model Human Processor (1983) is an abstract representation in database form of a general purpose user, having the attributes of memory and processors with values expressed as working memory, cognitive processor, etc. In contrast, Whitefield´s Blackboard Design Model (1984) is an abstract representation of the engineering designer, in hierarchic form having the attribute of design-related knowledge with values expressed as individual knowledge sources. This definition, then, should help to characterise models more completely and so reduce confusion by its greater discrimination.

2.3 A Format for Models

In addition to distinguishing the characteristics of the representations and their entities, there is a need to reflect the genesis and purpose of models. Whitefield (1986) suggests a useful format to this end and it is used here. Failure to specify a model´s genesis and purpose can lead to confusion concerning its status.

The characterisation is as follows:
A Creator $\underline{C}$ creates with a purpose $\underline{P}$ a model $\underline{M}$ expressed as a representation $\underline{R}$ of an entity $\underline{E}$ for a utiliser $\underline{U}$

In the case of users´ interaction models, the user him- or herself would be both the creator and the (model) user and the model would consist of a mental representation of an interaction entity such as the computer, for the purpose of understanding and predicting its behaviour. For example, a wordprocessor operator would create and use a model of the computer to understand and predict its behaviour under certain circumstances e.g. if a particular document stored on disc exceeded the capacity of the processor. In the case of agents´ models of the user, the agent and the (model) user are likely to be different people and the model would consist of an abstract representation of users for the purpose of designing a new interface. For example, a computer-aided design researcher would create a model of the engineering designer to be used by the system developer to help decide, for example, what library functions for prototypical objects might be made available by the program.

The format supports the development of a complete

classification and hence the operationalisation of the
profusion of models. It makes clear that the creator and
the user of the models may be different people and are
certainly different from the computer user (who in turn
will have his or her own interaction model). Further, the
important difference between descriptive, prescriptive and
predictive models can be reflected by the model's purpose.
The difference constitutes a major source of confusion
concerning user models. Last, even if a model is of the
exemplary kind, that is a prescriptive model for a user to
acquire, it must still be a model of some entity e.g. the
computer. Its entities, attributes and relationship then,
can all be questioned. In short, the format provides a
means for reflecting the genesis and purpose of user
models. Together with the distinction between user models
and their definition, this completes the framework.

3. Application of the Framework

The utility of the framework can now be briefly illustrated by applying it to the examples of confusion cited earlier. First, in terms of the framework, Hammond's "designer's user model" (1985) is an agent's model of the computer system for the purpose of being acquired by the user. As such, it is equivalent to Norman's "designers conceptual model" (1986) and, consistent with Norman's view, is neither a user's interaction model nor an agent's model of the user. However, neither Hammond nor Norman propose a model in terms of the definition offered here. Both authors propose only characteristics general to models, but not a specific model. Second, Whitefield's "user's model of the system" (1986) and Norman's "conceptualised user's model of the system" (1986) are in fact equivalent. In terms of the framework, both are agent's (researcher's or designer's) models of the user's interaction model of the system. Both are descriptive and the purpose of both is to aid system design. Third, the Keystroke-level Model of Card et al (1983) is not purely a developer's model of the user as claimed by Young (1985), since it includes the attribute of "system response operation". The framework supports Whitefield's view (1986) that the model is an agent's model of the system, where the system includes the user and the computer.

The framework, then, is consistent with the claim that it supports a complete classification of user models. It has been able to clarify a number of instances in which some confusion existed concerning the status of individual user models or the relations between different user models.

4. Future Development

The framework, however, constitutes no more than an initial proposal. More extensive application would provide a more adequate test than has been possible here. In particular, the framework needs testing for coherence, completeness and explicitness. The applications should include disambiguating apparently confusing instances of user model terminology; classifying user models of all types; and accommodating model taxonomies offered by others. Such applications will provide evidence of the framework's efficacy and suggest ways in which it can be usefully developed.

ACKNOWLEDGMENT

The author acknowledges the influence on this paper of the presentations and discussions which took place at the Alvey Joint IKBS/MMI User Modelling Workshop held at Abingdon in September 1985.

REFERENCES

Card, S., Moran, T., & Newell, A., 1983, The Psychology of Human-Computer Interaction. Erlbaum: Hillsdale N.J.

Hammond, N.V., 1985, The user's model of the system. In Human Factors of the User-System Interface, edited by B. Christie. Elsevier: North Holland.

Long, J., 1986, People and computers: designing for usability. In People and Computers, edited by M. Harrison and A. Monk. CUP: Cambridge.

Long, J. (in press), Cognitive Ergonomics and Human-Computer Interaction. In Psychology at Work, edited by P. Warr. Penguin: London.

Norman, D.A., 1986, Some observations on mental models. In User Centred System Design: New Perspectives on Human-Computer Interaction, edited by D.A. Norman and S.W. Draper. Erlbaum: Hillsdale N.J.

Whitefield, A., 1984, A model of the engineering design process derived from Hearsay-II. In Interact '84, edited by B. Shackel. North Holland: Amsterdam.

Whitefield, A., 1986, Constructing and applying a model of the user for computer system development: the case of computer aided design (Unpublished Ph.D. Dissertation: London University).

Young, R., 1985, User models as design tools for software engineers: Paper presented at Alvey Joint IKBS/MMI User Modelling Workshop.

**THE EVALUATION AND GENERATION OF ICONS
FOR A COMPUTER DRAWING PACKAGE**

R.M. BROWNE and R.B. STAMMERS

Applied Psychology Division
Aston University
Birmingham B4 7ET

This study evaluated icons used in the MacPaint™ package. In the first phase, icons were assessed in terms of comprehensibility. The second phase consisted of a group design exercise to generate alternative icons for six functions. Three variants of each redesign were then further assessed, leading to a final selection of six new icons. These were then tested along with the remaining icons. Results showed a significant improvement in five of the six new icons.

INTRODUCTION

The use of icons in human computer interaction is now common. What has not yet been established is the effectiveness of different icons, their comprehensibility and sound principles for icon design. There is however an increasing interest in these issues. This comes from both a general human factors perspective (eg, Gittings, 1986), and from a theorectical viewpoint (Jervell and Olsen, 1985), where there are such questions as what is an icon, and how does it differ from a symbol? There has also been a number of studies evaluating icons and relating design issues to such concepts as 'concreteness - absractness' (ie, Rogers, 1986). The area also benefits from the considerable amount of work that has already been carried out on the evaluation of symbols and signs (Easterby and Zwaga, 1984).
This paper reports on a straightforward evaluation of an icon set used in a commercial package. A procedure for generating new icons is then described, followed by a further evaluation of the new icons.

STUDY 1: INITIAL EVALUATION
Introduction

The aim of this experiment was to guage the comprehensibility of a set of icons in the commonly used MacPaint™ program which runs on the Apple® Macintosh™ computer. The program can be used in a variety of drawing roles. It has a set of twenty icons which appear on the left hand side of the screen when in use.

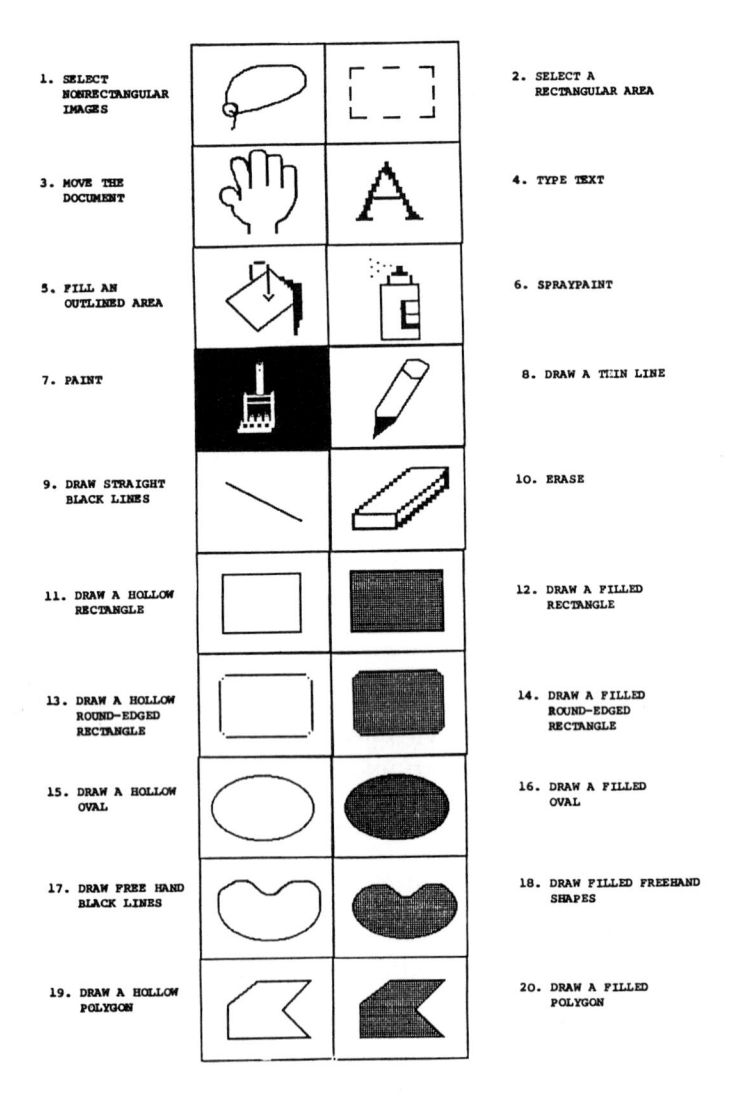

1. SELECT NONRECTANGULAR IMAGES			2. SELECT A RECTANGULAR AREA
3. MOVE THE DOCUMENT			4. TYPE TEXT
5. FILL AN OUTLINED AREA			6. SPRAYPAINT
7. PAINT			8. DRAW A THIN LINE
9. DRAW STRAIGHT BLACK LINES			10. ERASE
11. DRAW A HOLLOW RECTANGLE			12. DRAW A FILLED RECTANGLE
13. DRAW A HOLLOW ROUND-EDGED RECTANGLE			14. DRAW A FILLED ROUND-EDGED RECTANGLE
15. DRAW A HOLLOW OVAL			16. DRAW A FILLED OVAL
17. DRAW FREE HAND BLACK LINES			18. DRAW FILLED FREEHAND SHAPES
19. DRAW A HOLLOW POLYGON			20. DRAW A FILLED POLYGON

FIGURE 1 The original MacPaint™ icon set.

The icons represent a number of drawing functions or shapes that can be utilized in producing pictures on the main screen area. Typical interaction with the icons involves their selection with a mouse input device. In this study the icons were examined in a more simple mode.

Method

Individual subjects were tested on the time taken to indentify icons. They were each presented with a different random order of functions.

Subjects were presented with reproductions of the twenty icons (4.5 x 3.5cm, black figures on white). The icons were laid out in a similar way to how they appear on the screen, five different orders of icons were used, Figure 1 illustrates the original icon set (in this example the icons are labelled and one is shown as active, this was not done in the experiment).

The subjects were ten undergraduates, five female and five male, average age was 21.6 years.

Subjects who had used MacPaint before were not used. A practice trial was given to all subjects followed by a trial for each of the icons. Subjects were tested on the time taken to indentify by pointing to the icon which represented the function stated by the experimenter. They were each presented with a different random order of functions. Selection time was recorded by stop watch and any errors noted. If an incorrect icon was chosen the subject was told this, and had to continue until the correct one was chosen. Selection time included error selection time.

Results

Detailed results were not analysed at this stage. Mean selection time was 12.59 secs and the range was 1.22-70.97 secs. The six icons with the poorest performance were chosen as candidates for further study, these were (see Fig. 1) :-
    1. Select non-rectangular images
    4. Type text
    5. Fill-in an outlined area
    7. Paint
   10. Erase
   17. Draw a free hand black line.

STUDY 2 DESIGN OF NEW ICONS
Introduction

A design production session was organized to generate six new icons for those functions listed above.

Method

Four female and five male students took part in an initial brainstorming session, led by a female experimenter. The students came from a variety of backgrounds. The session lasted approximately one hour and involved a number of stages. The end-product of the session was a set of rough sketches of icons.

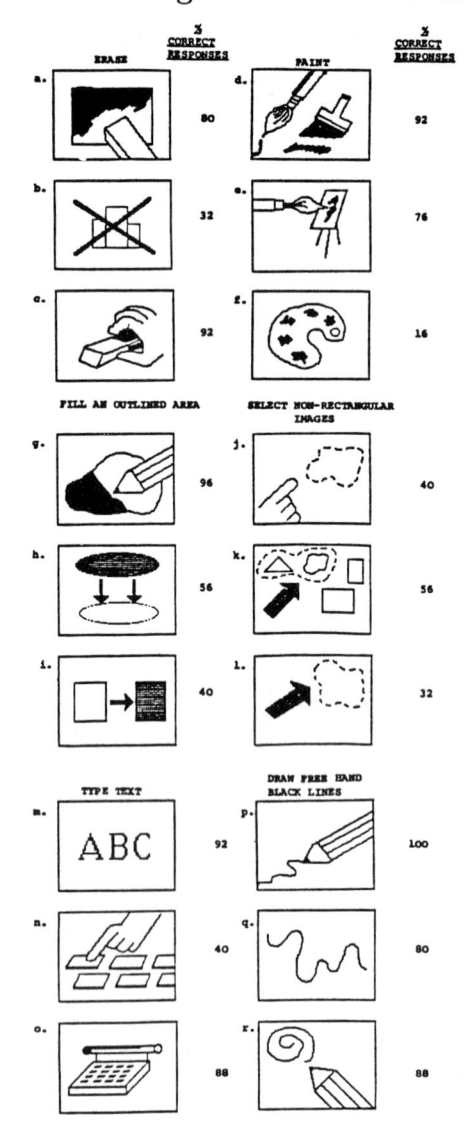

Figure 2 Icons derived from design exercise.

These sketches were then redrawn in a more comprehensible format and a copy sent to each participant. They were then asked to select the three icons for each function that they felt best represented its meaning. For each function the three icons receiving the most 'votes' were selected for further evaluation in the next study (Figure 2).

STUDY 3 SELECTION OF NEW ICONS
Introduction
The aim of this study was to determine which of the three variants produced in the session above was best understood by a new set of subjects.
Method
Individual subjects were presented with booklets in which each icon appeared on a separate page (sizes were as used before). Subjects were asked to name by writing the function to which the icon on each page referred. They were given one practice trial and required to go through the whole book. Twenty-five student subjects were used, thirteen females and twelve males, average age 21.08 years.
Three judges then assessed the responses independantly, using descriptions of the functions given in the program manual. Agreement by two of the three judges was required for the response to be given as correct.
Results
The percentage correct for each icon is given in Figure 2. Two thirds of the icons were found to yield scores of over 50%, the highest scoring icon for each function was chosen for the next stage of the study. Five out of the six icons had scores of over 66%, the ISO standard for acceptability of symbols.

STUDY 4 EVALUATION OF NEW ICON SET
Introduction
The aim of this experiment was to determine whether the redesigned icons, when placed back in the context of the original icon set, would yield improved performance. The original study was therefore replicated to give comparison data.
Method
The same procedure and methodology was used with the six new icons replacing the six originals in the set. Ten new subjects were recruited, five of each sex, average age 20.3 years.

Results

Selection times this experiment were compared with those collected in Study One, using analysis of variance. In a two-way analysis, with icons as a repeated measure and icon set as a between subject factor, a significant difference between sets was found (F=14.7; df 1, 18; p<0.01).

More interestingly, analyses of simple main effects showed significant differences only for five of the six redesigned icons and for no other comparisions. The icons that led to improved performance were :-

1. Select a non-rectangular area	p<0.001
4. Type text	p<0.05
5. Fill in an outlined area	p<0.05
10. Erase	p<0.01
17. Draw a free hand black line	P<0.001

The remaining new icon just failed to reach a significant difference at the 5% level, no differences approached significance for the original icons.

DISCUSSION

It has been demonstrated that variation in the degree of comprehensibility of an set of icons can exist. It has also been found that redesign of some of those icons led to improvement in performance.

Further work is now required on an analysis of all the sub-groups of icons to determine if their characteristics can be used to predict performance (in the manner of Rogers, 1986). Additionally it will be neccessary to study the different icon sets in the more realistic environment of active use in an interactive computer-based task.

REFERENCES

Easterby, R.S. & Zwaga, H.J.G. ,1984,eds., **Information Design.** (Chichester; Wiley).

Gittens, D. ,1986, Icon-based human-computer interaction. **International Journal of Man-Machine Studies, 24,** 519-543.

Jervell, H.R. & Olsen, K.A. ,1985, Icons in man-machine communications. **Behaviour and Information Technology, 4,** 249-254.

Rogers, Y. ,1986, Evaluating the meaningfulness of icon sets to represent command operations. In **People and Computers: Designing for Usability,** edited by M.D. Harrison & A.F. Monk (Cambridge University Press) pp.586-603.

Knowledge Acquisition

THE DESIGN, DEVELOPMENT AND EVALUATION OF
A CLIMATIC ERGONOMICS KNOWLEDGE BASED SYSTEM

T.A. SMITH and K.C. PARSONS

Department of Human Sciences
University of Technology, Loughborough
Leicestershire LE11 3TU

INTRODUCTION
    A computer system should be designed to meet the needs of its users. In the case of a Knowledge Based System (KBS) an additional problem is posed in that the differences between experts and non-experts may have to be recognised and an appropriate information structure created such that both can use and understand the system. With this in mind a prototype system was created such that an individual with a minimum level of climatic ergonomics knowledge could obtain expert information. The potential user group were those people who have undergone the basic course in climatic ergonomics given to undergraduate and postgraduate ergonomics students at Loughborough University. The system could also be used by Ergonomists outside the University.

AIMS
    The aims of the project were, in the context of assessing human response to thermal environments, :-

    1   to develop a mental model for the expression of
         the knowledge domain
    2   to produce a working system
    3   to evaluate the system with respect to three main
         aspects :
                    -    ease of use
                    -    appropriateness of design
                    -    utility provided
    4   to ensure that the design allowed the system to be
         updated.

DESIGN CRITERIA
    Academic tasks are often based upon information retrieval where a literature search is required in order to review a given topic area. However, climatic ergonomics by its nature demands practical experimentation to be conducted such as the thermal assessment of a room and recommendations made as result. The mathematical and empirical indices used can be time consuming and

laborious to calculate and therefore ideal for computerisation. Having derived information from a literature review and personal experimentation a conclusion will be drawn. If the results of similar experiments produce a common trend then this trend could be classified as knowledge and it is this facility that the indivivdual would benefit most from. A Task Analysis approach was used to identify the needs of the users and from this three main facilities were created : literature search, simulation, and knowledge.

As well as utility the system was intended to allow effective communication between man and machine. Therefore, when designing the system attention was paid to the knowledge of the user with respect to both their climatic ergonomics and computing experience. Due to the variety of work performed by the user group it was possible to predict that there would be infrequent use of the system. In keeping with the recommendations by Damodaran et al (1980) in such circumstances a menu driven system was created. However, the dialogue style was changed to a question and answer format for the use of the mathematical models and indices as a menu dialogue was impractical.

Although the factors listed above need consideration it is the issue of knowledge organisation which is of importance and as a consequence attention must be paid to the way information might be represented in memory and how the relationships amongst items in computer memory might be represented.

ERGONOMICS KNOWLEDGE

Climatic ergonomics draws mainly upon the pure disciplines of psychology, physics and physiology to ensure that human comfort, performance and well-being are assured. It is precisely this multi-disciplined approach and the nature of their interactions which makes climatic ergonomics a complex domain. If it is accepted that knowledge bears reference to known events, then the knowledge of an individual may be dependent upon memory. Barfield (1984) states that it is the ability to link appropriate information together and retrieve it that makes the expert distinct from a non-expert. As a consequence attention must be paid to the way information might be represented and the relationships between different items in memory.

One idea proposed by cognitive psychologists for knowledge representation is that of semantic networks (Collins and Quillian, 1969). The main feature that this theory offers is that if generic information is held then it is possible to infer about a particular concept in terms of generic information. From this it is obvious that proposition plays a critical role in the representation and acquisition of knowledge. To state that an individual becomes an expert in a given domain simply through exposure would omit a subtle but important process which occurs during exposure, the development of rules.

Experts recognise domain-specific "chunks" of information, possess rules relating the chunks and the ability to apply the rules to obtain relevant information (Barfield, 1984). For this partial problem of assessing human response to thermal environments, having established the contents of the knowledge domain, the expert and the potential user group a matrix structure was used to describe the common areas in the domain. Five main classes were formed ("areas of interest") and these are : thermoregulation ; physical performance ; mental performance ; indices ; and adverse effects.

This information was combined with the range of environmental conditions (hot, moderate and cold) and the facilities required and a series of three matrices were created to express the domain. The expert, a lecturer in climatic ergonomics with practical experience, was then used to apply his rules for the knowledge domain and place the appropriate information in each section of the matrix. This process of completing the matrices could be referred to as "mapping" the knowledge domain into the conceptual model of the user group, thus meeting their needs (figure 1).

The KBS was implemented on a Future FX30 micro-computer using Prolog, Basic and Concurrent CP/M.

EVALUATION

An experiment was conducted to evaluate the system with respect to usability (the ease with which the system could be used), appropriateness (whether knowledge of climatic ergonomics could be placed into a computer), and utility (that the design of the system allowed the desired task, or range of tasks to be performed).

Each experimental period consisted of five distinct stages:
Stage 1: a short questionnaire was given to determine computer and climatic ergonomics knowledge. Both these factors could have a significance upon the way the subject viewed the system and needed to be quantified
Stage 2: before using the system each subject was given introductory information about the system and a tree diagram to indicate the structure of the system (figure 2).
Stage 3: determination of ease of use requires all aspects of the system to be considered and hence written instruction was given about how to turn on the system and prepare the system for use.
Stage 4: once the system was ready the subject was given a set of five exercises to complete using the system. The first exercise was one which allowed the user to familiarise himself with the system and like the remaining four was goal related.
Stage 5: finally, a questionnaire was issued to determine subjective opinion using rating scales, YES/NO response and open-ended questions.

16 subjects were used for the experiment, 8 male and 8 female, and each had completed the basic course in climatic ergonomics. Their ages ranged from 19 to 43, with the majority of subjects in their early twenties.

RESULTS

The subject knowledge questionnaire revealed that the subjects were able to indentify the generic terms for particular examples. It was interesting to note that only two subjects failed to understand a given generic term and that thirteen subjects understood all five. The generic terms used were : physical performance ; mental performance; thermoregulation ; adverse effects ; and indices. Eight of the sixteen subjects stated that they disliked using computers and that they did not feel confident when using one.

From the results obtained from the experiment it is possible to state that the majority of subjects found the system was:easy to use (due to the use of menus); appropriately designed; (with a structure which was easily understood)provided utility; (through the provision of literature, knowledge and simulation facilities).

One important feature of the system which needs to be mentioned but is not referred to explicitly in the results is that all subjects were able to perform the exercises set. This system allowed the subjects to achieve task related goals. The semantic network model (Collins and Quillian, 1969) provided the most logical and easy to comprehend structure for the classification of the knowledge domain.

CONCLUSION

The major objective of this study was to provide utility to a particular user group through the creation of a knowledge based system. From the results this objective has been shown to have been achieved. The use of generic information made it possible to develop the matrices that allowed both the expert and non-expert to classify information in the same way. As a result of this commonality between knowledge structures the difference between expert and non-expert was bridged.

In conclusion the following aims were achieved :

1. A mental model of the user group was created which allowed the knowledge domain to be expressed in a form the user would understand.
2. A working system was produced
3. The evaluation procedure showed the system to be generally, easy to use, appropriately designed and that it provided utility.
4. The design of the system is such that it can be updated.

REFERENCES

Barfield. W., (1986). Expert-novice differences for software : implications for problem-solving and knowledge acquisition, Behaviour and Information Technology, Vol 5, No. 1, 15-29.

Collins, A.M. & Quillian, M.R. (1969), Retrieval time for semantic memory, Journal of verbal learning and verbal behaviour, 8 : 240-247.

Damodaran. L, Simpson. A.S., & Wilson. P.A., (1980) Designing systems for people , NCC Publications.

Area of Interest	Hot Environment	Moderate Environment	Cold Environment
INDICES	KERSLAKE McINTYRE LEITHEAD & LIND	FANGER	SIPLE & PASSEL BURTON & EDHOLM
THERMO-REGULATION	HENSEL STOLWIJK & HARDY	STOLWIJK & HARDY	KEATING STOLWIJK & HARDY
ADVERSE EFFECTS	LEITHEAD & LIND	NO INFORMATION	GOLDEN
MENTAL PERFORMANCE	PEPLER MEESE McINTYRE	McINTYRE	HORVATH & FREEDMAN
PHYSICAL PERFORMANCE	CHRENKO MEESE	McINTYRE	PARSONS & EGERTON FOX BURTON & EDHOLM

Figure 1 . Literature matrix.

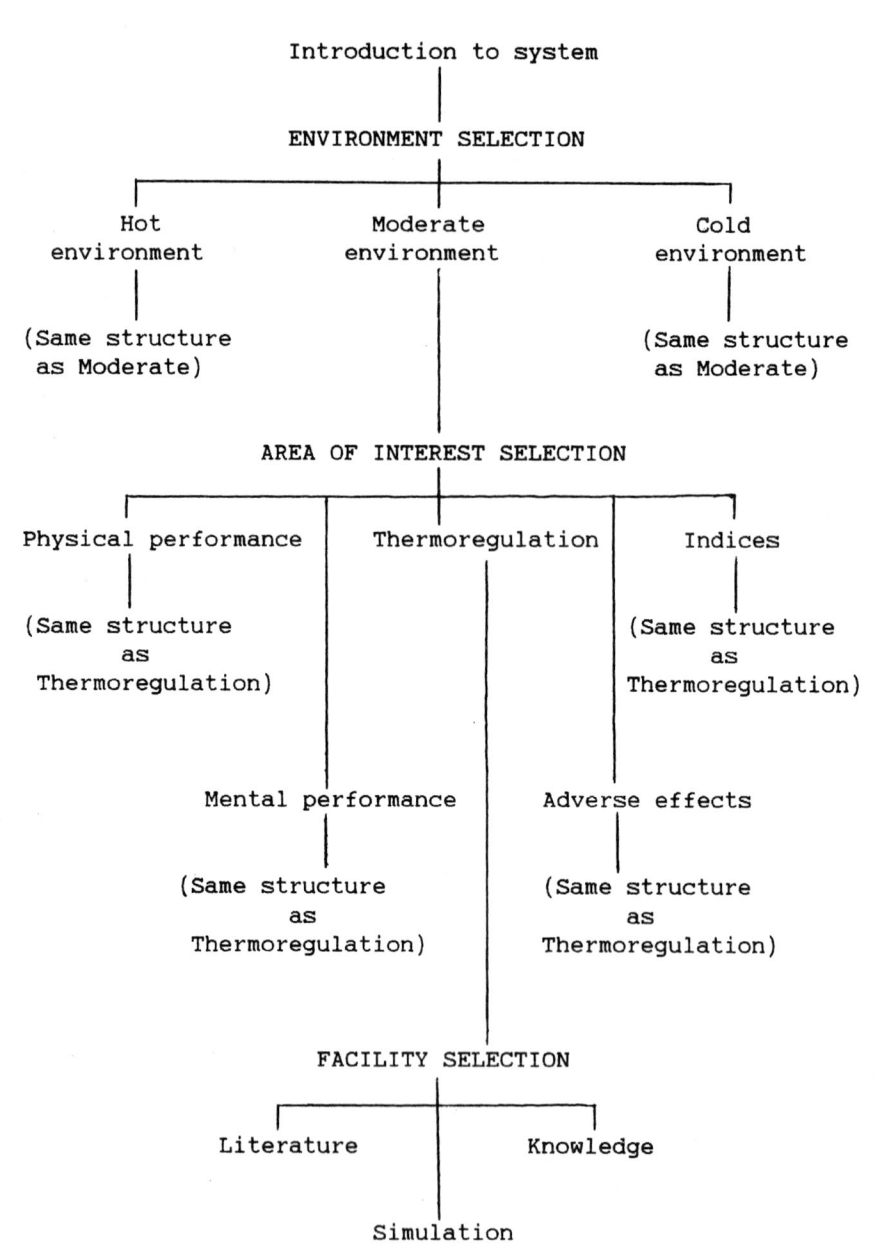

```
                    Introduction to system
                             |
                    ENVIRONMENT SELECTION
                             |
        ┌────────────────────┼────────────────────┐
       Hot                Moderate               Cold
    environment          environment          environment
        |                    |                    |
 (Same structure                            (Same structure
  as Moderate)                                as Moderate)
                             |
                   AREA OF INTEREST SELECTION
                             |
        ┌────────────────────┼────────────────────┐
 Physical performance   Thermoregulation        Indices
        |                    |                    |
 (Same structure                            (Same structure
       as                                         as
  Thermoregulation)                          Thermoregulation)

                      Mental performance    Adverse effects
                             |                    |
                    (Same structure       (Same structure
                          as                     as
                     Thermoregulation)    Thermoregulation)

                         FACILITY SELECTION
                             |
              ┌──────────────┼──────────────┐
          Literature                     Knowledge
                             |
                         Simulation
```

Figure 2 . System tree diagram.

PROBLEMS OF KNOWLEDGE ACQUISITION FOR EXPERT SYSTEMS

N.K. TAYLOR, E.N. CORLETT and M.R. SIMPSON

Department of Production Engineering and Production Management, University of Nottingham
Nottingham NG7 2RD

Probably the greatest, but least documented, of the problems encountered in the development of an expert system is that of knowledge acquisition. Whilst the validation of the information used is clearly an essential and exacting process, capturing the expertise in the first place is by no means simple and the choice of appropriate efficient techniques is not clear-cut. This paper reports on the problems encountered and the solutions applied, over the past three years, in building a variety of ergonomics knowledge bases for the ALFIE expert system.

INTRODUCTION
It is now over twenty years since work started on the DENDRAL system for mass spectrogram analysis. During this time much effort has been channelled into developing a wide range of 'expert systems'. A rather disturbing statistic, however, is the small number of such systems which are actually in use :
'As of September, 1984, the number of fully operational expert system applications in regular use under field conditions is probably no more than ten'. Reitman & Weischedel (1985).
Even allowing for the fact that this statement is two years old, given the fever of activity in expert systems development over the past five years, we are entitled to ask the question : Where are the fruits of all the labour? Reitman and Weischedel give us a hint when they proceed to inform us that there are between 100 and 200 other expert systems in 'some stage of development'. This suggests, and confirms our own experience whilst developing the ALFIE system (Taylor & Corlett 1987), that it is unusually common to under-estimate the resources

required to build a viable expert system. So just where are these errors of judgement being made and why?

SOME STUMBLING BLOCKS

The evolutionary nature of the development of an expert system is summarised by Buchanan et al. (1983) thus :
1. Identify the problem characteristics;
2. Find concepts to represent the knowledge;
3. Design a structure to organise the knowledge;
4. Formulate rules to embody the knowledge;
5. Validate the prototype system;
6. If need be start again from an earlier stage.

Stages 1 and 2, although they may take some time to accomplish, are not likely to result in any surprises so far as the amount of effort required to achieve them is concerned. Stage 3 is the point at which a proprietary expert system shell might be purchased and it is essential that such a shell is suited to the domain under consideration. A badly chosen shell may well turn stage 4 into a mammoth exercise in accommodation if the true nature of the domain is more conducive to another approach. It is probably worth noting here the existence of model-based approaches to expert systems as well as the more conventional rule-based approach. See Johnston (1986) for instance. Stage 5 should be, like stages 1 and 2, relatively problem free, even though it will require a significant proportion of the overall effort.

Stage 6 enshrines the evolution principle. It can in fact be applied at any time and will be applied quite frequently - even if the software designer, knowledge engineer and domain expert are one and the same person. The evolution principle is not unique to expert systems. It appears wherever something is being created or modified. Unfortunately, such unpredictable iterations are very hard to cost in terms of man-power requirements which may well be much greater than anticipated. The problem is compounded by the fact that a proprietary expert system shell may have to be thrown out as part of this iteration with a consequent increase in cost.
Perhaps some of those systems under development have been 'temporarily shelved' due to a lack of resources and are not under active development at all.

Throughout the above development process there is also the underlying problem of actually obtaining the expertise needed to 'arm' the system. Because knowledge acquisition is not mentioned explicitly, as is all too often the case in reports of expert system developments, the new-comer may well be lured into thinking that it is a relatively trivial problem compared to the others. This is far from being the case (Fox et al. 1985, Welbank 1983).

THE KNOWLEDGE ACQUISITION BOTTLENECK

Buchanan et al. (op. cit.) report the existence of a knowledge acquisition bottleneck between the knowledge source (domain experts and/or definitive works) and the expert system knowledge base. This will exist for at least as long as the representations used by the knowledge source cannot readily be mapped onto those used by the expert system. However, it is important that the knowledge engineer does not exacerbate this mapping problem through a lack of awareness of the subject matter or terminology employed.

Unless the knowledge engineer is himself the domain expert, the first stage in knowledge acquisition must be for him to familiarise himself with the domain under consideration. Grover (1983) proposes the production of a Domain Definition Handbook which should contain the following :

A general problem description;
A bibliography of reference documents;
A glossary of terms;
A list of domain experts;
A list of performance metrics;
Examples of reasoning scenarios.

As Grover points out, the production of the handbook can be speeded up by having access to an expert at this stage. However this must be set against the amount of time which the expert is prepared to devote to the project. Certainly, the expert will be invaluable in reducing the potential bibliography to those references which provide up to date, authoritative information on the domain but he cannot be expected to provide answers to each and every query which the knowledge engineer might encounter whilst studying the references. Having obtained a selective bibliography, a glossary of terms and a provisional list of domain experts should not be difficult to produce. Short communications with those people on the provisional list should enable the knowledge engineer to produce his working list.

Before proceeding however, we recommend that the knowledge engineer re-examines his problem description in the light of what he has learnt from studying the bibliography. The original proposal may now appear to be unfeasible given the state of the art of the domain itself. A less ambitious objective may still be a worthwhile subject for the expert system and, if necessary, this should be investigated. Once the knowledge engineer and domain experts have agreed on a feasible and useful domain for the system they can consider the performance criteria required of the system and produce some examples of its use.

KNOWLEDGE SOURCES

There are two major types of source from which knowledge of a domain might be extracted. These are the literature, comprising the bibliography mentioned earlier, and human experts. In general, the literature on a subject will be more structured than a transcript compiled from an interview with a human expert. It will also allow the knowledge engineer to acquaint himself with the basics of the domain without wasting the time of an expert. The literature is therefore a more suitable starting point for the knowledge acquisition process. The main drawbacks of the literature are that it is not interactive and it may well be incomplete. In order to fill in the gaps left by the literature the knowledge engineer will have to approach human experts.

Not all developments will require the use of both types of knowledge source. If the literature is sufficiently detailed and complete (but so diverse that an expert system is still required to collate the information into a usable package) then a human expert may not be needed until the validation stage. On the other hand, if the literature is too scarce or idealised then the human expert will be required from a very early stage.

The knowledge engineer should approach a human expert with as clear a set of objectives as is possible. This may seem obvious. However, the objectives should include an interviewing technique which has been carefully thought out to ensure that the interview does not go 'off the rails' into areas which are not of immediate concern to the knowledge engineer.

A number of knowledge acquisition techniques have been documented ranging from the study of textbooks through interviewing and questioning techniques to automated learning and text understanding systems. We do not have the space to describe all of these here and so a summary has been provided for some fifteen techniques which contains references to sources where full descriptions may be found. A large proportion of the approaches cited have been employed differently by different people and have consequently been given different labels by those people. We have attempted to show how these different labels equate with each other.

A Summary of Knowledge Acquisition Techniques

	Fox et al (1985)	Buchanan et al (1983)	Grover (1983)	Welbank (1983)
1		Textbooks		
2	Informal Interviews		Forward Scenario Simulation	Interviews
3	Protocol Analysis		Procedural Simulation	Observing Experts
4				Critical Incidents
5				Repertory Grids
6				Distinguishing Goals
7			Pure Reclassification	Reclassification
8				Systematic Symptom-to-Fault
9				Intermediate Reasoning Steps
10			Goal Decomposition	Dividing the Domain
11	Interactive Computer Techniques	Intelligent Editors		Questioning with Shells
12				Using Half-Built Systems to Elicit Further Knowledge
13	Rule Induction	Inference Systems		Induction Systems
14	Heuristic Discovery	Automated Learning		
15		Text Understanding Systems		

CONCLUDING REMARKS

We have shown that the problem of actually obtaining the knowledge required for an expert system is far from trivial. Not only this, but it may also necessitate the refinement of the initial goals of the development - perhaps to the point of making the resulting system unviable as a practical tool.

The process of continually refining one's goals to make them less ambitious can be very demoralising. It is hoped that this paper, by recognising the existence of the problem, will provide a morale booster to others engaged on work in this field as well as a warning to those about to embark on it.

We are currently investigating, in a joint experiment with our colleagues mentioned in the Acknowledgements section below, the applicability and efficiency of a number of techniques for eliciting knowledge from domain experts. The results of this work will be presented in a suitable organ in the near future.

ACKNOWLEDGEMENTS

We are enormously grateful to Dr. Richard Schweickert of Purdue University who led us (to our advantage) to devote more effort to the problems of knowledge acquisition than might otherwise have been the case. We are also grateful to Dr. Mike Burton and Mr. Andy Hedgecock of the Psychology Department at Nottingham University for useful discussions. The development of ALFIE is supported by the Science and Engineering Research Council.

REFERENCES

Buchanan, B.G., Barstow, D., Bechtal, R., Bennett, J., Clancey, W., Kulikowski, C., Mitchell, T., & Waterman, D.A., 1983, Constructing an Expert System. In *Building Expert Systems*, edited by F.Hayes-Roth, D.A.Waterman & D.B.Lenat, (Addison-Wesley), 127-167.

Fox, J., Myers, C.D., Greaves, M.F., & Pegram, S., 1985, Knowledge Acquisition for Expert Systems : Experience in Leukaemia Diagnosis, *Methods of Information in Medicine*, 24, 65-72.

Grover, M.D., 1983, A Pragmatic Knowledge Acquisition Methodology, *Proceedings of the 8th International Joint Conference on Artificial Intelligence*, 426-438.

Johnston, R., 1986, Breaking the Rules, *Expert Systems User*, 2 (2), 23-26.

Reitman, W., & Weischedel, R., 1985, Automated Information Management Technology : Considerations for a Technology Investment Strategy, Air Force Aerospace Medical Research Laboratory Report TR-85-042, 27-48.

Taylor, N.K., & Corlett, E.N., 1987, ALFIE - Auxiliary Logistics For Industrial Engineers, *International Journal of Industrial Ergonomics*, Forthcoming.

Welbank, M., 1983, A Review of Knowledge Acquisition Techniques for Expert Systems, British Telecom Research Laboratories Report.

ERGONOMICS AND MECHANICAL ENGINEERING DESIGN

P. JOHN

HUSAT Research Centre
The Elms, Elms Grove, Loughborough
Leicestershire LE11 1RG

Fieldwork has been conducted with mechanical engineers designing fuel injection pumps. This has been used together with a literature survey to propose a model of the design process. This model has been applied to specify the necessary tools for an intelligent knowledge based design support system for this domain. This work forms part of the Alvey large scale demonstrator project 'Design to Product'.

INTRODUCTION

The HUSAT team have five objectives within the Alvey Design to Product project : (1) to develop a model of the mechanical engineering design process (2) to apply a blend of methodologies to elicit mechanical engineering design knowledge (3) to specify a man-machine interface to an intelligent knowledge based design support environment (4) to prototype aspects of such an interface and develop such by extensive user trials and (5) to investigate and define the appropriate human roles for a design to product system. It is work towards the first and third objective that is described within this paper.

In order to formulate a model of the mechanical engineering design process two strands of work were brought together, fieldwork and literature studies. These two approaches are discussed separately and then integrated by the proposed model. Finally, the model is applied to specify the necessary tools for an intelligent knowledge based design support system.

FIELDWORK

Essentially, three phases of fieldwork have been conducted to date in order to explore the design process. All the fieldwork has been conducted with personnel of the design office at Lucas CAV Ltd, Gillingham and within the context of designing fuel injection pumps.

The first phase of fieldwork consisted of a top-down approach. The aims of this work were (1) to get a feel for the design process at the organisational level (2) to gain some basic understanding with regard to the design of fuel injection pumps (3) to understand the constraints imposed upon the designer by the organisational culture and other pressures and (4) to identify implications for a design support system. Relatively unstructured interviews were used to meet these aims, involving fourteen designers.

The second phase of fieldwork adopted a bottom-up approach. This time the research was conducted at the part/sub-assembly level. Six fuel injection pump parts were selected. Eight designers received relatively structured interviews with regard to a given pair of sub-assemblies. The designers verified and refined each other's responses in separate interviews. In each case they described (1) the rationale behind the design of each feature of the part and (2) the sequence of design stages to achieve that design. An understanding was gained of the low-level strategies at work and the relative commonality of design stages. Implicit rules underlying the designs were gleaned and implications for a design support system were again identified.

The third phase of fieldwork utilised a middle-out approach, this time operating at the product (fuel injection pump) level. A designer had kept 24 sketches representing a trace of the redesign of a fuel injection pump. An interview-and-teachback method, based on these sketches, was used to elucidate the thinking that had occurred. An in-depth feel for the conceptual design process was gained as well as a further list of implicit design rules. As usual, the outcomes of this exercisewere converted to design support system implications.

LITERATURE

In the first instance, it was hoped that studies of the design process literature would reveal an appropriate model that could be validated in the specific context of the fuel injection pump design domain. No such model was forthcoming. However, by studies of the literature, important insights were gained into the nature of design and the history of design process research. For the latter, essentially three generations emerge.

The first generation methods prescribed systematic design, which borrowed its scientific methodology from systems engineering techniques of military and space missions see Jones (1963), Archer (1965). Different methodologists selected different labels for the prescribed design process stages, but they are commonly known as analysis, synthesis and evaluation. The design process laid down by RIBA for architects reflects this model, see Lawson (1980) as does that for mechanical engineers, see Pahl & Beitz (1984). Systematic design methods are also currently actively advocated by Roth (1981), Hubka (1983). These methods strongly proposed that the 'designer knew best'.

The second generation methods prescribed participative design. These arose out of disillusionment with the structured methods. Rittel (1972) maintained that you cannot understand the problem without having a concept of the solution in mind, and that design is not so much a sequential process but one of juggling different tasks. Others, like Daley (1982), reacted to the structured methodologies, with an anti-expert stance, the "user knows best". It was felt that designers' roles should be to encourage other people to determine what they themselves want.

The third generation methods were proposed by Broadbent (1979) based on Popper's 'conjecture and refutation' model of scientific method. Darke (1979) elaborated the model of Hillier et al (1972) to include the concept of a primary generator ie the concept or objective that generates a solution, as a conjecture which is then either accepted or refuted. If refuted, new solutions are sought and the cycle continues until a solution is accepted. The role of the designer is thus seen to be to make expert design conjectures to be refuted by others until an irrefutable solution is arrived at. The term 'interactive design' is used to describe the third generation methods, as the conjecture-refutation process depends on interaction. (Note, this title is not used in the literature).

The evidence of the literature, alongside the fieldwork, led to the conclusion that none of these three models was in itself satisfactory as a model of the mechanical engineering design process. Consequently, the strengths and weaknesses of each of the three models was evaluated. Indispensable qualities of the design process were also established. All this was assessed in light of the fieldwork in proposing a new model as follows.

PROPOSED MODEL

Stated very briefly, the proferred model suggests that the design process, pictured as a cosine wave, oscillates between the systematic and interactive models throughout its entirety. This oscillation occurs at multiple levels throughout the organisation as pictured below.

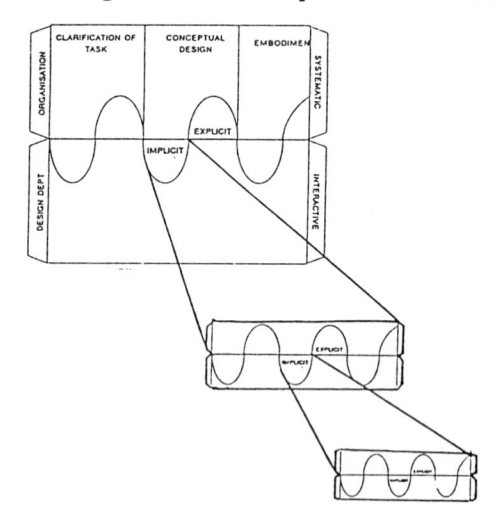

A horizontal line dissects the cosine wave. Above the line, systematic methods are most apparent, below the line interactive methods are most significant. This holds at each level of the organisation. Whilst three levels are described in the diagram, this is only illustrative. A second description is given to 'above' and 'below' the line in the diagram. This is the organisational dimension. The semi-circular areas above the line describe the explicit design process with respect to the particular organisational level in question. Those below the line describe the implicit design process with respect to the same level.

As can be seen from the diagram, that which is implicit at one level, has explicit and implicit components at the next level down, and so on.

From an organisational viewpoint, design appears to happen in a sequential, structured fashion (the systematic element). Controlling work to a deadline necessitates progress measured by explicit deliverables (product specifications, sketched solutions, layout drawings, etc). However, it is clear much goes on in between, and in order to achieve, the explicit deliverables. These implicit processes allow solutions to exist at the earliest stages prompted by the primary generators such as previous designs, discussion with other experts, new materials, etc. Conjectures are refuted by designers themselves, their colleagues, manufacturing experts, the customer, etc., until a satisfactory solution emerges explicitly (the interactive element). The model actively prescribes the customer and/or user to be involved in this interaction (the participative element).

ADVANTAGES OF THE MODEL

From an ergonomics viewpoint, the model offers the following (1) it provides a common framework for accommodating disparate design process views (2) it provides a framework for developing a CAD interface specification (3) it provides a framework for evaluating CAD systems in terms of task coverage and match (4) it provides an open enough structure to incorporate cognitive models of the design process eg blackboards for instance Whitefield (1986) (5) it provides an opportunity to define user oriented CAD systems as opposed to technology driven ones (6) it provides a framework for future research (7) it spans the gap between organisational design and individual design needs (8) it emphasises the need to build CAD systems that support unobservable human behaviour as well as that which is observable (providing for Computer Aided Design in contrast to Computer Aggravated Draughting) (9) it accommodates individuals with different design styles.The next vital stage in the process is to elaborate this model by identifying and specifying appropriate and usable tools to support the designer above and below the line, for each defined level of the organisation.

ACKNOWLEDGEMENTS

The work being carried out at the HUSAT Research Centre, Loughborough University of Technology, on the Design to Product project comes under the Alvey programme of research and is funded by H.M. Government, through the Science and Engineering Research Council. Additional funding for other parts of the project is being provided by GEC plc, Lucas CAV Ltd and the Department of Trade and Industry. I would like to thank the personnel of the design office in Lucas CAV Ltd, Gillingham, for giving their time so freely during the course of this fieldwork. I would also like to express my thanks to my colleagues in the HUSAT Design to Product team: Murray Sinclair, Fran Marquis, Enda Fallon, Reza Lajevardi and Sue Sutton, for their help and advice in writing this paper.

REFERENCES

Archer, L.B., 1965, Systematic Method for Designers, in Cross, op.cit.
Broadbent, G., 1979, The Development of Design Methods, Design Methods and Theories, 13(1), 41-45.
Cross, N., (ed), Developments in Design Methodology, John Wiley
Daley, J., 1982, Design Creativity and the Understanding of Objects, Design Studies, 3(3), 133-137.
Darke, J., 1979, The Primary Generator and The Design Process, Design Studies, 1(1), 36-44.
Hillier, W., Musgrove, J., O'Sullivan, P., 1972, Knowledge and Design, in Cross, op.cit.
Hubka, V., 1983, Design Tactics = Methods + Working Principles for Design Engineers, Design Studies, 4(3), 188-195.
Jones, J.D., 1963, A Method of Systematic Design, in Cross, op.cit.
Lawson, B.R., 1979, Cognitive Strategies in Architectural Design, Ergonomics, 22(1), 59-68.
Pahl, G., Beitz, W., 1984, Engineering Design, (ed.Wallace, K), Pitman Press, 38-44.
Rittel, H.W.J., 1972, Second Generation Design Methods, in Cross, op.cit.
Roth, K.H., 1981, Foundation of Methodical Procedures in Design, Design Studies, 2(2), 107-115.
Whitefield, A., 1986, An Analysis and Comparison of Knowledge Use in Designing with and without CAD, CAD 86 proceedings, 89-97.

Physiological Stress

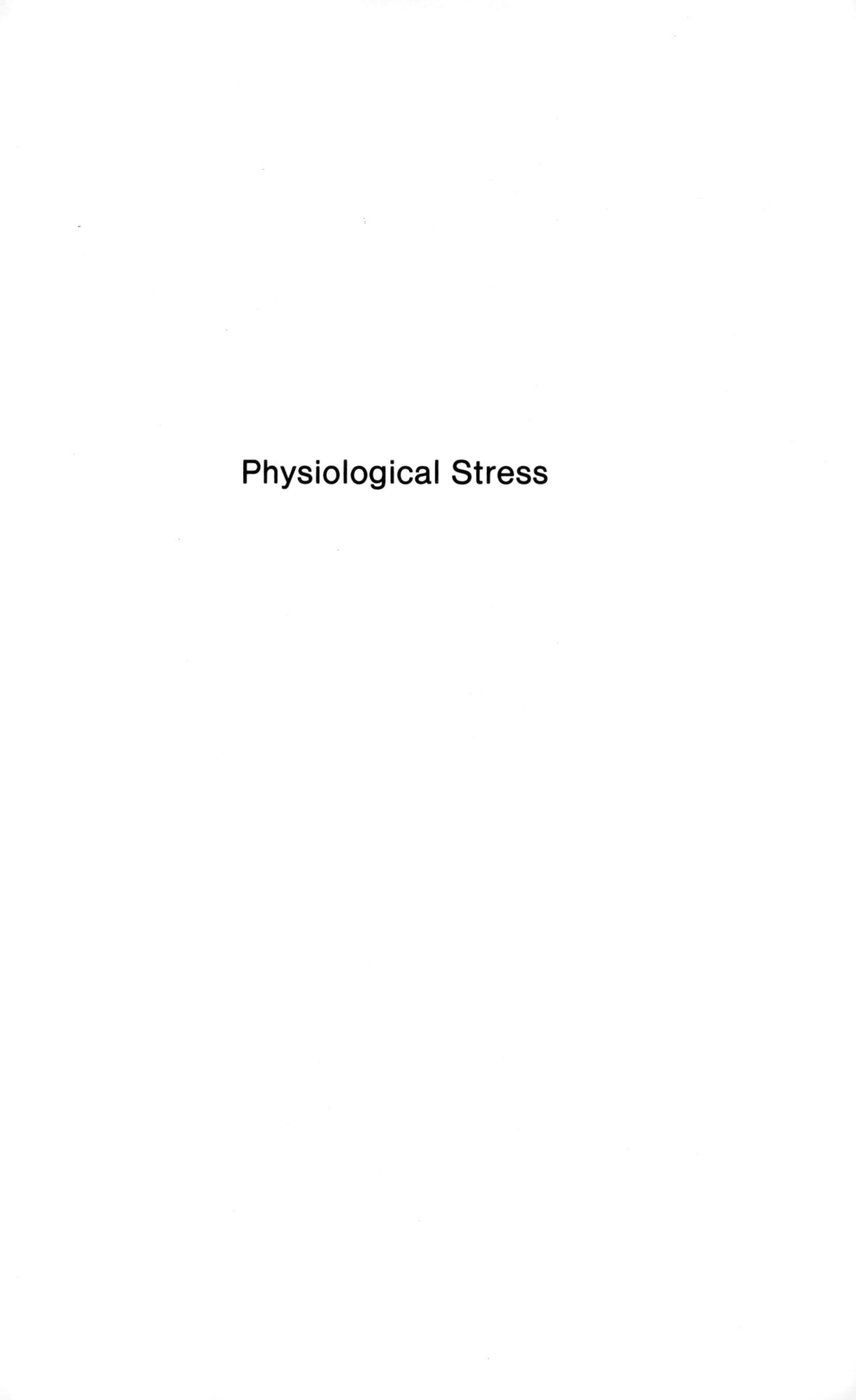

A MODEL OF HUMAN THERMOREGULATION: THE EFFECTS OF BODY SIZE ON THE ACCURACY OF PREDICTION

S.L. COOPER, R.A. HASLAM and K.C. PARSONS

Department of Human Sciences
University of Technology, Loughborough
Leicestershire LE11 3TU

The study investigated the effects of body size, in terms of mass, volume, surface area and heat capacity, on the accuracy of a predictive model of human thermoregulation. Values for the factors were found for eight male subjects and were substituted in the model in place of values for an 'average' man. Original and revised temperature predictions were compared with the subjects' actual body temperatures for exposure to a hot and a cold environment. The results show that the predictions based on data for an 'average' man were more accurate than those based on data for the actual subjects. Possible reasons for this are discussed.

INTRODUCTION
    The thermal environment influences a person's ability to work or function effectively. Mathematical models can predict human response to the thermal environment, but most are based on the response of an 'average' man. One such model is that proposed by Stolwijk and Hardy (1977). The model consists of two basic components, the controlled system (the human body) and the controlling system (temperature sensors and activators). The controlled system (see Fig. 1) is represented by cylinders (the trunk, arms, hands, legs and feet), and a sphere (the head). Each is divided into four concentric layers or compartments representing the core, muscle, fat and skin. The blood is represented as a separate compartment. Each compartment is assigned a mass, volume and heat capacity. These values relate to an 'average' male weighing 74.4 kg and with a surface area of 1.89 m^2.

Figure 1. The Stolwijk and Hardy (1977) Representation of the Human Thermoregulatory Controlled System.

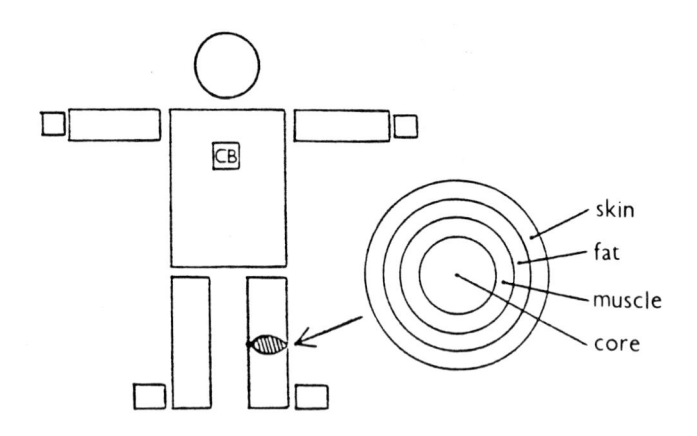

AIM

The aim was to compare predictions made using the data for an individual's body size with those made using the data for the 'average' man of the Stolwijk and Hardy model.

METHOD

Eight adult male subjects of varying body size and composition were chosen to take part in the experiment. The total surface area of each subject was calculated from an equation by Jones et al. (1985) and from this the surface area of each segment was determined according to DuBois & DuBois (1915). The heat capacity of each compartment was found from the subject's body volume and density using a table by Scammon (1953), volume and density being determined from the principle of water displacement (Rahn et al., 1949) using a body volumeter.

To produce data with which to assess the accuracy of the new predictions, each subject, minimally clothed and resting, was exposed to a hot environment (40° C, 60% R.H.) and a cold environment (5° C, 50% R.H.), each for 75 minutes. For both environments the mean radiant temperature was equal to the air temperature and the air movement remained below 0.1 m/s. The subject's response was measured by means of an aural thermistor and seven skin thermistors.

Using both the original and the revised data for surface area and heat capacity, predictions were obtained for the two environmental conditions. Temperature/time graphs of core, hand and mean skin temperature in the cold environment and core and mean skin temperature in the hot environment were plotted and compared.

RESULTS

In most instances the original predictions were closer to the observed data than were the revised predictions. An example is given in Fig.2. Where the accuracy of the predictions was improved, the improvement was small compared with the difference between the predicted and observed data. There appeared to be no correlation between body size and composition and either the predicted or observed results.

Figure 2.

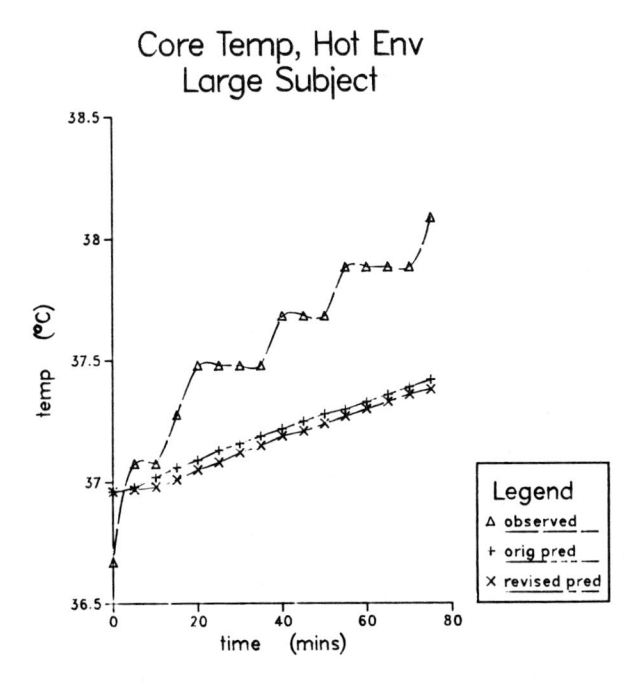

DISCUSSION

There are several possible reasons why, in general, there was no improvement in the accuracy of the model's predictions when using the revised figures for body size.

Firstly, the model's representation of the controlled system (the human body) is a gross simplification of the real system. It is possible that variables other than those of surface area and heat capacity are of equal or greater importance in determining an individual's response to the thermal environment. An example is the distribution of body tissues. When determining the compartmental heat capacities for each subject the tabulated values (Scammon, 1953) were based on a total body fat of 15% and a muscle content of 43%. However, the percentages of fat and muscle were known to vary greatly between subjects; for example, body fat ranged from 8% to 29% of the total body mass. Therefore, although the mass of each body segment was taken into consideration, the percentage distribution of the body tissues (skin, fat, muscle and core) within the segments was not considered. The values for heat capacity substituted in the model do not therefore reflect individual differences in the distribution of body tissue.

A second possible reason for the lack of improvement in the predictions concerns the controlling system (temperature sensors and activators). Stolwijk and Hardy have simplified this highly complex system and it is therefore likely that inaccuracies have been introduced. It is possible that the simplified representation of the controlling system has masked the effects of changing data for the controlled system (the human body).

Thirdly, only the controlled system of human thermo-regulation has been investigated in this study. It is possible that an efficient controlling system has the capacity to counteract the effects of body size on a person's response to the thermal environment. This may account in part for the lack of correlation between body size and composition and the observed results.

CONCLUSIONS

To be of practical use, the model must be able to take into account variations that exist between individuals' responses to the thermal environment. This study has shown that taking account of a person's body size, in terms of surface areas and heat capacities, has had no significant effect on improving the model's predictions.

It is likely that there are other variables relating to both the controlled and the controlling systems of human thermoregulation that influence an individual's response to the thermal environment. These variables must be identified, and the differences between individuals taken into account, if the model is to predict with greater accuracy an individual's response to the thermal environment.

REFERENCES

DuBois, D. & DuBois, E.F., 1915, Clinical calorimetry: 5th paper, The Measurement of Surface Area of Man, Archives of Internal Medicine, 15, 868-888.

Jones, P.R.M., Wilkinson, S. & Davies, P.S.W., 1985, A Revision of Body Surface Area Estimations, European Journal of Applied Physiology, 53, 376-379.

Rahn, H., Fenn, W.O. & Otis, A.B., 1949, Daily Variations of Vital Capacity, Residual Air, and Expiratory Reserve Including a Study of the Residual Air Method, Journal of Applied Physiology, 1, 725-736.

Scammon, R.E., 1953, Developmental Anatomy. In: Morris' Human Anatomy, 11th edn, edited by: J.P. Schaeffer (McGraw-Hill, New York).

Stolwijk, J.A.J. & Hardy, J.D., 1977, Control of Body Temperature. In: Handbook of Physiology, Section 9: Reactions to Environmental Agents (American Physiological Society).

ACKNOWLEDGEMENTS

The authors wish to thank Dr. P.R.M. Jones for his help and guidance with the collection of the anthropometric data.

LIMITING HEAT STRAIN: CAN WE RELY ON THE SUBJECT?

L.A. GAY and N.T. THOMAS

Ergonomics Unit, Department of Mechanical and
Production Engineering, The Polytechnic of Wales
Pontypridd, Mid-Glamorgan CF37 1DL

Are protective clothing wearers able to regulate their heat strain through perceived discomfort? Laboratory and field data which were gathered from wearers of protective assemblies during work may provide the answers. Most subjects were accurate in judging which assemblies kept them cooler. However, other factors received excessive perceived importance, so that less physiologically acceptable garments were preferred. Less fit subjects were generally unaware of their imposed strain. Designing protection for 'comfort' may increase thermal work exposure, where workers control this subjectively.

INTRODUCTION
Of all the factors which impose a heavy burden and limit the amount of work which can be tackled, heat stress is probably the heaviest. At first this statement may be questioned. Surely hot and humid working environments are decreasing and rare, now that hostile conditions are encountered more by machines than by human beings? Although this is probably true for regular working periods, there are still many occasions when heat stress may be critical.
New hazardous environments have to be entered for maintenance, inspection and other duties. Often layers of clothing and equipment surround the head and body. Some physically active 'leisure' sports require similar layers between sports people and their sport. From the skin of a wearer to these protective layers, stressful temperatures and humidities can build up quickly. The deaths of, and accidents to wearers in industry and sport have been attributed to excessive heat strain (e.g. Ramsey et al,

1983). This is particularly critical when activity levels are high and a covering fairly impermeable to sweat is worn.

What can be done to minimise such fatalities and risks to safety and health? This paper examines whether the wearers of protective clothing and equipment are able to regulate their exposure to heat stress within acceptable limits by self assessment of their heat strain. 'Strain' in this context is the physiological and psychological response to the combined effects of work load, elevated temperature and humidity.

METHODS

Recent laboratory and field data have been gathered from wearers of fire fighting assemblies (Tattersall & Thomas, 1985; Mawby et al, 1987) and from experiments with foundry workers' garments. The results of comparable experiments with different assemblies used in 'nuclear' environments (Thomas et al, 1976) have also been re-examined. These data are being studied to assess whether the relationship between heat stress and heat strain can be applied in the field for safer design and more healthy organisation of work.

A range of work loads, generally in the moderate to heavy categories, has been monitored mainly in the laboratory, at slightly above normal working temperatures (e.g. 25°c) and at 50% to 60% relative humidity. Tasks, similar to those usually undertaken in the appropriate clothing and equipment, were simulated in the laboratory. Follow-up studies have also taken place in the field (e.g. Mawby et al, 1987).

Physiological responses monitored continuously included heart rate, aural temperature and skin temperature. Heart rate is generally considered to be the most appropriate index of overall strain. Subjective data was gathered by questionnaire and structured interview.

RESULTS

The general picture emerging from the above studies was that most subjects were indeed able to judge accurately which protective clothing assemblies imposed the least and the greatest thermal loads. Figure 1 and Table 1 are representative, showing close relationships between subjective and physiological assessments for six 'nuclear' assemblies (Figure 1) and five foundry garments (Table 1). Comparisons based on heart rate, aural temperature and predictive working times gave similar results.

Figure 1. Mean objective and subjective responses to simulated work in six types of 'nuclear' assemblies (O, H, C, V, P & N). a) relative rate of rise in aural temperature. b) relative subjective rating. Replotted from Thomas et al, 1976.

[Figure 1: Two bar charts. (a) RELATIVE RATE OF RISE IN AURAL TEMP vs CLOTHING ASSEMBLY (O, H, C, V, P, N). (b) INCREASING DISCOMFORT vs CLOTHING ASSEMBLY (O, H, C, V, P, N).]

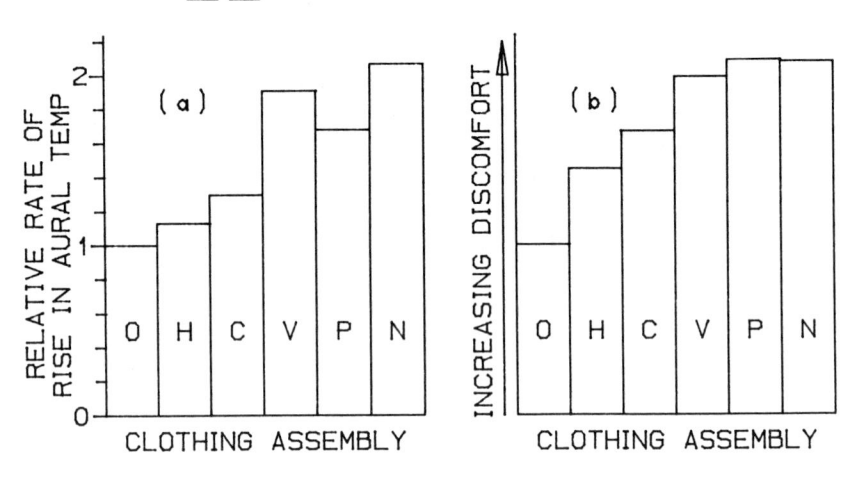

Table 1 indicates that the cotton based foundry protective suit (Cn) was generally a more acceptable garment from a physiological and subjective viewpoint than the heaviest of the wool suits (W). As expected, ordinary clothing (O) produced least strain. Overall, it was not possible to distinguish between two woollen garments made from the same fabric, one in coverall (w1) and the other in jacket and trouser design (w2).

Apart from thermal sensation; garment fit, design, weight, bulkiness and texture have all been identified by subjects as sources of persistent impressions. Where these caused excessive perceived effects, less physiologically acceptable garments were preferred. The overall perception could be a deception. For example, a subjective preference for non-belted styles accounted for the anomalously superior ranking of the woollen coverall (w1) by subject 4 (Table 1). Three subjects based a preference for coverall (w1) or jacket and trousers (w2) on the design they were accustomed to wear for occasional maintenance work.

In the series of experiments, the least fit subjects completed less work because they reached preset limits for heart rate (around 160/170 b.p.m.). Limits for aural temperature increase (+1.5°C) were rarely exceeded. These 'high risk' subjects were apparently unaware of heart rates indicating the approach of unsafe conditions. The fittest subjects wore most of the assemblies without a significant increase in thermal strain (e.g. Table 1, S1): heart rates were usually less than 100 b.p.m.

TABLE 1. Rank order of foundry clothing assemblies in four subjects. (Subj = subjective ranking; Obj = objective ranking based on increase in heart rate during manual task - lifting and placing bricks)

S	Assessment	Assemblies	(decreasing	acceptability	→	)
1	obj	[O	Cn	W	w1]	w2
	subj	O	Cn	[W	w1]	w2
2	obj	O	Cn	[w1	w2]	W
	subj	O	Cn	[w1	w2]	W
3	obj	O	[w2	W]	Cn	w1
	subj	O	[Cn	W]	w2	w1
4	obj	O	[w2	Cn]	[w1	W]
	subj	O	w1	Cn	[w2	W]

Brackets indicate no significant difference in heart rate (5% level) or same subjective ranking. (Details in Gay & Thomas 1986).

DISCUSSION

The most disturbing sign from the recent studies was that those who appear to be vulnerable to the first effects of heat stress - the individuals who have the lowest levels of physical fitness - seem unaware of the risks they are taking. Consequently attention may be diverted towards fitter workers, who seem able both to cope, and to predict limits to their ability to cope with strain which ironically, occurs less frequently for them. This distraction accentuates the problem. Ergonomists are very much aware of the dangers of designing work for average and above average subjects. Those workers in the greatest need of health protection have to be the prime targets for optimum design and organisation of work.

Rarely in industry is physiological monitoring available, apart from the worker's internal feedback, to control thermal stress exposure. Almost invariably workers regulate their own exposures, with little or no external information or guidance.

Many calls for improved physiological assessments have been made (e.g. Roantree, 1951; Brouha, 1967; NIOSH, 1986). Oral temperature checks are recommended during severe heat exposures (Dukes-Dobos, 1981). Little correlation was found between climatic comfort and effective temperature, radiant heat or skin temperature, in experiments at the workplace in the iron and steel industry (Hettinger et al, 1984).

Although laboratory experiments enjoy advantages of controlled and predictable conditions, these known conditions could be the main causes of poor validation between laboratory and field. The frequent and sudden changes which occur during real industrial tasks, cannot be reproduced accurately in the laboratory.

A prolonged study of thermal stress and strain within industry is recommended. This should consider a range of physical fitness levels, in hot weather conditions and/or hot industrial work. Such a study could take into account other factors, such as acclimatisation and the effects of conditioned clothing.

A broad aim merely to minimise discomfort from protection, could cause increased exposure beyond heat strain limits. This danger should be assessed during the ergonomic field study. Instead of relying upon the 'subject' we may then answer the question; 'What about the worker?'

ACKNOWLEDGEMENTS

The authors wish to express their gratitude to the Central Electricity Generating Board and to the Health and Safety Executive for their support of various heat stress investigations. Personnel employed by these organisations and The Polytechnic of Wales have been particularly helpful in developing these investigations.

REFERENCES

Brouha, L.A., 1967, Physiology in Industry, Pergamon, Oxford.

Dukes-Dobos, F.N., 1981, Hazards of heat exposure: a review, Scandinavian Journal of Work and Environmental Health, 7, 73-83.

Gay, L.A. & Thomas, N.T., 1986, Foundry workers garments: a pilot study to assess physiological and subjective responses from wearers of protective assemblies, Unpublished report for H.S.E.

Hettinger, T., Eissing, G., Hertting, R. & Steinhaus, I., 1984, Strain and demand through the wearing of personal protective clothing (in German), Bundesanstalt Fur Arbeitsschutz, Dortmund, Research Report, 392, Pt. 1.

Mawby, F.D., Street, P.J. & Norman, C.J., 1987, Laboratory and field evaluation of selected fire fighting assemblies, Contemporary Ergonomics, E. Megaw, Ed., Taylor & Francis.

NIOSH, 1986, Occupational Exposure to Hot Environments, Revised Criteria, p. 112.

Ramsey, J.D., Burford, C.L., Beshir, M.Y. & Jensen, R.C., 1983, Effects of workplace thermal conditions on safe work behaviour, Journal of Safety Research, 14, 105-114.

Roantree, W.B., 1951, Work in high air temperatures in a fire in Mysore Mine, Kolar Gold Field, Transactions of the Institute of Mining Metallurgy, 60, 513-539.

Tattersall, A.J. & Thomas, N.T., 1985, Flames and sweat - a physiological assessment of fire-fighting apparel, Contemporary Ergonomics, D.J. Oborne, Ed., 171-180, Taylor & Francis.

Thomas, N.T., Spencer, J. & Davies, B.T., 1976, A comparison of reactions to protective clothing, Annals of Occupational Hygiene, 19, 259-268.

PREDICTING THE METABOLIC COST OF INTERMITTENT LOAD CARRIAGE IN THE ARMS

I.P.M. RANDLE

Ergonomics Research Unit, Robens Institute
University of Surrey, Guildford
Surrey GU2 5XH

Four metabolic rate prediction models for load carriage in the arms were validated by direct comparison with measured data. A total of 52 observations on 20 subjects involving intermittent load carriage tasks were made. The actual metabolic rate was then directly compared with the individually predicted value given by each of the four models. The prediction formulae generally over-estimated metabolic rate, by between 13 and 50%. The results indicate that the prediction models studied were not appropriate for use with intermittent load carrying. A revised model is proposed which is more accurate at predicting the metabolic cost of this type of work.

INTRODUCTION
Despite widespread mechanisation, the manual carriage of loads remains a feature of many industrial jobs. Problems due to over-exertion and fatigue may be minimised by matching the metabolic demands of these tasks to the abilities of the workers. Therefore models which can predict the metabolic cost of load carriage tasks may be useful ergonomic aids in the design and analysis of many jobs. A few metabolic rate prediction models have been produced specifically for load carriage in the arms, which is by far the most common way loads are carried in industry (Drury et al 1982). Givoni & Goldman (1971), Garg et al (1978) and Morrissey & Liou (1984) have all produced such prediction formulae, but

each is based on the responses to steady state continuous load carriage. This involves subjects carrying loads for ten minutes or so, without pausing or releasing the load. Task analyses have shown that most loads in industry are carried intermittently, for much shorter durations (Drury et al 1982). The applicability of these models to realistic situations in which loads are carried intermittently may therefore by limited. The aim of this study is therefore to compare the accuracy with which these prediction formulae, based on steady continuous load carriage, can describe the metabolic cost of a more realistic intermittent load carriage task.

METHODS

Metabolic rate data were collected from two previously unpublished load carriage studies performed by the author. Both consisted of subjects walking on a treadmill carrying a load in both arms in front of the body. In each case it was carried for 30 seconds, and then the subject walked without it for 30 seconds, in repeating cycles. This was designed to simulate a loading or unloading task. In study I the subjects carried 25% body weight (18-23kg) in a steel lifting pallet (25 x 50cm). This was performed at two workrates, equating 25% and 50% of the subject's aerobic capacity. The desired workrate was calculated by manipulating the treadmill speed ($0.83-1.67 m.s^{-1}$) and its gradient (6-11%) in each condition. Subjects worked for ten minutes to allow sufficient equilibration time. In study II a fixed load of 25kg was carried for 15 minutes in each of three conditions. In one the steel pallet was used and in the other two the subjects carried a cardboard box (36 x 28 x 32cm). The treadmill was level, and walking speed was $1 m.s^{-1}$ in each condition.

Eight subjects performing two conditions each were used in study I, and 12 subjects performing three conditions were used in study II. Thus a total of 52 observations of intermittent load carriage were used in the analysis.

Models studied

Four metabolic rate prediction models were studied:

Model I (G & G) was that proposed by Givoni & Goldman (1971) with the additional '+M' factor for loads carried in the hands.
M=$[0.015L^2v^2]$+n(W+L)$[2.3+0.32(v-2.5)^{1.65}$+G$(0.2+0.07(v-2.5))]$
Where M = metabolic rate (kcal.hr^{-1}); v = walking speed (km.hr^{-1}); L = load weight (kg); W = body weight (kg); G = treadmill gradient (%); n = terrain factor (treadmill = 1)
Model II (GARG) Garg et al (1978)
M=$0.024W+10^{-2}$ $[68+2.54WV^2+4.08LV^2+11.4L+0.379(L+W)GL]$
Units as above except V = walking speed in m.s^{-1}
Model III (M & L) Morrissey & Liou (1984)
m=$-75.14+3.11W+V^2(2.72L+87.75)+13.36(W+L)(L/W)^2$
Units as above except m = metabolic rate in watts
Model IV (M & L + W) Morrissey & Liou (1984) with box width factor
m=$312.94+V[2.39(W+L)-481.62]+V^2(218.3+0.36Z)+17.35(W+L)(L/W)^2$
Units as above except Z = box width in cm.

Metabolic rate was measured for each subject in all conditions. This was then directly compared with the individually predicted metabolic rate for that subject in that condition, using the above four formulae. Paired t-tests were used in all statistical comparisons.

Metabolic rate was measured by collecting the subject's expired air in douglas bags, for the last two minutes of work in each condition. Gas volume was measured with a dry gas meter, and gas composition using Beckman analysers. Oxygen uptake was then calculated using the Weir formula, and metabolic rate in watts by multiplying $\dot{V}O_2$(l.min^{-1}) by 293.02.

The physical characteristics of the male subjects are as follows:

	Mean	S.D.		Mean	S.D.
STUDY I			STUDY II		
n=8			n=12		
Age (yrs)	24.5	2.3		30.3	7.1
Height (cm)	173.6	6.2		178.4	7.2
Weight (kg)	71.78	3.01		72.0	11.3

RESULTS

Table 1 gives the results of the comparison between the observed and predicted metabolic rates from study I. The difference between the observed and predicted values was highly significant (p<0.001) for each of the four models.

Table 1. Observed and predicted metabolic rates (watts) from study I - mean of 16 values (8 subjects x 2 conditions).

Observed		Predicted			
		G&G	GARG	M&L	M&L+W
Mean	493	601	738	387	405
S.D.	117	388	144	53.7	45.9
p<		0.001	0.001	0.001	0.001
% diff		22	50	-21	-18

The results from study II are given in table 2. Again there were significant differences between the measured values and those predicted by all four models.

Table 2. Observed and predicted metabolic rates (watts) from study II - means of 36 values (12 subjects x 3 conditions)

Observed		Predicted			
		G&G	GARG	M&L	M&L+W
Mean	430	486	565	471	500
S.D.	82.6	38.3	37.1	5.94	20.3
p<		0.001	0.001	0.01	0.01
% diff		13	32	10	16

The data from study I were used to develop a new model to predict metabolic rate from walking speed, body weight, gradient and load weight. Stepwise multiple regression analysis yielded the following equation:

$m = [(GV)^2 1.25] + 0.1[(47.33G) + (87.75L) + (21.96W) + (197.51V)]$

All units as in previous models.

This model accounts for nearly 98% of the variance in the data (r = 0.975). It was validated in two ways. Firstly, the data from one subject were removed from the main data set, and the model reconstructed using the remaining data. The measured metabolic rate for that subject was then compared to the predicted value from the model. This was done for each subject in turn. No significant differences were found between the measured and predicted values. The second phase of the validation involved testing the model against other load carriage data. Previously unpublished data from this laboratory, again involving walking and carrying in 30 second cycles (n=8) were compared to the values predicted by the model. Again no significant difference was found, although the model did underpredict the measured metabolic rate by around 6%.

DISCUSSION

The results show that none of the four models studied could accurately describe the metabolic cost of intermittent load carriage. The equation of Garg et al (1978) appeared least accurate, overpredicting by up to 50%. This may be explained in part by the fact that their subjects walked on a concrete floor, not a treadmill, which incurs a slightly greater energy expenditure. The Morrisey & Liou (1984) models contain no gradient factor, and hence underpredicted the values in study 1, which used an inclined treadmill. In all other cases however, each model grossly overpredicted the metabolic rate. This error in prediction was quite consistent for the Morrisey & Liou equations, however those of Givoni & Goldman, and Garg et al produced some massive inaccuracies under some conditions. In three out of the 16 cases in study I these models predicted values in excess of 1000 watts, when the measured values were around 650. In one other case the Givoni & Goldman equation predicted 86 watts, when the observed was 346. In each of these cases the values of speed, load and gradient were within those used by the authors in their original articles. It appears that the range of conditions in which these models can be used may be narrower than the authors suggest.

Table 3 shows the results of a comparison between the measured and predicted metabolic rates in an additional condition where no load was carried. As can be seen, three out of the four models gave reasonably accurate predictions.

This suggests that the differences apparent in the other conditions, where loads were carried, was due to the intermittent nature of the load carriage, rather than inherant inaccuracies in the predictive abilities of the models.

Table 3. Observed and predicted metabolic rates for walking on a level treadmill (n=12)

Observed		Predicted		
	G&G	GARG	M&L	M&L+W
Mean 226	240	282	224	212
S.D. 45.0	31.2	30.5	27.4	21.3
p<	NS	0.001	NS	NS
% diff	6	24	-1	-6

The model proposed in this paper provides a means of predicting metabolic rate for an intermittent load carriage task. This offers an alternative to the other models studied, which tend to overestimate the cost of this task. However, the small range of conditions on which the model is based limits its usefulness. Further development is obviously required to extend its accurate range.

REFERENCES

Drury, C.G., Chau-Hing, L.A. & Pawenski, C.S. 1982, A survey of industrial box handling. Human Factors, 24, 553-565.

Garg, A., Chaffin, D.B. & Herrin, G.D., 1978, Prediction of metabolic rates for manual materials handling jobs. American Industrial Hygiene Association Journal, 39, 661-674.

Givoni, B. & Goldman, R.F., 1971, Predicting metabolic energy cost. Journal of Applied Physiology, 30, 429-433.

Morrisey, S.J.& Liou, Y.H., 1984, Metabolic costs of load carriage with different container sizes. Ergonomics, 27, 847-853.

A COMPARISON OF METHODS FOR MEASURING BODY 'CORE' TEMPERATURE IN ERGONOMICS APPLICATIONS

A.P. PAYNE and K.C. PARSONS

Department of Human Sciences
University of Technology, Loughborough
Leicestershire LE11 3TU

The study evaluated methods and techniques for measuring body core temperature, in ergonomics applications, with respect to the accuracy of measurement and discomfort and strain placed on both subject and experimenter. Twelve male subjects were exposed to a hot ($40^{\circ}C$ 60% rh) and a cold ($0^{\circ}C$) environment. Rectal, aural (zero gradient and ear plug) and oral temperatures were compared. It was found that the most suitable method will depend upon the environment. At least two methods of measurement should be used. Urine, eye (using infra-red), EAR ear plug aural and the use of a radio pill were also considered.

INTRODUCTION
It is often convenient to consider two types of body temperature, core (deep body) temperature and shell (surface) temperature. In hot or cold environments core (or shell) temperature can provide an indication of the strain on the person exposed. In industrial application the measure of core temperature should be accurate and should also be acceptable to workers. The idea of core temperature is simplistic and in practice deep body temperature can vary depending upon measurement site. Usually the most important site is the brain, however main body mass temperature may also be of interest.

The aim of this study was to compare methods of measuring human deep body (core) temperature with respect to ergonomics applications.

METHOD

There are a number of traditionally used measurement sites on the body which can be used for estimating body core temperature. These include measurement of oral, aural, tympanic, oesophageal, urine, rectal and visceral temperatures. Some of the methods can provide accurate estimates of 'core' temperature however it is likely that for practical reasons (discomfort and pain for workers, cumbersome for investigators etc) that they would not be acceptable for many industrial applications. For this reason an experiment was conducted to compare primarily oral, aural and rectal measures with respect to accuracy and acceptability for use by subjects and investigators.

Twelve male subjects were exposed to a hot ($40°C$ 60% rh) and a cold ($0°C$) environment. Rectal, oral and aural (4-zero gradient and 8-aural plug) measures were taken during each experimental session. Subjects' temperatures were stabalized in a thermo-neutral room ($28°C$ 50% rh approx) for thirty minutes before entering an environmental chamber for one hour. During the time in the chamber subjects rested for thirty minutes, performed a step test (200mm high, 12 steps per minute) for fifteen minutes and rested for fifteen minutes. Subjective measures concerning comfort and convenience of the measures were taken in the thermo-neutral room and in the chamber. Subjects wore briefs and shorts only.

Rectal temperature was measured with a Grant's rectal probe thermistor (accuracy $\pm 0.2°C$). Subjects applied the probe themselves in a private room. Aural (plug) temperature was measured using a Grant's aural probe ($\pm 0.2°C$), a probe was placed in each ear and a cotton wool pad and tape insulated the ears. A zero gradient thermistor was used with headphones which included a heating element to warm the outer ear to reduce effects of the external environment on measurement. Oral temperature was measured with a mercury in glass thermometer placed under the tongue for at least three minutes with the mouth closed. After sixty minutes in the chamber the subject was asked to give a urine sample the temperature of which was measured by a thermistor in a vacuum flask.

RESULTS

Figures 1 and 2 show the mean responses of subjects exposed to the hot and cold environments respectively. In the neutral room (0-30 minutes) rectal temperature was highest and stable. Oral temperature was also fairly stable however both forms of aural temperature rose steadily to about 36.7 °C on average. Oral temperature was around 37.0 °C and rectal temperature was above 37 °C. The results of the subjective measures were that the zero gradient aural method was uncomfortable and the experimenters found the equipment bulky and expensive. Aural temperature plugs were found to be uncomfortable and experimenters need practice to fit them correctly. Oral temperature was inconvenient for the subjects but not uncomfortable or painful. Rectal probes were disliked by the subjects however once applied they were found to be comfortable and painless.

DISCUSSION

For a full discussion of the experiment and results the reader is referred to Payne(1986) on which this paper is based. A comparison of methods is also given by Mairiaux et al(1983) and ISO(1986). For methods to be used in industry they should be acceptable to workers and provide accurate results. The concept of core temperature is vague and has not been defined in terms of a measuring point within the body. It is useful however to have an estimate of brain temperature (including rate of change) and body mass temperature. Measurement of rectal temperature and of aural and oral temperature if correctly administered will provide satisfactory estimates. One measure alone will not provide complete information however.

Acceptability to the workforce will depend upon the attitude of the workers to the job and on how the measurement requirement is presented. This should be considered carefully by the investigators. Aural temperature can provide discomfort. Rectal temperature may initially be unacceptable, however it will provide a comfortable unobtrusive method and 'accurate' results.

Subjects generally found the idea of swallowing a radio pill unacceptable. The use of EAR plugs

Figure 1. Responses to the hot environment

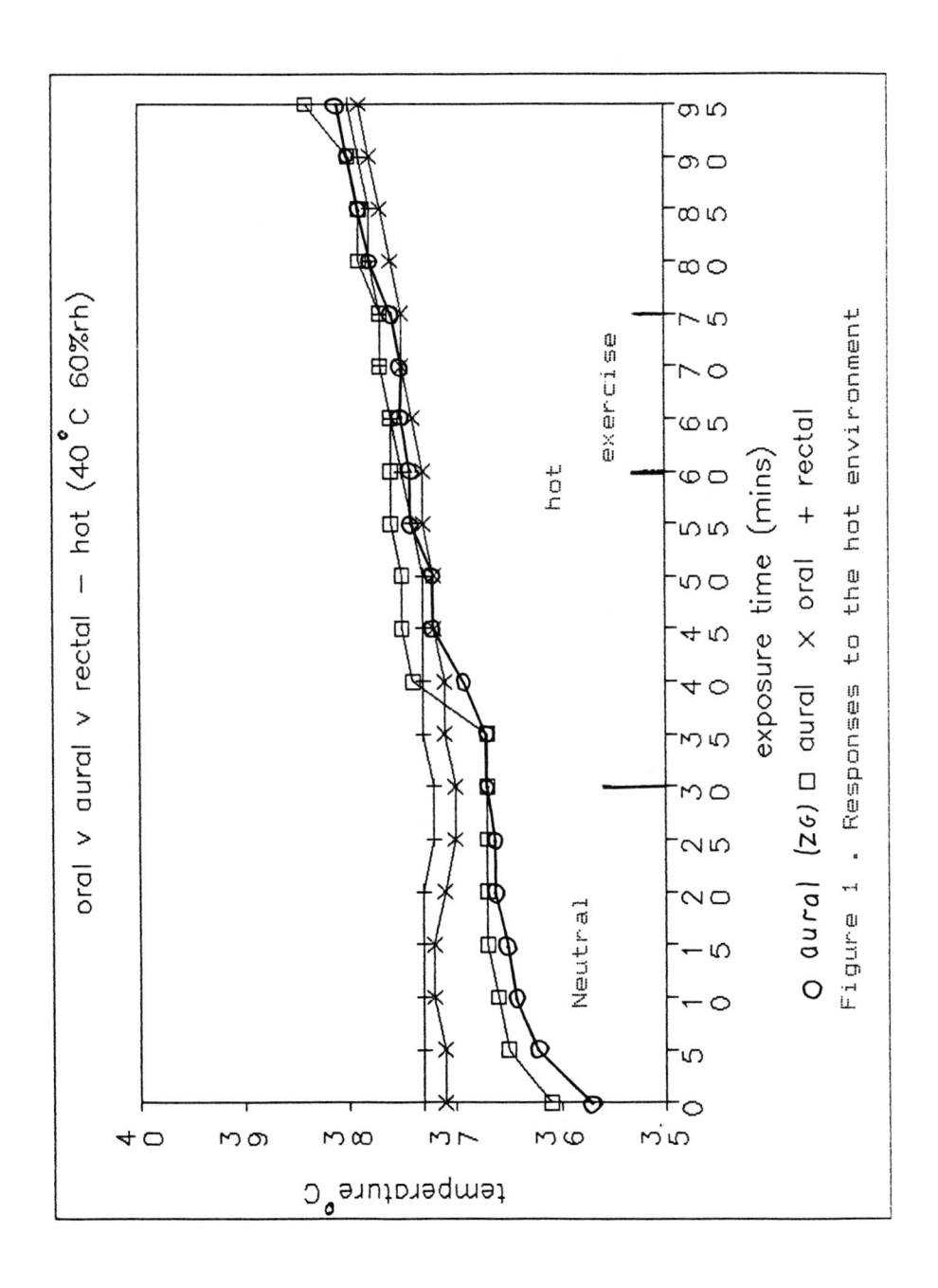

Figure 2. Responses to the cold environment

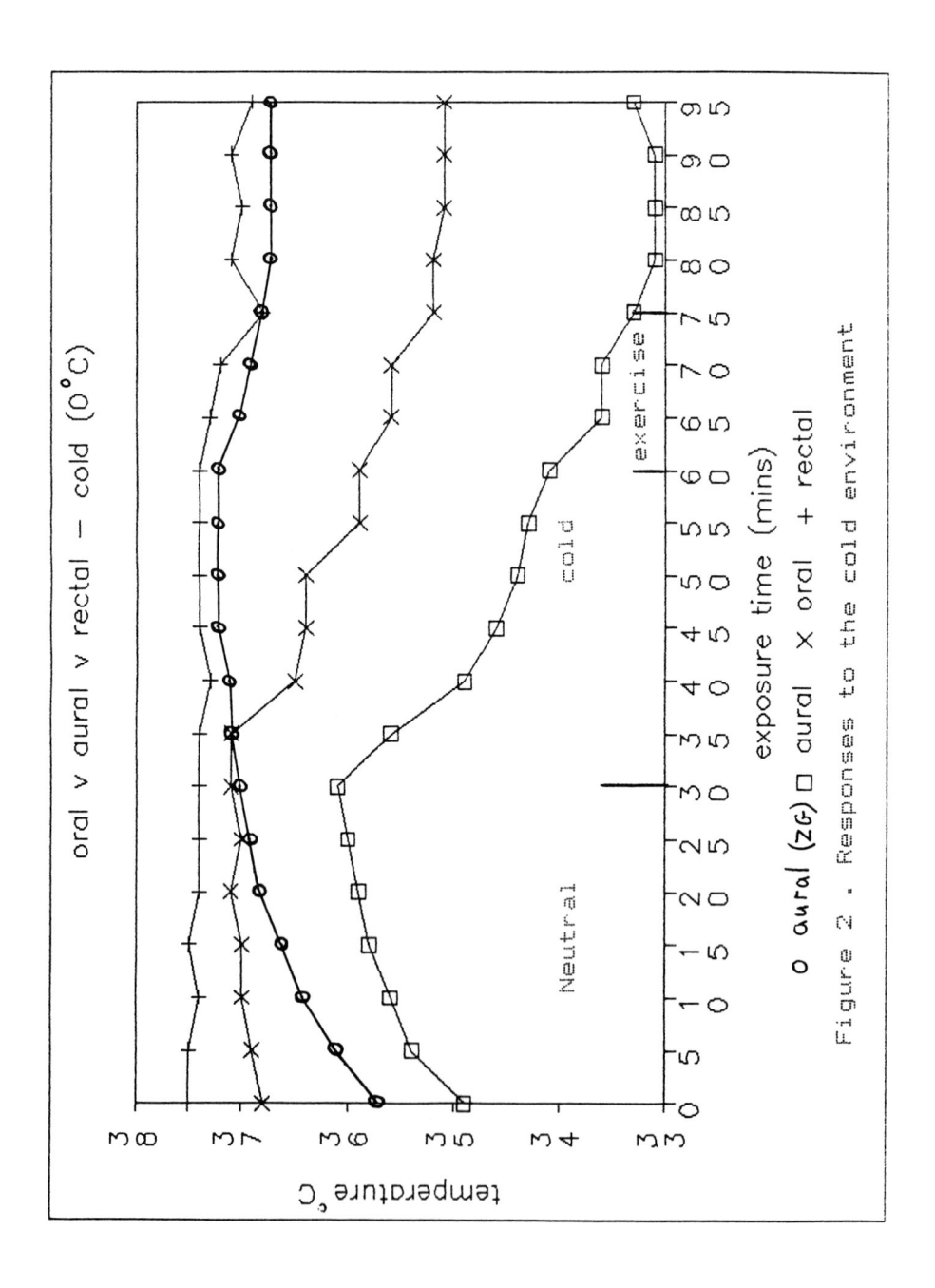

instead of plastic ear plugs was promising (more comfortable to subjects) however it was more difficult to judge distance of insertion into the ear canal. Urine temperature was acceptable however it depends upon the subjects capability and only infrequent measures can be obtained. Infra-red measurement of eye temperature was acceptable however the sensor used in this experiment 'averaged' the eye socket temperature. A measurement system with greater resolution should be investigated.

REFERENCES

ISO (1986) Assessment of thermal strain by physiological measurements. ISO TC 159/SC5/WG1 working document - Malchaire.

MAIRIAUX P H., SAGOT J C and CANDAS V (1983) Oral temperature as an index of core temperature during heat transients. Eur J Appl physiol. Vol. 30. pp331-341

PAYNE A P (1986) An investigation into methods for measuring body core temperature in ergonomics application. Ergonomics final year project report Loughborough university of technology

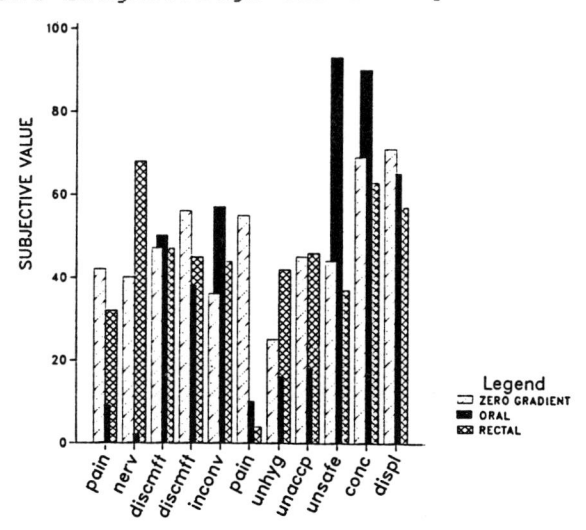

Table 1 . Subjective ratings

REDUCING THE EFFECTS OF VIBRATION THROUGH THE CONTROL OF WORKPLACE AMBIENT TEMPERATURE

N.K. AKINMAYOWA

Ergonomics Unit, Department of Psychology
University of Lagos
Akoka, Nigeria

This paper describes a method of reducing the effect of vibration on the hand-arm system. The finger temperature was altered by warm and cold air and by a warmed handle and then vibrated. The observed changes in skin temperature during and after vibration were found to be dependent on the air temperature because the warmed air and handle appeared to suppress the effects of the vibration. It is indicated how these findings could be used to reduce the effects of vibration with reference to ambient conditions.

INTRODUCTION

Vibration White Finger (VWF), a disease of occupational origin, is extensively discussed by Taylor and Pelmear (1975). The factors which influence the susceptibility of the operator are discussed in the ISO (1979) guidelines including the vibration characteristics, work practices and ambient environmental conditions.

The multidimensional nature and interaction of these factors in any working situation make the specification of vibration conditions where injuries to the hand will be eliminated, a very intricate task. Davies et al (1957) indicated that in a warm climate the disease develops slowly. Damage to the blood vessels does not generally appear and the vascular spasm which is provoked by the vibration and accentuated by cold exposure or the handling of cold objects, can be eliminated by warming the affected parts or whole of the body.

Because the prevailing finger and ambient temperatures are identified as important factors influencing vibration effect, it becomes important to investigate changes in the hand-arm system using warm or cold temperature and vibration conditions where the response of finger temperature to vibration exposure can be controlled. The influence of hand-grip was also investigated.

METHOD

Subjects were industrial workers including both grinders and fettlers (N = 20). Those with VWF (N = 10) had a mean age of 33.6 years, +/- 1.9 years (S.D.) and a mean weight of 73 kg, +/- 1.4kg. Their mean time of exposure to vibration conditions was 9.9 years, +/- 1.4 years. Those without VWF (N = 10) had a mean age of 33.9 years, +/- 1.7 years, and a mean weight of 70.8 kg, +/- 1.2kg. Their mean time of exposure was 11.4 years +/- 1.3 years. 16 subjects (students), randomly selected, acted as controls. Their mean age was 26.1 years +/- 1.8 years and with a mean weight of 65.8kg, +/-1.8kg. They had no history of vibration exposure.

The handle gripped by the subjects was warmed to 42 - 45 ^{o}C and three air temperature conditions were used to alter the thermal conditions, 42 - 45°C, 32 - 35°C and 15 - 17°C. The laboratory temperature was kept at 31 - 35°C. When the skin temperature was observed to have stabilised to the temperature condition, it was vibrated at 100Hz at an acceleration vibration of 4.5g (rms), produced by standard B & K Vibration Equipment. The skin temperature response at previbration, vibration and postvibration periods was recorded with a Comark Thermometer (Type 1604) equipped with a copper constant thermocouple. The latter was attached to the index finger in such a way that the skin temperature nearest the blood temperature was registered and subsequently recorded on a Rackal Thermionic Tape Recorder.

The effects of two types of gripping, commonly used by industrial workers, were also investigated. In the palmar grip the hand clenched the vibrated handle, whilst in the fingergrip the tips of all the fingers held the handle along its side. Grip force was maintained at 2kg throughout the experimental trials by a device designed by Akinmayowa (1982).

For the air temperature exposure of 42 - 45°C and 32 - 35°C an adjustable hair drier was used, while for air at 15 - 17°C compressed air was cooled via a nozzle. For the warmed air handle condition, the handle was heated using the hair drier attached to an inlet. An exhaust attached to the outlet enabled expired air to pass outside the laboratory.

The various conditions were allocated in a random order

to the subjects with trial intervals of 30min. During the experimental stages, subjects were asked to identify symptoms associated with the hand-arm system resulting from the vibration.

RESULTS

The analogue signals recorded on the tape recorder were reproduced as skin temperature-time-curves with an X-Y pen recorder and digitised with a package on a PDP 11 computer.

From the temperature curves (N = 224) it was shown that with the 32 - 35°C air exposure condition, the skin temperature increased during and after vibration. This increase was greater for the industrial group particularly when the palmar grip was used and greater than the increase in temperature which accompanied air and vibration exposure at 42 - 45°C. Under the 15 - 17°C condition, there was an increase in the finger temperature during vibration which was followed by an initial decrease in temperature when the vibration ceased, followed by a further increase in temperature. The increase in the skin temperature with the cold air was highly significant particularly in the case of the industrial group while the drop in temperature after vibration under the cold air condition was more pronounced for the VWF group, particularly with the fingergrip.

There was no significant difference in the skin temperature response when the warm air and handle were used by the control group.

DISCUSSION

The drop in skin temperature which is generally observed during vibration exposure (Sakurai, 1977) and is attributed to a disturbance of the vasomotor nerves leading to decrease in blood flow was not observed in this study. It was shown that the skin temperature rose during vibration exposure above the initial temperature and that this was accentuated by the lower air temperature. It could be argued that the warm air condition permitted a greater percentage of skin blood flow during vibration.

The increase in skin temperature during cold air exposure may have arisen from the passive warming of the hand by the energy from the vibration. The increase was higher for the fingergrip condition. This may have been a result of the difference associated with the dissipation of vibration energy between the two types of grip. The impedance of soft tissue is firmer with the fingergrip. Therefore, it can be argued that the gripping pressure during fettling

should be minimised if the fingergrip is being used.
Pain was reported after vibration under the 15 -17°C condition (VWF, N = 5; Non-VWF, N = 7; controls, N = 12). But under the warm air conditions no symptoms of pain were reported. Akinmayowa (1986) observed that although tests of nail press, of grasping power and of finger sensitivity could differentiate between controls and subjects who used vibrating hand tools in warm air conditions, symptoms indicating the presence of vibration syndrome in the operators were not confirmed. Whether the symptoms manifested in vibration disease under cold conditions are different from those found under warmer conditions needs to be further investigated. The present study does show that it is possible to reduce the effects of vibration by modifying the exposure conditions since the application of heat alleviates temperature loss during vibration exposure.

Whether the apparent encouragement of the blood flow compensates in any way for the effects of the vibration other than to reduce the evidence of the temperature drop is not known. Further research will be required to investigate this phenomenon. The application of the present technique to job and equipment design appears simple and economical.

REFERENCES

Akinmayowa, N.K., 1982, Factors influencing vibration effects. Unpublished Ph.D. Thesis, Department of Engineering Production, University of Birmingham.

Akinmayowa, N.K., 1986, Techniques for improving occupational health in industry. Paper presented at a seminar on Health and Safety at Work organised by the Nigerian Airport Authority Medical Department, Lagos.

Davies, A., Glasser, M. and Collins, P., 1957, Absence of Raynaud's phenomenon in workers using vibration tools in a warm climate. Bulletin of Hygeine, 32, 860.

International Standards Organisation ISO,1979, Principle forthe measurement and the evaluation of human exposure tovibration transmitted to the hand. ISOPublication ISO/DIS 5349.

Sakurai, T., 1977, Vibration effects on the hand-arm system: Observations on skin temperature, Part 2. Industrial Health, 15, 59-66.

Taylor, W. and Pelmear, P.L., 1975, Vibration White Finger in Industry. London: Academic Press.

Lighting

LIGHTING FOR THE PARTIALLY SIGHTED

P.T. STONE

Department of Human Sciences
University of Technology, Loughborough
Leicestershire LE11 3TU

A small percentage of people registered as blind have no sight at all. Most possess a degree of visual capacity which can be helped by improving environmental cues and especially by providing suitable illumination. Visual disabilities occur throughout all ages and in both sexes, but the greater proportion are found among the elderly. Various types of eye disorder impose certain constraints on lighting design and factors such as intensity, glare and contrast have to be given appropriate consideration to ensure that illumination does not increase the disability of partial-sightedness.

THE GENERAL PROBLEM
    Partial sightedness arises from a variety of conditions which commonly involve neural (e.g. optic atrophy), ocular (e.g. high myopia, cataract) or retinal (e.g. macular degeneration, retinitis pigmentosa) mechanisms. In general, low vision implies reduced visual acuity which may also be accompanied by a visual field loss. Such afflictions are found throughout the age ranges of both sexes but it is elderly people who constitute the larger proportion of the population with a sight loss, cataract, macular degeneration, and glaucoma being among the most frequent conditions. These will be additional to the loss of transmission of light through the eye arising naturally from the ageing process.
    It is not possible to find medical remedies for all conditions of partial sightedness, but many people have a residual visual capacity that can often be usefully assisted by optical aids, by special treatment of the environment, and by lighting. The loss of acuity which characterises these various conditions would suggest that more light should be of prime importance, but the problem

is, how much light? It must be pointed out that some people are very sensitive to disability or discomfort glare or both. The former is a particular problem for the cataractous eye because of the scattering of light in the opacity in the lens, and discomfort glare is a severe problem for those with senile macular degeneration, albinism and nystagmus. Thus without due care, increasing the illumination may also increase the individual's disability. Gilkes has pointed out (1978) that elderly people often live in a gloomy environment with the curtains drawn in order to offset the effect of glare from windows. Many elderly people live in an unnaturally dim environment, as has been shown in surveys conducted by Cullinan et al. (1979), Levitt (1980) and Simpson and Tarrant (1983). The illuminances found in homes were frequently in the range 20 to 70 lux. A recommended value for domestic environments is between 150 and 300 lux, and under these circumstances dramatic improvements in acuity would be attained.

Apart from considerations of intensity and glare there is also the important factor of contrast sensitivity loss which occurs with low vision. Southall and Stone (1986) investigated 50 partially sighted children all of whom had varying amounts of loss of contrast at different spatial frequencies. The reading performance of these children was significantly affected by print contrast, especially at low light intensities. Thus the preservation and enhancement of contrast both for work tasks and within the environment generally is a critical consideration. Another finding that emerged from the above study was that a high proportion of the children exhibited colour vision losses, in the red, green, and blue colour zones.

LIGHTING RECOMMENDATIONS
(a) Illuminance

It is difficult to give a definitive prescription for illuminance since individuals vary widely in their needs. However recent investigations by Boyce (1986), Julian (1984), Gazeley and Stone (1981), Southall and Stone (1986), all suggest that the range between 600 to 1000 lux will meet the requirements of many young and elderly partially sighted people for visual tasks. The most flexible arrangement is to have adjustable illuminances achieved either by dimmer or switching controls, or by providing different light areas for various tasks and people. All researchers have shown very wide differences in performance among partially sighted subjects for the same level of illumination and therefore, the requirement for intensity is a personal one. It is important that good general illumination is provided in rooms and passageways in order that the physical boundaries of the environment are perceived for purposes of orientation.

(b) Glare

Disability glare is caused by light from bright sources being scattered internally within the optic media of the eye. Such an effect creates a "mist" of light through which images of the external world have to be seen. This results in reduced contrast sensitivity down to a total obliteration of detailed vision when the viewer is aware only of a sensation of light. Such conditions arise naturally with age but particularly, with cataracts. Those sources of light that are especially offensive are a direct view of the sun, car headlights, spotlights, directional reflections from glossy surfaces such as paper, water, or metal, and point sources of light. It is therefore important that these sources are avoided. Spotlights and other highly directional light sources are in common use today and should not be installed in homes where cataract patients are likely to reside unless the source is always shielded from direct viewing. Discomfort glare is a daily commonplace experience and although not interfering with seeing, it can be intensely unpleasant or at the least, distracting. The control of glare from light sources can be achieved with directional louvred reflectors or with plastic diffusers, or, on table lamps with suitable shades. Dimmer controls are always very useful, especially for home reading lamps and for individual room lighting in residential homes for partially sighted people.

Windows are frequently a source of glare, generating discomfort, and sometimes, disability glare. The problem apart from direct sun penetration, arises from bright skylight contrasting sharply with dark surroundings to the window, something that occurs often at the end of a corridor. Good ambient internal light or light reflecting surfaces are required to offset such effects but roller blinds may also be necessary. Especial care must be taken to reduce glare from a window that faces on to a staircase as critical visual cues may easily be lost.

(c) Contrast

The differential reflection of light from objects and their backgrounds produces differences in contrast, a highly important parameter for vision. A high contrast black and white text is vital for many partially sighted people who need to read. But so also is the contrast of a cup or a plate or cutlery on the table, and especially glassware. Also, the very clear marking of door frames, lighting switches, electrical sockets, skirting boards, is just as important as an aid for seeing as is high intensity light. Glossy finishes tend to interfere because reflections arising from them cause veiling reflections which reduce contrast. Thus texts and working surfaces need to have a matt finish. On the other hand it is possible to have excessive contrast, which may arise from striped patterns such as from wallpapers or carpets, or, it may also be caused by highly directional light fittings which create pools of brightness and deep shadows. This type of light-dark effect on floor surfaces creates perception problems, when shadows may be seen as obstacles by people with low vision.

(d) Local illumination

Many partially sighted people find that a local light source can assist their vision, but they may often be disappointed following a purchase, since the particular item chosen has not met their needs. Careful selection is required but unfortunately, local shops rarely have a full range of the available brands to enable a person to make comparisons. The desirable characteristics that a lamp should provide are a widespread distribution e.g. to cover at least, an A3 size sheet; an even intensity without "hot spots", a variable intensity control. Another important requirement is freedom from glare, which may be achieved with modern desk lamps by fitting a small skirt to extend the sides enclosing the lamp. Modern desk lamps now use miniature fluorescent lamps instead of incandescent lamps and these are excellent light sources but again, glare control is required. This is a matter that the author is looking into at the present time. With the traditional domestic table lamp the requirement is a large drum shade, so that a 150 watt lamp may be used in conjunction with a dimmer control. This will give flexibility for a lot of people. Some partially sighted people need to place the lampshade of a desk lamp against their face in order to carefully track the object of regard, such as print, and a small wattage tungsten halogen lamp with dichroic reflectors (to keep the shade cool) will be useful in such cases.

CONCLUSIONS

With the increasing population of elderly people there will be a concomitant increase in the number of partially sighted people. Many of them have useful residual ability and it is suggested that lighting and aids for enhancing the environment can assist their vision. These aids include, a variable intensity of light for individual work places, well illuminated surroundings, promotion of good visual contrast in objects in the environment, careful control of glare and provision of suitable task lights. The important factor to recognise is that lighting must be flexible as it is a matter of satisfying individual needs which are not necessarily predicted from diagnostic categoires and vision tests.

REFERENCES

Boyce, P.R., 1986, Lighting for the partially sighted: some observations in a residential home, Capenhurst Research Memorandum; ECRC/M1980, January, (The Electricity Council, Chester).

Cullinan, T.R., et. al. 1979, Visual disability and home lighting, The Lancet, March 24, 642-644.

Gazely, D. & Stone, P.T., 1981, Visual capacity, lighting and task requirements of partially sighted school children, Light and Low Vision Report, No.3, Department of Human Sciences, University of Technology, Loughborough.

Gilkes, M., 1980, Let there be light, Proceedings of the Light for Low Vision Symposium (1978) London, pages 103-107, (Partially Sighted Society, Doncaster).

Julian, W.G., 1984, The design of buildings for the aged partially sighted, Lighting in Australia, December, 17-22.

Levitt, J., 1980, Lighting for the elderly: an optician's view, Proceedings of the Light for Low Vision Symposium (1978), London, pages 55-61, (Partially Sighted Society, Doncaster).

Simpson, J. & Tarrant, A., 1983, A study of lighting in the home, <u>Lighting Research & Technology</u>, 15, 1, 1-8.

Southall, D. & Stone, P.T., 1986, Light for Low Vision research at Loughborough University. <u>Proceedings of the Light for Low Vision Symposium</u>, (1986), London. (Partially Sighted Society, Doncaster), (In press).

LIGHTING FOR CONTROL ROOMS

J. WOOD

Communications Complex Design Ltd
76 Church Street, Weybridge
Surrey, KT13 8DL

Lighting is a particular problem for ergonomists designing control rooms. Control of daylight and design guidelines for uplighting are described. Measured levels of artificial illumination in controls are generally found to be lower than those currently published. Recommended levels of control room lighting are presented and the concept of "diurnal variation" in artificial lighting explained.

Special Requirements for Control Room Lighting

The design of the visual environment in control rooms poses a number of special problems for the human factors engineer. Twenty-four hour operation and changing workload levels seem to be reflected in demands by operators for different quantities of light at various times. In a typical control centre, an operator will have to deal with radio, telephones, intercoms, VDUs and on the wall around him, maps and electronically driven displays. It is almost certain that he will also be involved with some paper-based tasks which may vary from looking at directories to jotting down telephone numbers and addresses. The control equipment being used will probably have been supplied by different manufacturers giving rise to the common problem of demanding that an operator copes with a light-emitting display, say an LED, alongside a light reflecting display say an LCD.

Although electronic equipment has become more

sophisticated and compact, display technology has often not kept pace, from a human factors point of view - compare the visibility of an LED indicator on a modern telephone unit with the tungsten bulb which it replaces.

Finally one must consider the way in which information is brought to the attention of operators. Voice communications are likely to be an absolutely crucial link and there may be neither the time nor the opportunity to get repeats of poor, noisy and largely indecipherable calls. In order to conserve the auditory environment the human factors engineer will attempt to replace the various auditory alarms with visual ones - once again emphasising the importance of optimising the visual environment.

Natural Lighting

Very few people find that a totally windowless environment is acceptable and the control of daylight and provision of windows is a particular problem faced by the ergonomist. In some cases windows are an essential feature of the control room function, such as ground movement control at an airport or in a plant process control centre.

The debilitating effects of the rising sun over VDUs can to some extent be avoided by carefully considering the location and orientation of control towers during the planning stage or by use of solar film on the windows. The underlying conflict, however, between the constant and relatively weak electronic displays and the enormous variation between the midday to dusk illumination of the plant remains.

Experimental Trials on the Artificial Lighting of Controls

Various lighting solutions have been tried over the years for the illumination of electricity grid control centres - none with any great success. It was against this background that the CEGB asked CCD to conduct some comparative tests on alternative lighting for the future generation of these centres.

It was suggested to the Board than an existing

full scale mock-up of an entire grid control centre be used such that alternative direct and indirect lighting systems could be assessed for both room and wall displays.

The main variables examined were as follows:-
- downlighting versus uplighting of the main room
- high pressure sodium versus fluorescent sources
- preferred levels of lighting on worksurfaces
- direct versus indirect lighting for mimic displays
- preferred lighting levels for mimic displays
- colour temperature preferences
- interaction between different sources and daylight

Representative control engineers from all over the country participated in the trials.

Fig 1. Lighting Trials for Grid Control Centres

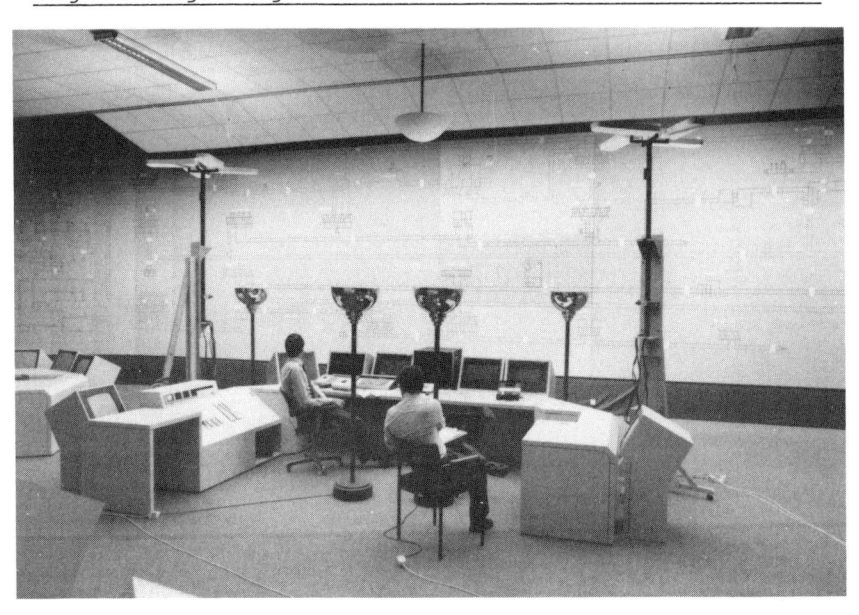

The trials ran through various working scenarios during which assessments were made about alternative lighting conditions. To help the engineers in their judgements some working VDUs were introduced on to the workstation, telephone line indicators could be activated and, on the mimic display, LED indicators could be switched on and off.

Alternative means of illuminating the wall mimic, direct versus indirect lighting, were tested and subjects were asked to use the simulated graphics, LED displays and analogue meters in their judgements. In addition, alternative colour temperatures and levels of illumination were tested by paired comparisons.

Some of the conclusions derived from this study, see below, were subsequently incorporated into the specification of a new lighting system, which is currently being installed.

- an uplit lighting scheme was preferred for room lighting with an average level of illumination on the worksurfaces of 250 - 300 lux.
- there was some evidence that a slightly lower level of illumination was acceptable with an uplighter with a fluorescent source compared to an uplighter with a high pressure sodium one
- an indirect wall mimic lighting system was preferred to the traditional direct lighting solutions.
- cooler diagram lighting was preferred when some daylight was present.
- dimmable, flicker-free fluorescent sources were preferred for the room uplighters.

The Use of Uplighters in Controls

The luminaires should not be suspended closer than a metre from the ceiling, otherwise, the sought after result of an even illumination will be marred by "hot spots" which themselves could create unacceptable reflections off VDU screens. When this constraint is added to those of minimum false floor and ceiling voids together with the height ofa99th percentile male it can be seen that slab to slab distances become a concern.

Since the object of uplighting is to create a glare free, softly lit environment in which there are no hard shadows it is all to easy to create a bland and uninteresting interior. CCD's experience is that it may be necessary to use lighter and fresher colours in the interior scheme to compensate for the lack of shadow and flatness created by indirect lighting, Table 1.

Table 1. Interior Finishes for Uplit Schemes

	SELECT	AVOID
Floors	light tones to reflect light	dark colours - will make room gloomy
Ceilings	white finishes; either plaster or a suspended ceiling with tiles. Repaint when off colour	strong architectural features such as beams, aluminium grids, large nos. of dark heating & ventilating outlets.
Walls	light to mid tones	dark colours - will make room gloomy
Windows	Vertical louvre blinds, light to mid-tone, supported both top & bottom for pos -itive adjustment	Venetian blinds, Vertical louvre blinds supported only at top, South orientated windows completely
Worktops	light to medium tones to reduce contrast with white paper	white, black or saturated hues
Equipment Finishes	medium tone, neutral colours with textured finishes to mini -mise reflections	chrome fixing screws, brushed aluminium bezels, shiny finishes or saturated hues

Recommended Levels of Illumination

Visits to a number of different control centres over the years has revealed a somewhat surprising but common pattern. There is a "daytime" level of illumination largely indicated by the amount of fenestration. During these daylight hours the artificial lighting may sometimes be on as well- when it is, this is as often as not due to nobody bothering to switch it off. From dusk to around 2 am, the artificial lighting tends to be fully on - "evening" illumination.

When things have got quieter and the pace of activity wound down the lights also tend to be turned down to a "night-time" level of illumination which can be as low as 10 lux on the worksurface, Table 2. Even in windowless environments operators appear to wish to vary the quantity of artificial light to synchronise with workload levels and hours of darkness.

Table 2. Diurnal Variation in Control Room Lighting

" Day-time"	400 lux
" Evening-time"	100 - 300 lux
" Night-time"	10 - 50 lux

Generally recommended levels of illumination at VDU workstations seem to be too high for control centres, HSE (1983), Wood J. (1986). An indication of the type of values which may be more appropriate for these installations is summarised in Table 3.

Table 3. Recommended Levels of Illumination for Control Rooms

TASK/EQUIPMENT	ILLUMINATION (lux)	
	Vertical	Horizontal
VDU screen	20 - 50	
VDU keyboard	-	150
Radio & telephone equipment incorporating LEDs etc	-	50 - 150
Writing areas	-	200 - 250
Walkways/Ambient Lighting	-	50 - 100

References

1. Health & Safety Executive 1983, 'Visual Display Units', Her Majesty's Stationery Office.

2. Wood, J., VDU Lighting in Control Environments National Lighting Conference, Nottingham, 1986.

Acknowledgements

The author would like to thank the CEGB for permission to publish trials results reported in this paper.

EMERGENCY LIGHTING AND MOVEMENT THROUGH CORRIDORS AND STAIRWAYS

G.M.B. WEBBER and P.J. HALLMAN

Department of the Environment
Building Research Establishment, Garston, Watford
Hertfordshire WD2 7JR

This study investigated human movement on a corridor and a stairway in different emergency visual conditions, which included the illuminance recommended by British Standard BS5266, illuminances above and below the recommended level, non-uniform illuminance and a totally different approach based on photoluminescent marking of the route. Findings for speed of movement and subjects' opinions are presented. Speed with photoluminescent markings was comparable to that under standard emergency lighting.

INTRODUCTION
    The study aimed to examine the requirements for escape route lighting - that which is provided when the normal power supply fails and is to enable occupants to evacuate quickly and safely from a building. Performance measurements were made under the current British Standard BS5266 (1975) recommendations for escape route lighting (minimum of 0.2 1x on the centre line of the floor of the escape route) and under alternative provisions, including photoluminescent material. This material can be activated by the uv spectrum of either natural or artificial light. It emits very low levels of light but has high visibility when normal lighting is switched off. The initial luminance of the material depends upon the illuminance on the material produced by the exciting source. Though it decays continuously in the dark, because the human eye is also adapting at a rate roughly matching this decay, the material continues to be visible over a period of several tens of minutes.
    The study took place in a special experimental simulation facility at the Building Research Station (BRS). It was

equipped with infra-red floodlights and instrumented with low-light-sensitive video cameras to find out how well people can find their way about in a typical building when lit by different levels and types of low-intensity lighting. The facility comprises a long L-shaped corridor (length 24 m), incorporating a single step change of level, and a separate domestic stair leading to a first floor landing. The wood stair consisted of a straight flight of 42° pitch, rise 200 mm and going 222 mm. The distance along the pitch line from the nosing of the top landing to the nosing of the bottom tread was 2.6 m.

The study involved 84 subjects, with an equal number of men and women. All were staff from BRS. None of the subjects was previously familiar with the building or its escape routes. Subjects were selected to fill the following age groups: under 30 years, 30 to under 40 years, 40 to under 50 years, and 50 years and over. None of the subjects suffered any movement disability.

TEST PROCEDURE

In the corridor tests, subjects had to find their way from a desk, down to the end of the L-shaped corridor, turn round and find their way back to the desk. Subjects were first light-adapted to normal room lighting (approximately 500 1x) at the desk. The normal lights went out, simulating a power failure and the subject had to negotiate the escape route in a particular emergency visual condition. Each test run was video-recorded with a digital time superimposed. The time taken for the outward and return journey (each 24 m) was analysed from the video tape.

In the stair tests, subjects had to find their way up and down a single flight of stairs. Again subjects were light adapted and their movement in particular emergency conditions was viewed and recorded through a one-way screen.

Five different emergency visual conditions were used, see Table 1. In the corridor, the uniform lighting on the route

Table 1. Emergency visual conditions.

Condition	Corridor	Stairway
1	Nominally 1.0 1x uniform	
2	Nominally 0.2 1x uniform (as per BS5266)	
3	Nominally 0.02 1x uniform	
4	Non-uniform, single source of 1 1x at end of corridor and 0.02 1x at desk	Non-uniform, single source of 1 1x at top of stair
5	Photoluminescent material	

was provided by 12 luminaires fitted with tungsten lamps which were controlled by dimmers; the photoluminescent markings consisted of painted strips of plywood of 100 mm width which could be fixed temporarily against the skirting board and riser of the single step. On the stairway, this material marked the risers, wall string and skirting board, top of handrail and balustrade. Photoluminescent signs were also placed on the wall to indicate the upward and downward direction of the start of the stair.

Each subject experienced just two of the visual conditions on the corridor and stair: all experienced condition 2 and one of the other conditions. In order to investigate 'familiarity' or 'learning effects', half of the subjects did condition 2 on their first run and the other half on their second.

FINDINGS FOR CORRIDOR

Mean speed of movement varied with visual condition, as shown in Figure 1a and 1b. Mean speed of subjects undergoing condition 3 was markedly slower than under condition 2, as found by Simmons (1975) and Boyce (1985) for

Figure 1. Mean speed of movement on corridor in different visual conditions (a) comparison of outward and return journeys (1st and 2nd runs combined), (b) comparison of 1st and 2nd runs (outward and return journeys combined).

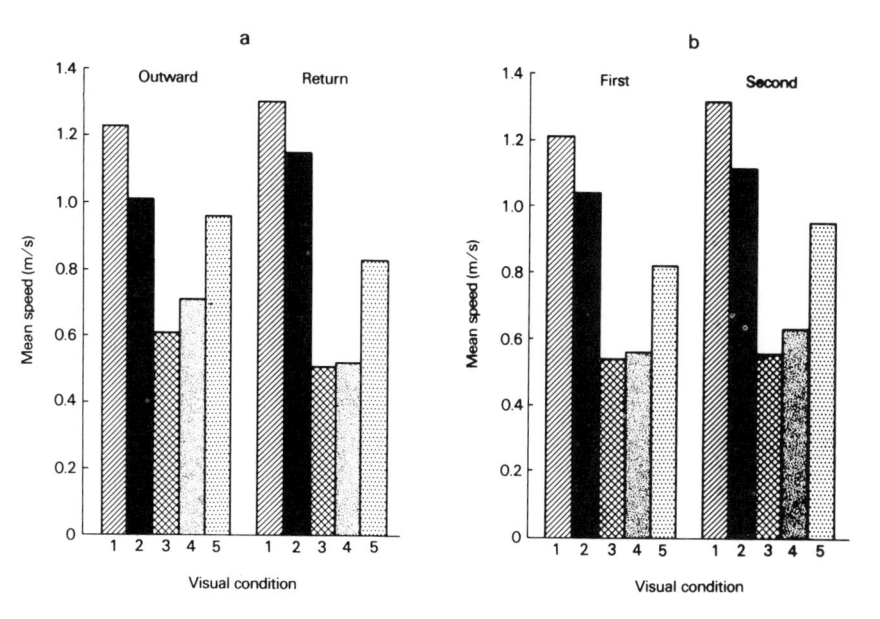

movement in an office area. Under uniform illuminance conditions 1 and 2, the return journey was quicker than the outward journey, but the reverse applied to the other three conditions. Part of the return journey involved a less well defined path (width exceeded 2 m as compared to 1 m in narrow corridor section) where subjects were required to turn a corner. Some subjects lost their way and some used their hands to detect the presence of the corner. For conditions 1 and 2 there was less difficulty with the abrupt change of direction, suggesting that visual conditions/cues were adequate. With condition 5 the markings had been positioned principally with the outward journey in mind.

The speed of subjects using the total route for the second time is compared to those using it for the first time in Figure 1b to indicate the effect of familiarity. in all five conditions, the 'familiar' subjects were quicker, eg by 7% and by 16% under 0.2 1x and photoluminescent material respectively.

The mean speed for females over the total route (48 m) was found to be slower than for males in all five conditions eg by 18% under condition 2. The fastest group were the under 30 year old males, otherwise there was little difference with age within each sex.

Subjects' opinions were obtained by means of a questionnaire completed after all the tests. They were asked to rate (on seven point scales) how difficult it was to see where they were going, and how satisfactory they considered the emergency visual condition experienced would be in evacuating the building quickly. For visual condition 1 subjects reported it easy to see and were moderately satisfied. There was greater difficulty and more dissatisfaction with conditions 2 and 5, approximately 1 and 2 scale points respectively higher than condition 1.

FINDINGS FOR STAIRWAY

The speed of movement for ascent and descent was calculated along the slope (pitch) of the stair. Figure 2 shows the mean speed of ascent and descent in different visual conditions. For visual condition 1 descent was faster than ascent - Fruin et al (1984) found that this is generally the case for stair movement under normal lighting conditions. For conditions 2, 3 and 4 descent was slower than ascent. With condition 5 the speeds of descent and ascent were very similar, and both were closely comparable to that under condition 2. Figure 3 shows a subject descending under condition 5.

Figure 2. Mean stair speed in different conditions.

Figure 3. Photograph of condition 5.

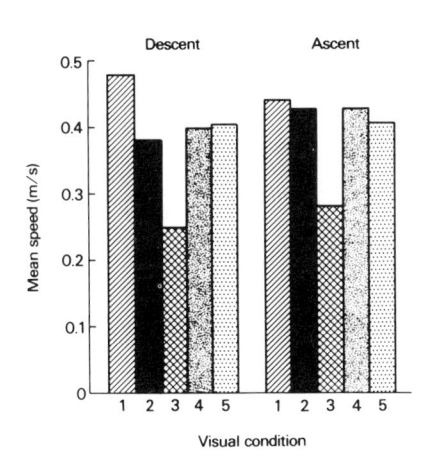

Comparison of mean speeds for ascent and descent for first and second runs showed that only under condition 1 was the second run quicker (by about 20%). Under the other conditions there was little difference in speeds, in ascent and descent, between the two runs.

Subjects' opinions indicated that in terms of ease and satisfaction with visual conditions, condition 1 was ranked first, followed next by condition 5. Condition 2 was ranked fourth.

On comparing mean speeds of ascent and descent, females were generally slower than males, eg by about 20% and 30% respectively under condition 2. Male subjects aged under 30 years were quickest.

DISCUSSION

Overall the findings for level walking through a corridor under uniform illumination ranging from 0.02 to 1 1x are similar to those of Boyce (1985). Namely, under the illuminance standard (0.2 1x) subjects experienced a little difficulty on some parts of the route; under a mean illuminance of 1 1x, people moved more smoothly at a speed typical for movement under normal lighting. When visual cues such as photoluminescent markings were provided for key elements on the route, speed was comparable to that under the traditional approach of uniform lighting of 0.2 1x. A limited amount of this material was used in the corridor in this study and movement under this alternative might be improved by increased areas or different positioning.

The findings for the stairway indicate that movement under 1 1x was faster, easier and of less risk of a stumble or fall incident than under 0.2 1x, suggesting that an increase in mean illuminance to about 1 1x or more on the stair treads would be beneficial. Comparison of performance between the photoluminescent material and 0.2 1x indicated that the mean speeds on the stair itself were very similar. However, approaching and leaving the stair was slightly quicker with photoluminescent material. Of these two conditions, the photoluminescent was preferred.

The use of photoluminescent material as an alternative or supplementary provision on stairways and corridors warrants further serious consideration. It is anticipated that use of these materials would involve a relatively smaller resource cost (sum of material, installation, running and maintenance costs) compared with traditional emergency lighting.

ACKNOWLEDGEMENT

The work described has been carried out as part of the research programme of the Building Research Establishment of the Department of the Environment and this paper is published by permission of the Director.

REFERENCES

Boyce, P.R., 1985, Movement under emergency lighting: the effect of illuminance, Lighting Research and Technology, 17, 51-71.

British Standard BS5266, Part 1, 1975, Code of Practice for emergency lighting of premises, BSI.

Fruin, J.J., Guha, D.K., and Marshall, R.F., 1984, Pedestrian falling accidents in transit terminals, Port Authority of New York and New Jersey, New York, USA.

Simmons, R.C., 1975, Illuminance, diversity and disability glare in emergency lighting, Lighting Research and Technology, 7, 125-132.

Management of Change

MANAGING CHANGE IN THE LOCAL ECONOMY: INFORMATION TECHNOLOGY SUPPORT AND THE NEW ENTREPRENEURS

C. BROTHERTON, M. ALDRIDGE, S. CHARLETON and P. LEATHER

Department of Psychology
University of Nottingham
Nottingham NG7 2RD

The paper outlines research on the changing structure of Nottingham's economy and presents the central findings of an in-depth evaluation of the role of information technology and telecommunications in supporting newly created companies in Nottingham. Data is presented indicating the role information technology and telecommunications play in the decision styles of new companies. The paper indicates some of the necessary precursors to the development of appropriate software for the small company and raises some issues which are vital for the appropriate design of training.

In the past decade major structural changes have occurred within Nottingham's economy. Although unemployment has remained at or around the national average there has been a clear trend away from manufacturing. Major job losses have occurred in every manufacturing sector. For example, T.I. Raleigh, Nottingham's largest engineering company, has reduced its workforce by over 70% from its 1971 level and is still making people redundant. Comparable reductions have also occurred in the numbers employed at John Player's Tobacco Factory. New company formation is occurring but this is being vastly outstripped by the number of closures. Much of the manufacturing sector of the city is dependent either upon its links with the major companies or upon selling its products to domestic householders (see Aldridge et al (1986) for a fuller discussion). The changes occurring in the structure of Nottingham's local economy are reflected in situations being faced by every city in Britain (see for example, Boddy (1986), Buck (1986), Cooke (1986), Spencer (1986). Neither is there anything particularly unique about the way Nottingham is facing the issues of change (Aldridge & Brotherton, in press). The issues being tackled in Nottingham may then have some generality beyond the immediate locality.

As in many cities facing dramatic structural change, Nottingham has embarked upon many schemes to promote and support the creation of new jobs through self-employment. These include a range of business advisory and promotion services; flexibly licensed managed workshops providing a range of wood and metal working equipment, textile machinery and practical advice on a common service basis; a managed innovation centre; a science park and an Advanced Business Centre - which provides centrally serviced information technology and telecommunications facilities. The data presented in this paper has been gathered from in-depth interviews conducted in 70 new businesses, many of which are located in these facilities. All of the companies have been established within the past two years, and represent a broad cross-section of manufacturing and commercial activities. Only a fragment of the data can be hinted at here.

The first point that we would draw attention to here is the great diversity to be found in the behavioural dynamics of those owning and managing newly formed companies. Elsewhere, (Charleton et al 1986) we have argued that there are important differences within new entrepreneurs as a group. These differences are found to be particularly salient in terms of the route into entrepreneurship, the orientation to business growth and development, the value put upon the various rewards of self-employment, and the resultant strategies used in business management. There is also considerable gender variation in motivation for the entry into self-employment as expressed by the majority of male and female entrepreneurs within our sample. 27% of our women respondents gave as their main reason for their entry into self-employment the fact that working in one's own company allowed them a better accommodation to their personal and domestic situation than did formal employment. 38% of the men in our sample cited real or anticipated job loss as a general reason for their entry into self-employment, whilst no woman gave this reason directly. (We have called this group the 'pushed' entrepreneurs in comparison with those we have called the 'elective' entrepreneurs.) 30% of the men, but no women, in our sample gave the anticipation of improved financial reward as a reason for entering self-employment (Charleton et al 1986). These different orientations are strongly correlated with different styles of decision making and business management in general.

We have found it useful to distinguish between those

companies handling their information and decision processes strategically and those handling them operationally. 76% of our companies are generally operational, by which we mean they adopt a reactive and non-predictive mode of information tracking and decision-making. 24% we have categorised as being strategic, that is they are proactive and predictive in their approach to decision. 80% of the femal owner-managers in our sample are strategic, whilst only 22% of the male owner-managers are so classified (Brotherton et al, in press).

We raise these points because it becomes clear that we can no longer rely upon descriptions of entrepreneurship that are available in a literature which predates the recent large scale restructuring that Britain is experiencing. These variations in motivation have not been reported in earlier studies. More difficult still, entrepreneurship is regarded in the exisiting literature as being an abstract and highly general concept, necessitated either by the broad assumptions of economic theory or by those concerned with the contrast between the 'entrepreneural spirit' and the impulsion towards conservatism characteristic of large scale formal organisations. Neither of these positions assists the understanding of the present problems at the level of human factors being created by the restructuring of the economy (Brotherton et al 1986). Since there are few lessons to be learned from the past, this raises sharply in our view, the issue of how support, especially that support which is provided by the new technologies, might best be designed for new small companies.

The central problem, from the point of view of systems design, is that, firstly, as our data shows, there is a good deal of separation of the process of financial accounting from the ongoing financial decisions being made by the majority of small businesses with whom we have spoken. This separation is sometimes physical - the majority of our operational businesses have their books done for them either by a professional accountant or by another member of the family, and on a basis which is less than frequent. Secondly, there is an emphasis upon external reasons for book-keeping, such as producing accounts for scrutiny by the Inland Revenue, rather than for internal reasons, such as the desire to know how the business is performing. Thirdly, there is little use of accounting information to both understand the present position of the business and, more importantly, to predict future action. Many of these problems are actually encouraged by the majority of training texts currently available to small businesses (Charleton et al 1985). We also feel that the lack of prediction and search in the cognitive strategies of the majority of our respondents might well explain why so many small businesses in general fail as a result of cash-flow difficulties.

These are particularly difficult issues for the designers of soft-ware to come to grips with. Jensen (1983) tells us that "Every user of a system has a subjective mental representation of the task being carried out and the nature of the system which he or she is confronted. This applies both to the novice and the sophisticated user. An understanding of these perceptions is an important ingredient in making a suitable fit between the user and the system. Messages presented to the users should be in a form consistent with the users' perception of the activity.". In our view, Jensen does not go far enough since these perceptions and mental representations need to be conceptualised on two levels. Firstly, those representations concerned with actually using hardware and software. Secondly, those representations concerned with the particular business and managerial issues and areas which are themselves addressed by the application of information technology. Given our data base, it is clear that the interface between the small business user and information technology must be flexible and sensitive enough to take adequate account of individual user and business profiles and work through requisite change or development at the information technology level simultaneously with change or development at the general management level (Leather et al 1985).

There is considerable scope for information technology to increase its potential in supporting small businesses. At present, the general picture is that there is a lack of appreciation, on the part of the designers of software, of the actual decision strategies that are possible for small businesses. Instead, there is an overwhelming tendency for the programmer to simply scale-down the software designed for a large company to that considered appropriate to a small company. What is required is that the base-line from which the company is actually working be taken as the point of departure for design.

Without such sensitivity in design there seems a danger that systems will continue to be under-utilized. The East Midlands has the lowest uptake of information technology in the country (DoE 1981). It may be that the structure of many traditional industries has not lent itself to implementation. The initiatives aimed at changing this through the economic development of Nottingham have so far made but a small impact in this respect. There is a very marked tendency in our data for the companies we interviewed both those housed in centres with IT facilities and those based in more conventional common support developments, for many telecommunication and computer devices to be seen as inappropriate. Very often, when some advantage was taken of the facilities, it was the image, rather than the actuality, which first engaged the company in IT utilization. The issue, generally, was one of lack of perceived applicability rather than ignorance (Leather et al 1986). Our data have been fed back to the policy makers and there are changes in hand but the process of learning is, we fear, a long one.

Many of these points apply not only to the design of software for that proportion of the small business population that may actually come to grips with using information technology but also to the design of training. Sparrow (in press) makes the point that as a whole Britain's managers make very little use of formal training. He sees the issue as being one of attitude. However, we feel that there is again likely to be a problem of lack of perceived applicability. Our new entrepreneurs are very much caught up in the here and now of the running of their business - they need an immediate return on their time investment if they are to be thoroughly rewarded for undertaking training. On the whole, they report to us a level of generality and technicality in the language of training which makes it difficult for them to relate their own immediate problems to that of formalised presentations. To bridge the gap, trainers have to be prepared to enter the discourse at a level of understanding of the business - not of that of the text (most of which, as we observed above, seems to lead the reader astray in any event!).

A concrete example may assist here. Many training texts will present formulae for analysing the ratio of fixed to variable costs. Most, but not all, of our 'pushed' entrepreneurs are tempted, when faced with matters so formally presented, to leave the questions to their advisors. However, our data suggest that if the same point is made by way of a series of grounded questions, such as 'What rent do you pay on your unit?', 'What are your material costs?' and so on, then understanding not only follows, but is acted upon day by day. Adult learning is enhanced, we argue, if presentation begins at the level of immediate understanding and experience and builds gradually by way of allowing the testing of hypotheses about knowledge relationships in such a way that interactions and linkages between events are encouraged and so maximized. Formalised sequencing often cuts across these processes (Brotherton et al 1987).

We cannot yet tell what impact this wide variety of

support schemes is having upon the Nottingham economy. The initiatives are so diverse that direct measurement through the traditional indicators is impossible. There is, though, some evidence that the presence of some of the initiatives within the city is beginning to attract new developments in the employment field. For example, the Advanced Business Centre has now attracted a Training Access Point project which will provide data banks of training information to enquirers from new small businesses. The Science Park is raising enquiries from potential clients with High Tech businesses which are too large to be housed in the Park itself. Beyond that it is only possible to guess. Neither is it possible through national figures to see what is happening to the general level of self-employment - official records are simply not kept in a way which would provide an answer (Curran et al 1986). Although observation tells us that the pace of new job creation lies considerably behind the processes of employment destruction, what is clear is that Local Authorities are very much at the margin in their attempts at helping cities cope with large scale changes that are themselves generated by the international economy (Aldridge et al 1986, Aldridge & Brotherton, in press). The pace of structural change means that policies have to be set, with all too limited a resource, and then the lessons learned during the course of implementation. The research which we outline briefly here is aiding that learning process at both the policy level and at the level of the participants. However, whilst the management of change in the local economy is being pursued with vigour, economic change is occuring at a rate which is furious.

REFERENCES

Aldridge, M.E., Brotherton, C.J., Gillingwater, D. & Totterdill, P, 1986. In Global Restructuring - Local Response, edited by P. Cooke (Economic and Social Research Council).

Aldridge, M.E. & Brotherton, C.J., in press, Being a Programme Authority: Is it worthwhile? Inner city policy making in Nottingham, Journal of Social Policy.

Boddy, M., Lovering, J. & Bassett, K., 1986, Sunbelt City? A Study of Economic Change in Britain's M4 Growth Corridor, (Clarendon Press: Oxford).

Brotherton, C.J., Charleton, S. & Leather, P., in press, Job Creation - New Work for Women?, Work and Stress, Women and Work, 1,3.

Brotherton, C.J., Leather, P. & Charleton, S., 1986, Social Psychological dimensions of Job Creation. Paper delivered to the Occupational Psychology Conference of

the British Psychological Society.
Buck, N., Gordon, I. & Young, K., 1986, The London Equipment Problem, (Clarendon Press: Oxford).
Charleton, S., Brotherton, C.J. & Leather, P., 1986, Management decision and management education: searching for synchronicity. Paper delivered to the London Conference of the British Psychological Society.
Charleton, S., Brotherton, C.J. & Leather, P., 1986, The Small Business Boom: Social Context and Social Construction. Paper given to the Social Psychology Conference of the British Psychological Society.
Cooke, P., 1986, op cit.
Curran, J., Stanworth, J. & Watkins, D., 1986, The Survival of the Small Firm, Vols. 1 & 2, (Gower Press: London).
Department of Employment, 1981, Information Technology Uptake in the UK, (DoE, London).
Jensen, s., 1983, Software and User Satisfaction. In New Office Technology: Human and Organizational Aspects, edited by H. Otway & M. Peltu, (Francis Pinter: London).
Leather, P, Charleton, S. & Brotherton, C.J., 1985, Sounds OK for Interpol - but what about us? Paper delivered to the Occupational Psychology Conference of the British Psychological Society.
Sparrow, P. & Pettigrew, A., in press, Britain's Training Problems: The Search for Strategic Human Resources Management Approach, Human Resources Management Journal.

ORGANISATIONAL CHANGE AND MANAGEMENT TRAINING

A. DAVIS and T. COX

Department of Psychology
University of Nottingham
Nottingham NG7 2RD

This paper describes the development and evaluation of a management training programme concerned with the "management of change" in a large engineering and shipbuilding company. Its aim was to facilitate the introduction of major changes in work methods. The evaluation data suggested that the programme met its agreed objectives, and produced changes both in the understanding of change and of the skills required to manage it successfully.

The company on which this study is based is at present subject to major structural and organisational changes resulting from privatisation and the introduction of new work technology (Chirrito 1983). This project is one of several initiatives commissioned to facilitate these changes.

TRAINING NEEDS ANALYSIS : "ATTITUDES TO CHANGE"
The necessary training needs analysis took the form of an employee survey (Bailey 1983). This not only provided objective data on which to base the development of training materials, but also allowed company employees to become involved in the project at an early stage. An "Attitudes to Change" questionnaire was designed to assess a) employee knowledge of the planned changes, b) how they had obtained that information,   c) the attitudes towards potential problem areas and d) the degree of support for change.

The respondent group consisted of 51 employees involved in a submarine build project which had already felt some of the effects of the planned changes. Respondents were selected from as many different levels within this project area as possible.

The survey (Davis & Cox 1986a) highlighted the fact that most of the managers within the company suffered from a lack of training in management skills having simply been promoted from first class engineers. Four particular areas were identified in which skills could usefully be developed. These were in relation to: 1) the support given by management during the planning and implementation of change; 2) the structure and nature of communication and interpersonal communication skills; 3) the anxieties of the workforce concerning change, in particular relating to issues of job security and safety, personal development, and group and shift working, and 4) problem solving during the period of change and after (cf action learning).

DESIGN OF TRAINING WORKSHOPS

It was agreed that the most practical strategy for improving management knowledge, attitudes and skills would be a training course based on a series of linked workshops (Cox 1985). The workshops were designed to examine the process of change itself and in doing so, develop managers' knowledge and skills in relation to the management of change, communication within a large organisation, and problem solving (Davis & Cox 1986b, 1986c). In addition, an attempt was made to develop a more proactive organisational culture.

The **first workshop** was designed to improve management understanding and skills in relation to person management during the process of change and to examine problems arising from poor communication. Exercises and case studies were used to highlight the points made during the workshop. The link between the two workshops was provided by **project work** in which managers were encouraged to examine their own problem solving skills. It was based on participants' use of the Critical Incident technique (Flanagan 1954) to record and analyse their own problem solving activities. The **second workshop** began by examining the project reports and then discussed strategies of effective problem solving. Finally, the principles of Action Learning (Revans 1984) were outlined and **"action learning teams"** were established. These teams were to look at the progress of change in their members' various areas and discuss how members attempted to solve the problems that arose and what might be learned from the exercise. The teams also provide peer support for their members.

Groups of 16 managers attended the two workshops for a total of three days, including two evening sessions. The evaluation studies were conducted by Stress Research in conjunction with the Training Department.

EVALUATION OF THE WORKSHOPS

A "Knowledge, Attitudes and Skills" framework was used as

a basis for the evaluation of the immediate effectiveness of the workshops (see Warr et al 1970; Davis & Cox 1986d). Evaluation took the form of an attitude questionnaire, a knowledge test of the information presented (both administered pre and post workshop), an assessment of how much participants understood and would use the managerial skills under consideration, and a course assessment form (both post workshop only).

The Attitude Questionnaire

One of the aims of the course was to promote a positive attitude towards the planned changes. A semantic differential procedure was used for the attitude enquiry. The form consisted of three sets of statements defining important attitude areas: a) the current changes in work practices, b) likely problems resulting from change, and c) communication in general within the company.

Overall, there were no marked changes in managers' attitudes, although the problems arising from change were seen to be somewhat less threatening and important after the workshops. Generally change was seen as somewhat complex but highly necessary, beneficial and challenging. The problems arising from change were described as reasonably frequent in occurence, important, and on balance detrimental to productivity, personal and reasonably avoidable. Communication within the company was seen as fairly poor, somewhat bureaucratic, rather complicated and slow.

The Knowledge Test

The data obtained from this enquiry indicated that a marked improvement in participants' knowledge of the change process and effective communication had occured. Statistical analysis using Student's t-test (matched pairs) indicated that this difference was highly significant (t=3.58, df 9, p=0.006). The increase in participants' knowledge was related to their original level of understanding (r=0.76; p<0.01).

The Skill Assessment

Participants were asked how much their understanding of management skills had increased as a result of attending the two workshops with respect to: a) the management of change, b) effective communication, c) problem solving, and d) action learning. Five point rating scales were used, from 1 (not at all) to 5 (complete understanding).

In all four cases participants' responses indicated that they had gained from the course. The mean ratings (modes) were: the management of change 3.6 (4), effective

SOCIO-TECHNICAL CHANGE IN TRUCK MANUFACTURING

P. LEATHER[*] and C. BROWN[**]

[*] Department of Psychology
University of Nottingham, Nottingham NG7 2RD

[**] Department of Psychology
University of Lancaster, Bailrigg, Lancashire

This paper considers the lessons to be learned from an applied intervention in a major British truck manufacturing plant, where the introduction of a new mass-produced lightweight vehicle necessitated major changes in systems of work organisation. A variety of indicators of organisational performance are considered, together with a number of alternative strategies for the management of change deriving from different attributional perceptions of this data. The paper emphasises the need to develop socio-technical prescriptions for change rather than purely technological ones.

THE CASE STUDY : BACKGROUND AND CONTEXT

The Plant, completed in 1979, was originally designed for the assembly of very heavy rigid lorries and articulated tractor units. However, adverse economic and market changes since then demanded that the plant now be used to manufacture a much wider range of vehicles than was originally envisaged. As other manufacturing operations within the Group were reduced or closed, the Plant was called upon to make an increasing contribution to the Company's total production.

These adverse market conditions led the Company to introduce a new product, a mass-produced lightweight vehicle, the TW1. The success of this "innovatory" vehicle was seen as crucial to the Company's survival. Survival, however, depended not only upon success in the market-place, but equally upon the extent to which appropriate social and technical systems could be developed within The Plant to achieve the efficient production of all the vehicle ranges to be manufactured there.

Much depended, then, upon the degree of flexibility with

which the plant's manufacturing facilities could adapt and respond to the quite radical changes in production demand imposed by the variable model mix. Central amongst these facilities were twin conveyorized assembly tracks, computer-assisted material control and warehousing, and computer-assisted vehicle-testing facilities.

With the need to manufacture both heavyweight (premium) and lightweight trucks within the plant, the Company's basic intention was that one of the two production lines would continue to be used for the assembly of premium vehicles - high specification models requiring a greater variety of work content - while the other would increasingly be used for the volume production of the lightweight vehicle.

Despite the recognition that the efficiency of an assembly line system is mix sensitive, the Project Team responsible for the design of The Plant nevertheless considered organisational responsiveness and flexibility to be important attributes of its chosen technological arrangements:

> "The factory can react more quickly to changes - for instance new demands from customers or lack of supplies of particular types of products - simply because it has more information about what is happening on the assembly line" (New Scientist)

Significantly, the movement towards the centralized assembly of mass-produced vehicles, symbolized in both the advent of The Plant and the TW1, encapsulated a fundamental change in organisational identity:

> "We were manufacturers of trucks in the old days, now we just assemble them. We're not really truck manufacturers any more. The idea in the Plant is that it is a purpose-built plant. Then, with the TW1, the idea is that we have a purpose-built truck for a purpose-built plant"
> (Senior Manufacturing and Engineering Manager)

Interacting closely with these changing beliefs about organisational identity and product choice were those about how things should be done within the organisation. In this respect, The Plant marked the advent of a new managerial philosophy: the decline of custom and practice and the ascendancy of scientific management as the governing rationale for work and organisational design. As one member of the now disbanded project team put it, "The Company had gone along for a long while with custom and practice traditions, but this was a deliberate attempt to regulate things more, to scientifically determine them." "When the doors of The Plant opened," another added, "we had already mapped out an industrial engineering strategy of how the work was to be done. Before The Plant, this was unknown. We never even sent rate-fixers down to the old works. But with The Plant we said 'No' to this state of affairs. We said we were going to measure and get standard times."

The consequences of these changes for the de-skilling of many shop-floor jobs were readily acknowledged by plant management, as were their likely negative effects upon motivation and morale:

"Much of it is woman's work now. I hate to admit defeat, but I believe that in de-skilling the jobs, we have not kept the motivation level of the men at the levels it used to be"
(Industrial Engineering and Manufacturing Manager)
"The Plant is not man motivated but machine motivated. You don't need thinking people for this kind of work. Our operatives' skills and experience are not matched by the tasks they have to do. They are predominantly fairly skilled people, whereas The Plant suits non-thinking people"
(Industrial Engineering Manager)

These various decisions surrounding the organisational design of The Plant manifest an attempt by the Company to employ a purely technological prescription for the management of change. However, empirical data of the plants actual productive performance argues emphatically against the efficacy of such a limited technological prescription.

INDICATORS OF PLANT PERFORMANCE
a) <u>Proportion of Completed Vehicles Off Track</u>
1. During the six month period over which the lines were monitored, little over 42% of <u>all</u> chassis leaving the conveyors were complete enough to go straight to test, i.e. contained all those necessary parts which must be fitted before the vehicle can be tested.
2. Over this <u>whole</u> period, the heavyweight track significantly outperformed the lightweight track in terms of its ability to produce a complete chassis (47% compared with 40% $X^2 = 15.539$, df = 1, p < .001).
3. Pre volume production of the TW1, the heavyweight track outperforms the lightweight track by 13% (48% : 35% $X^2 = 24.688$, df = 1, p < .001).
4. Post volume production of the TW1, this advantage is reduced to 2% (46% (premium) : 44% (lightweight) $X^2 = 0.408$, df = 1, n.s.).
5. Although volume production of the TW1 leads to a significant improvement in the performance of the lightweight track, from 35% to 44% ($X^2 = 15.543$, df = 1,

$p < .001$), this improvement is offset by (a) the fact that, simultaneously, there is a slight, but not statistically significant, decline, 2%, in the performance of the heavyweight track; and (b) the likely 'halo' effect associated with the introduction of the new vehicle.

Amongst plant personnel, attributions of the causes of this "incompleteness" problem highlighted two principal sources: (1) parts simply not being available at the right assembly stage at the right time, or (2) the fitting of "slave" or temporary parts which allow a chassis to progress further down the line, but which later need to be replaced before the vehicle can be tested.

b) <u>Track Stoppage Time : Cab Lines</u>
1. During a three-month <u>monitoring</u> period, both cab lines suffered badly from <u>unplanned</u> track stoppages, with the lightweight line <u>doing significantly</u> worse than the heavyweight (21.4% of total possible production time, and 10.01%, respectively).
2. Absenteeism either directly or indirectly accounted for 35.69% of this unplanned stoppage time on the lightweight track, and 24.23% on the premium track (9.02% and 2.77% of total possible production time).
3. Material supply and workflow problems account for 48.99% (lightweight) and 63.36% (heavyweight) (12.38% and 7.24% of total possible production time).

c) <u>Absenteeism</u>
For much of the period that the plant was being monitored absenteeism (hours lost measure) regularly fluctuated between 10% and 14%, although in some areas, e.g. final assembly, it rose as high as 25%. This meant that up to a quarter of the hours scheduled to be worked were lost to unplanned absences, including short-term sickness.

DISCUSSION : GROUP PROCESSES IN ORGANISATIONAL DESIGN

The unavoidable conclusion that results from this data is not only that the plant's productive performance is extremely poor, but also that this sub-optimal performance raises fundamental questions against the true flexibility of The Plant's present technologically determined work organisation.

While there was unanimous concensus amongst the different interest groups within the plant over <u>how</u> badly it was performing, there were marked variations, both within and between management and shop floor groups, in the attributed reasons as to <u>why</u> this was the case, and how it could be remedied.

Production managers, for example, emphasized further technological developments, e.g. the transfer of more sub-assembly tasks onto the line, or the proposed introduction of a new assembly line into the rectification area where, currently, almost 20% of the plant's direct labour was deployed.

A second "traditionalist" group within management considered the efficacy of purely technological solutions to be severely limited due to both (a) the "bespoke" nature of much of the truck market, and (b) the problematic effects of any new working practices on a workforce having strong craft traditions.

The workforce, for their part, tended to feel either that a basically malevolent management was deliberately ignoring their needs and aspirations, or that management was fundamentally ineffectual.

The foremen were the most alienated group within the plant. They felt particularly vulnerable on two counts: (1) they felt their work organisation and task skills to be taken away by the deskilling and new role relationships imposed by the technology, and (2) they considered their man-management skills to be eroded or "by-passed" by centralized management control and decision-making.

Based upon (1) the poor performance data, (2) the need to overcome these clear inter-group conflicts, and (3) the view that excessive model variance undermines the assumption of volume demand for a standardized product upon which the effectiveness of an assembly line system rests (Sabel 1982), the recommendation made to the Company was that The Plant required a socio-technical prescription for the management of change rather than a purely technological one.

Originating in the work of Trist and Bamforth (1951) on technical change in the Durham coal mines, the central idea in socio-technical theory is that joint optimization of both the social and technical systems must be sought. If either system is maximized at the expense of the other, then sub-optimal organisational performance will result. As a specific means of operationalizing the theory in the present context, a number of techniques were suggested including (1) the use of autonomous work groups moving with the line over a number of task-determined stages (rather than "static" line manning), (2) job rotation, and (3) a greater involvement of shop floor employees in aspects of product planning and development.

Three lines of evidence supported the case for such a socio-technical strategy. Firstly, the weekly quality audits undertaken by plant personnel seldom indicated the variance between actual performance and desired organisational goal to be controlled at source. Secondly, regardless of its technological sophistication, the plant's material supply system necessarily depended upon the intelligent and responsible co-operation of assembly-line workers in such areas as stock control and the maintenance of an adequate organisation of supplies at strategic points along the lines.

Thirdly, by open admission amongst senior personnel, The Plant's chosen technology was a duplicate of that successfully operating in a Swedish truck plant with similar manufacturing problems, e.g. variable model mix. However, as is reliably reported in Aguren (1973), Gregory (1978) and Davis and Cherns (1975), this same Swedish plant attributed much of its manufacturing success to a programme of socio-technical innovations similar to those being recommended to the British firm. Significantly, Plant personnel had visited the Swedish factory on a number of occasions, but - their "technological" philosophy informing their subsequent perceptions and cognitions - had failed to see any differences between the two plants' social systems, save that "the Swedish plant had a few nice little library areas on the shop floor" (Project Team Member).

As a prescriptive measure, the suggested use of the group as the fundamental unit of analysis and action is built upon a functional theory of group dynamics (Sherif 1961). That is, according to socio-technical theory, inter-group conflict within the organisation is largely the result of those competitive goals and divisions brought about by a scientific management organisational design. With the realization of the superordinate goal of "joint optimization", wherein organisational performance and job satisfaction are to be improved simultaneously, then, the theory predicts, such inter-group conflict should decrease if not entirely disappear.

In addition to these positive social, psychological and economic effects, the use of autonomous work groups would also serve to establish a direct feedback loop between the design and execution stages of the performance process, thereby enabling variance between performance and goal to be monitored and controlled at source - with an obvious and immediate increase in organisational flexibility and responsiveness to change.

Despite its theoretical relevance, its success in the Swedish plant, and the unequivocal condemnation over present performance levels, this socio-technical prescription met with only piece-meal approval within the Company. Rather than serving as a point of integration for the different group positions, it was instead perceived, interpreted, evaluated and responded to in terms of how well or badly it fitted with each group's pre-existing biases and judgements.

In other words, the socio-technical promise of joint-optimization did not reduce inter-group conflict in the manner the theory predicted. Exactly the same can be said of the introduction of the new vehicle, the TW1. The crucial need for its successful manufacture was clearly appreciated by all the interest groups within the plant. In this sense it constituted a clearly recognized superordinate goal. Yet its introduction did little to lessen the characteristic inter-group conflicts within the plant.

As a prescription for the management of organisational change, the "lesson" of this failure would seem to be that group processes need to be understood and taken into account not only in terms of their differentiation around objective task and work role structures, but equally in terms of their interdependent relationship with social and professional values, e.g. different group beliefs about organisational identity, product choices and "appropriate" systems of work organisation; "professional" changes in the structure and composition of "management"; social choices embodied in technological change; and even wider socio-cultural beliefs about appropriate strategies for the management of scarce resources.

REFERENCES

Aguren, S. & Norstedt, J., 1973, The Saab-Scania Report. (Stockholm: The Swedish Employers' Confederation).

Davis, L.E. & Cherns, A.B., 1975, The Quality of Working Life, 2 vols. (New York: Free Press).

Gregory, D. (ed), 1978, Work Organisation: Swedish Experience and British Context. (London: Social Science Research Council).

Sable, C.F., 1982, Work and Politics. (Cambridge University Press).

Sherif, M. et al, 1961, Intergroup Conflict and Co-operation (University of Oklahoma).

Trist, E.L. & Bamforth, K., 1951, Some social and psychological consequences of the Longwell method of coal getting, Human Relations, 4, 3-39.

MANAGING INDIVIDUAL INNOVATION

M. WEST and N. KING

MRC/ERSC Social and Applied Psychology Unit
University of Sheffield
Sheffield S10 2TN

Innovation is broadly defined and research projects exploring innovation at the individual, group and organisational levels are briefly described. It is argued that a focus on individual innovation offers much promise in this relatively unproductive research area. A model of individual innovation at work, integrating factors suggested by previous research to be facilitative of innovation, is described. It is argued that the use of such models provides a firmer conceptual base for future research and applied interventions.

Despite the publication of over 4,000 research articles on technological innovation alone little consensus has emerged about factors to be considered in managing innovation successfully. Perhaps one reason for this is the wide range of operational and conceptual definitions employed by researchers. Some define innovation in terms of profitability while others display a pro-innovation bias (West et al., 1986). Some definitions focus exclusively upon technological innovation while others include only administrative or managerial innovation. To avoid fragmentation it would seem most valuable to adopt a definition encompassing innovation from the individual to the organisational and even to the societal level, including innovation in all types of work organisations. Innovation may therefore be defined as '... the intentional introduction and application within a role, group or organisation of new and different ideas, processes, products or procedures, designed to significantly benefit role performance, the group, the organisation or the wider society' (West et al., 1986, p. 4). This definition would therefore include as an innovation the introduction of 'watchdog' community members onto the management teams of nuclear processing plants even though the move might not be designed to benefit the organisation. Innovation would also not be confined to technological developments but would include, for example, the community nurse's attempts to improve her liaison with other professionals by visiting them rather than telephoning or writing. We have operationally defined individual innovation as consisting in the implementation of new and improved work targets/objectives; new and improved methods used to achieve targets/objectives; new and improved working relationships (who one deals with or how one deals with them); and new and improved procedures or information systems (West, 1986).

Recent and current research projects at the Social and Applied Psychology Unit in Sheffield have been exploring innovation in the world of work at three levels: individual, group and organisational. In particular because of the neglect of the study of individual innovation, effort is being directed at examining the relationship between individual innovation and other work attitudes and experiences, as well as the nature of individual innovations. Projects include a longitudinal study among 1,100 British managers examining the individual and perceived situational predictors of role innovation (West, 1987); a survey of 30 managers and their supervisors validating a measure of role innovation (West, 1986); studies of reports of innovation among those in low discretion work roles (145 student nurses) and those in high discretion, high workload roles (100 health visitors). Additionally, three ongoing research studies in elderly care institutions are examining the management of, and staff reactions to innovation. The innovation research team has also completed a study of the relationships between career development policies, work attitudes, organisational culture, and individual and organisational effectiveness across a number of different organisations in the British wool textile industry (Brooks-Rooney and West, 1987).

Van De Ven (1986) has pointed out that **'People** develop, carry, react to, and modify ideas' and it is our belief that much can be learned from a focus on individual innovation, and that a psychological perspective is particularly helpful. We are therefore currently developing a model which identifies the major psychological and organisational factors which appear to facilitate individual innovation in organisations and offer it as a starting point for psychological research on the management of innovation. Figure 1 summarises the model.

Figure 1. A model of individual innovation at work

Facilitators of innovation

> Intrinsic to job; Group factors; Relationships at work; Role factors; Organizational factors

Individual characteristics

> Confidence/dominance; Growth need strength; Internal work motivation; Creativity; Specific self efficacy; Previous innovation experience

Innovations

> Implementation of new and different: Objectives; Methods; Procedures; Working relationships

FACILITATORS OF INNOVATION

Factors intrinsic to the job

Challenge posed by the specific task or job may be an important factor since it is an independent predictor of reports of innovation in both our research and Amabile's (1984). Task characteristics may include unpredictability (identified in our management survey) and novelty. Financial and technical resources are also mentioned by many we have interviewed as being vital to successful innovation attempts (Brooks-Rooney and West, 1987).

Group factors

There are indications in our data that the work group will influence innovation in a number of ways. Firstly the extent to which the group reinforces risk-taking and attempts at innovation will influence individual behaviour. Secondly, the cohesiveness of the group will determine the extent to which the individual believes that he or she can introduce and attempt to apply a new idea without personal censure. Nystrom (1979) has argued that high group cohesiveness may be desirable since it could motivate members to be creative by increasing feelings of psychological safety and self-actualisation. At the same time however, European social psychologists (Nemeth and Wachtler, 1983) have found that conflict producing minority influence can lead to majorities reviewing problems and finding creative solutions (cf. Janis, 1972).

Relationships at work

In both organisational and individual level studies of innovation the nature of the leadership relationship appears to be of importance in influencing innovation. Support from her superior emerges as a significant independent predictor of the health visitor's reports of innovation (Brooks-Rooney and West, 1987). Kanter (1983) criticises the 'elevator mentality' of organisations which are dominated by restrictive vertical relationships and by 'top down dictate'. Both Kanter and Peters and Waterman (1982) suggest that innovation is most likely to occur where the leadership style is participative and collaborative. Our recent findings in elderly care institutions show clearly that positive reactions to innovation are associated with perceptions that staff were consulted about and participated in decisions about the introduction of innovations (Brooks-Rooney and West, 1987).

Constructive feedback and recognition for appropriately innovative work is mentioned by both Peters and Waterman (1982) and Kanter (1983) and is singled out as an important facilitator in Amabile's (1984) study. It is generally acknowledged that feedback and knowledge of results provide mechanisms for improving employee performance and specific feedback for innovation might be expected to have a similar effect upon innovative performance.

Role factors

Both our research (West, 1987; Nicholson and West, 1987) and other research on creativity and innovation at work (Amabile, 1984; Peters and Waterman, 1982) consistently shows that perceived freedom or discretion is of major significance in predicting innovation and creativity in organisations. Freedom includes the extent to which individuals feel they have the freedom to act independently of their superiors; the extent to which they can set their own work targets and objectives; how far they can choose the methods for achieving work targets and objectives; and the extent to which they can choose who they deal with in order to carry out their work duties. Another important aspect of freedom is the extent to which the individual has latitude in relation to the use of time (Amabile, 1984).

Organisational factors

Researchers have focused on a variety of organisational characteristics correlated with innovative performance but Kimberly (1981) found no consistent pattern of results in relation to organisation size, visibility of consequences of performance, organisational complexity, and competition. He does conclude however that administrators in organisations who are professional, committed to innovation, who have participative leadership styles and who are open to change tend to facilitate innovation. Other researchers have looked at the influence of organisational culture and climate and Kanter (1983) has proposed that a culture of pride, good lateral communication, few unnecessary layers of hierarchy and large access to power tools for innovative problem-solving are relevant attributes of innovative organisations.

Recent European and American research (Kanter, 1983; Nomme, 1986) converge in identifying the following characteristics of the environments of innovative organisations: organisational consensus on goals and vision; an organisational environment which is empowering; an egalitarian culture which manifests a concern for human growth and is visibly pro-innovation; localness of decision-making and accountability for results; and stress on performance.

INDIVIDUAL CHARACTERISTICS

Most studies of individual innovation are located within the individual creativity tradition, which focuses on intellectual and personality factors in the creative individual. Our recent research has identified confidence and high needs for growth and development opportunities as predictors of role innovation (West, 1987). In addition to individual growth needs we also posit a specific motivation which relates to the job the individual is doing - internal work motivation. Internal work motivation is the degree to which the individual is self-motivated to perform effectively in his or her particular job and the extent to which the individual experiences positive feelings when doing the job well and negative reactions when doing it poorly. Specific skills in relation to the task in hand are cited by those we interview as being important in the introduction of innovations (Brooks-Rooney and West, 1987) and previous experience of successful innovation is also implied by our longitudinal findings on role innovation and transitions to newly created jobs (West et al., 1987).

The model we describe here represents an early stage in our thinking about a psychological approach to understanding innovation, but it has clear implications for those concerned with the management of innovation. We have yet to specify in detail the relationships between these factors or the precise nature of some of the recursive loops that undoubtedly exist. Neither do we specify which factors will have linear or curvilinear relationships with innovative performance. Nevertheless, the value of such a model is that it provides a theoretical basis for deriving predictions about the management of innovation and testing them in field settings.

REFERENCES

Amabile, T.M. (1984). Creativity motivation in research and development. Unpublished manuscript, Department of Psychology, Brandeis University.

Brooks-Rooney, A. and West, M.A. (1987). Innovation at work. A symposium convened at the Annual Conference of the Occupational Psychology Section and Division of the BPS, University of Hull, 5-7 January.

Janis, I.L. (1972). Victims of Group Think: A Psychological Study of Foreign-Policy Decisions and Fiascos. Boston, Houghton.

Kanter, R.M. (1983). The Change Masters. New York, Simon and Schuster.

Kimberly, J.R. (1981). Managerial innovation. In P.C. Nystrom and W.H. Starbuck (eds.), Handbook of Organizational Design. Oxford, Oxford University Press.

Nemeth, C.J. and Wachtler, J. (1983). Creative problem solving as a result of majority vs. minority influence. European Journal of Social Psychology, 13, 45-55.

Nicholson, N. and West, M.A. (1987). Managerial Job Change: Men and Women in Transition. Cambridge, Cambridge University Press.

Nomme, R. (1986). Towards innovation: A model of organizational change. Paper presented at the 94th Annual Convention of the APA, Washington DC, August.

Nystrom, H. (1979). Creativity and Innovation. Chichester, UK, John Wiley.

Peters, T.J. and Waterman, R.H. (1982). In Search of Excellence: Lessons from America's Best Run Companies. New York, Harper and Row.

Van De Ven (1986). Central problems in the management of innovation. Management Science, 32, 5, 590-607.

West, M.A. (1986). A measure of role innovation. British Journal of Social Psychology, in press.

West, M.A. (1987). Role innovation in the world of work. MRC/ESRC Social and Applied Psychology Unit, University of Sheffield, Memo No. 670.

West, M.A., Farr, J.L. and King, N. (1986). Innovation at work: Definitional and theoretical issues. Paper presented at the 94th Annual Convention of the APA, Washington DC, August.

West, M.A., Nicholson, N. and Rees, A. (1987). Transitions into newly created jobs. Journal of Occupational Psychology, in press.

THE ORGANISATIONAL REQUIREMENTS OF MODERN, AUTOMATED MANUFACTURING SYSTEMS

S. JOYNER

Department of Human Sciences
University of Technology, Loughborough
Leicestershire LE11 3TU

During the development of advanced manufacturing systems a decision must be made on the right organisational structure to support and control the technology. In 1984 a large engineering company set up a small site to develop an automated manufacturing system. This study examines the effectiveness of the flat organisational structure chosen to support the project. It looks at the perceived strengths and weaknesses of the flat hierarchy and recommends areas requiring further action. The study also highlights issues to be considered before a full-scale factory is set up.

Background

In 1984 a large British engineering company set up a greenfield factory in south-east England to investigate automated manufacturing technology whilst making pre-production samples of a new product. Fifteen highly qualified engineers under a site manager are permanently employed at the factory with responsibility for these development and production functions.

As part of the project an organisational experiment was set up 'to determine whether professional engineers can satisfactorily operate a plant without the need for specialisation, demarcation or supervision'. A number of criteria were identified to facilitate this overall aim. These included the need for a flat hierarchy, team building, no job demarcation, operation of the factory on a rota basis and the automation or sub-contracting of unrewarding jobs.

With the exception of the manager everyone is considered equal with authority based on knowledge rather than position. For five weeks out of six the engineers work a normal five-day week, concentrating on their own projects and areas of responsibility. During the sixth week they are allocated to a duty group and spend seven days (Sunday to Saturday) on the factory floor fulfilling and managing the production role. There are thus two or three personnel running the factory seven days a week on a roster basis.

Initial analysis

At the start of this study informal interviews were conducted with personnel in the factory and the product design team. (The latter are based on a nearby site and fulfil the role of customers, receiving all pre-production samples manufactured at the factory). Comments made during these interviews formed the basis of a questionnaire which was broken down into three sections:
1. In section one subjects used a five-point scale to rate their strength of agreement with 200 statements taken from the informal interviews. The areas covered included the structure, teamwork, aims, goals, priorities, training, duty group working, accountability, quality, passage of information and the future of the site.
2. A section of open-ended questions hoped to measure personal understanding of individual and factory goals and priorities. Critical incident questions were also used to find the perceived best and worst features of the site.
3. The final section asked subjects to complete measures of overall job satisfaction, general satisfaction, alienation, supervision and tension. For the purposes of control and comparison these will be given to production engineers of similar experience on other company sites.

The questionnaire was completed by each of the fifteen engineers on site and the results form the basis of the following discussion on various organisational issues.

Goals and priorities

The main concern of engineers on the site was the perceived lack of goals and priorities. This is supported by several authors who have identified the need to set clear, time-bound goals during the implementation of new technology (eg Gerwin and Leung, 1980; Dalzeil, 1986). Long and shorter-term goals were set down in October 1985 and have not been revised since that time although they are now out of date. The rapidly developing understanding of the technology involved and the changing economic climate has apparently made it difficult to decide on fixed long-term goals. It is suggested, however, that short-term goals be set with frequent revision by consultation.

The main short-term goals presently identified include the development of an effective automated manufacturing system, component production and build, outside contract work, and support functions such as training. The diversity of these goals and a scarcity of resources suggests that priorities must be set. This constitutes a real problem as the questionnaire and subsequent discussions suggest there is little agreement on priorities at this time.

The structure

There was concern from management as to the effectiveness of the very flat hierarchy and whether changes should be made. Empirically there is little evidence as to the most effective structure for a given technology. Various studies have produced guidelines based on, for example, the process, the manufacturing system, the external environment, the needs of the workforce etc.. Examination of these suggests that the present structure is probably correct for the site given its diverse roles, the small number of highly qualified people and the variety of complex functions to be performed.

Although the structure looks flat on paper, in reality there is an unofficial pecking order based on knowledge, personality and previous experience. This lack of complete equality was probably inevitable given the type of people employed and their supervisory and managerial backgrounds. Without exception personnel feel this perceived lack of equality to be right; in an environment where authority is based on knowledge it is natural that ability and experience will affect personal standing.

The organisation effectively means that the personnel form one large work group performing a range of tasks from the simple to the very complex. Although there is little formal guidance on the size of effective work groups the engineers generally consider fifteen to be near the maximum that could exist without a more defined structure.

To support the work group the management style practised is consultative rather than directive with decision-making encouraged at the lowest level. This type of supportive management is generally recommended for new manufacturing technology (eg Zylstra, 1986). Most personnel, however, feel there should be stronger leadership with increased discipline; this expectation of management may be due to the traditional, hierarchical sites most had previously worked in. It may also indicate a lack of communication about the style of management used resulting in perceived confusion over the extent of personal responsibility.

The question of discipline highlights the very fine line between guiding people towards maximum effectiveness and showing a perceived lack of trust in their actions. The real issue is the degree of self-discipline needed to work as a team; without the enforced discipline of a structured environment self-discipline becomes very important.

Teamwork

A team-based approach is normally recommended to operate the complex control systems used in advanced manufacturing technology. Effective team building was thus identified as a major organisational goal. On the present site it allows pooling of knowledge and shares out the less popular jobs.

With hindsight management feel that an initial team building exercise would have speeded up the development of the good team spirit now present. It is also obvious that the team was only seen to consist of the fifteen engineers on site. A greater effort to include the product designers in the team would have meant more co-operation and less of a 'them and us' attitude. Secrecy over the site and the new product also fuelled union suspicion in a company which is shedding jobs in an effort to become more competitive. The reasoning behind such secrecy should be carefully balanced against the longer-term damage caused.

Several of the positive and negative points identified during this study have been recognised as inherent within flat organisational structures. The advantages are normally intrinsic to the work itself with the increased variety, flexibility and responsibility. This was reflected in the open-ended section of the questionnaire where the freedom, team spirit, working environment and variety were considered the best features of the organisation. The results of the general measures also showed the freedom, responsibility, variety, fellow workers and workplace to give the most satisfaction.

The disadvantages of a flat structure are often peripheral to the main task and usually include problems with passage of information, the reward systems, training, documentation, outside relations, promotion and a lack of direction. The open-ended questions showed the latter to be considered the greatest problem on the site along with the lack of future certainty and a perceived lack of company support. The general measures showed the sources of most dissatisfaction to be pay, promotion and job security.

A prolonged pay dispute within the company postponed the use of the general measures on other sites but this should be rectified shortly.

Future work

Those areas identified above should form the focus of future action by building on the good points and improving areas which are perceived as weak. This will benefit the present site and provide foundations for the final full-scale factory. In addition to those issues previously mentioned the following areas also require further work.

Passage of information within the site is very informal with weekly planning meetings the only regular source of communication. The small number of people in the factory usually ensures that information is picked up through informal daily contact - in a larger facility the lack of proper communication flows would cause serious problems.

Although they simulate the form that will be used once computerised, the present factory procedures require paperwork. Possibly as a result of the automated environment individuals seem reluctant to use paper. This means that little has been documented including the history of the site. A compromise is needed between wasting time writing and allowing information to remain in individual heads. At present no-one has left the site but this will change; without knowledge capture the loss is likely to be great.

One source of criticism is the present reward system. In addition to the annual pay rise there is an extra increment based on individual performance. This is generally perceived to be unfair with no objective scale of measurement and no incentive to operate as a team.

Multi-skilling through training was one of the original aims of the organisational experiment. In 1980 Gerwin and Leung noted the difficulty of training people in an environment where work pressures are high. This is apparent in the factory and a systematic training plan is needed to provide the impetus to train. To assist in its implementation it is necessary to investigate the most effective and efficient ways of transferring knowledge. These must also cater for the intermittent user as many skills are only required whilst on duty, one week in six.

Unlike some companies who train workers to be flexible one problem is the complexity of the tasks to be performed. Many tasks require several months to master effectively; if possible expert knowledge of these processes should be captured and made available to the casual user. Other issues raised by the complexity of these tasks include the degree to which people can become multi-skilled, the background skills required and the implications for selection. Do you recruit people who are expert in one area or those who have the potential to become generalists?

Summary and recommendations

This study into one company's experience of implementing a flat structure identified areas of interest for the present site, the proposed full-scale factory, and other companies who decide on this form of organisation. The following summary highlights the main issues discussed in this paper and makes recommendations for the present site.

During organisational change certain important areas may be overlooked. These include documentation, information flows, personnel functions, training and setting goals. The latter helps to ensure that everyone has the same concept of where the organisation is heading.

Criteria such as the management style and reward system chosen should be considered carefully as they are difficult to change once implemented. Accurate decisions are also needed on work groups including their size, selection, team building, incentives, discipline and the inclusion of outsiders. The recognised benefits of flat structures such as freedom, variety and responsibility should be encouraged.

As a result of this study the following recommendations have been made for the present site:
1. Goals and priorities should be identified and set down.
2. The capture and efficient transfer of knowledge should be investigated for documentation and training purposes.
3. The roles and skills needed should be identified and implemented into a systematic training plan.
4. Information flows and procedures need improvement.
5. The present team spirit should be further encouraged and extended to include customers and suppliers.

Some of these recommendations need action from management whilst others will form the basis of future work. Hopefully knowledge gained now can be used on the existing site and eventually ease the implementation of a full-scale factory.

References

Dalzeil, M., 1986, Diagnosing potential obstacles to the introduction of automation, Proceedings of the 1986 Congress on the Human Aspects of Automation, 1, 99-116, (Los Angeles, September 1986).

Gerwin, D. & Leung, T., 1980, The organisational impacts of FMS, Human Systems Management, 1, 237-246.

Zylstra, K., 1986, CIM Implementation - The Human Aspects, Proceedings of the 1986 Congress on the Human Aspects of Automation, 1, 155-162, (Los Angeles, September 1986).

Trade Unions and Ergonomics

ON THE TILL: WORKERS' PERCEPTIONS OF HEALTH AND SAFETY IN SUPERMARKET CHECKOUT DESIGN

C. THORNE and D. RUSSELL

USDAW
188 Wilmslow Road, Fallowfield
Manchester M14 6LJ

This paper outlines some of the implications that supermarket checkout design has for the health of checkout operators and reports on a trade union initiative to organise a members' research project to investigate the design problems.

Introduction

The job of checkout operator in a busy supermarket is particularly fatiguing and stressful. Staff regularly spend 5-6 hours seated, often in inadequate chairs, making rapid repetitive entries into a keypad with the right-hand, while handling shoppers' goods with the left. Though few individual items are particularly heavy, the worker can handle up to 5 tons of goods per day dealing with 300-600 customers in the process. There is considerable variation in check-out design, from trolley-to-trolley systems to more sophisticated arrangements of moving belts to carry items to and from the operator's reach area. Recently, there have been major changes in the technology involved, with the introduction of Electonic Point of Sale (EPOS) systems linked to in-store computers and increasingly associated with laser scanners at the checkout which 'read' the bar codes on goods. Today, 95% of all goods passing through supermarket checkouts are bar coded 'at source' and by 1990 it is estimated that laser-scanning electronic tills will be installed at 75% of all main checkouts operated by the leading supermarket chains. The traditional checkout faces further change with the Electronic Funds Transfer at the Point of Sale (EPTPOS) schemes.(1)
 New point of sales technology makes the checkout the central location for data registration allowing rapid access

to a wide range of information: volume and type of sales, variations in buying patterns, customer frequency, accuracy and speed of cashier, etc. The new technology extends the range of the supermarket checkout function and increases the importance of the checkout operator.(2)

Problems of Design

As the largest trade union representing shopworkers in Britain, the Union of Shop, Distributive and Allied Workers (Usdaw) is uniquely placed to assess the impact of EPOS systems on operators. Our impression is that the new technology has not resulted in significant improvement in the working environment and working conditions of operators. Complaints from members about the physical and mental stress of checkout work are increasing and a recent pilot survey of employees in one retail chain indicated that 100% of checkout workers were dissatisfied with their seating arrangements. 90% suffered from back and neck muscular problems and 50% experienced migraine.(3) Wilson and Grey's 1984 study of laser-scanner checkout systems also observed that for workers using these systems "the discomfort felt in the neck and arms in particular was worse than for cashiers in conventional supermarkets".(4) Other research studies suggest that there are three key problem areas and this is confirmed by preliminary discussions with our members.(5)

1. The work itself which leads to fatigue and muscular complaints due to the manual handling of such large quantities, often in constrained space with inadequate furniture. Cervico brachial disorders of the right side of the body appear to have reduced with the introduction of electronic keypads, but low back pain, pain in shoulders and the left arm, and repetitive strain injuries in the left arm are still common.

2. The work environment. The most common complaint is of cold temperature and draughts, but artificial lighting and glare could be contributory to visual fatigue and headaches and noise is often a problem too.

3. Job organisation is also a major cause of stress. Interaction with customers, particularly in a busy store where long queues develop, is an obvious problem. Other factors such as boredom, lack of autonomy, isolation from fellow workers, and the need to remember a large number of changing prices are also important.

Supermarket checkouts would appear to embody few ergonomic design principles and certainly fail to suit the physical and mental aptitudes of the worker. In the words of one checkout operator, an Usdaw member, "We sometimes sit for eight to twelve hours a day in chairs where the backs are broken, they do not swivel. We are in an awful situation ..... We are stretching and pulling; we have queues miles long. It is just not on - something has got to be done".(6)

Developing a Strategy for Change

The problem facing the Union is how to develop a strategic programme that will bring about effective change at the workplace level. Over the next five years employers will be investing over £200 million in EPOS systems. Little of this will be used to involve workers in decisions about the design of the systems nor for research into the better adaption of the checkout to the physical and mental aptitudes of the operator. The Union lacks the financial resources of the employer but it does have access to one key resource for identifying design problems; the members themselves.

So Usdaw is embarking on a major research project of its own, investigating the health problems associated with supermarket checkout design. The primary aim is not to produce an academic piece of work but to involve members in the research process as a way of clarifying the issues and producing recommendations which can be used in bargaining for better conditions. Central to the project is the firm belief that our members who work as checkout operators have detailed and unrivalled knowledge of the problems of checkout design. The plan is to provide these members with the support and skills necessary to investigate the conditions within their own workplaces. It is hoped that this kind of 'worker investigation' approach(7) will develop trade union activity around work-design issues and lead to effective action for change within the workplace.

The role of the Professional Ergonomist

As the 'electronic revolution in-store' evolves, the need for an improved checkout design will become all the more pressing and it is an area full of potential for ergonomic research. How useful and acceptable this research will be to workers is a matter of some concern to trade unions. Researchers may have experience of negotiating with employers, both for sponsorship of research and for access to the workplace and workforce, but the workers may feel themselves unwitting guinea pigs in the research. A better protocol would start with consultation with representatives of the workers as well as employers and would entail a commitment to communication with and involvement of the workers. While it may at first seem a more unwieldy task, there would be considerable benefits - less resources wasted on misdirected research, better co-operation from the workers and ultimately better working conditions and a healthier, less fatigued workforce.

References

(1) Partly because of the massive investment involved, British retailers have been relatively slow to introduce EPOS system and scanner devices. It is predicted, however, that the next five years will see an electronic shopping revolution that will result in 80% of all packaged goods sold handled electronically. See, for example, Retail Business, March 1986, No. 337, pp. 15-17, Retail & Distribution Management, November/December 1986, pp. 29-40, The Electronic Revolution in Store, Marketing Services Department, Ogilvy and Mather, 1986, Electronics in Supermarkets, Post News, Somerset, 1986.
(2) New Technology in Supermarkets, Consolidated Report, European Foundation for the Improvement of Living and Working Conditions, 1985.
(3) Usdaw's 1986 Annual Delegate meeting (ADM) saw a major debate on stress and shopwork in which checkout operators voiced numerous complaints about their working environment.
(4) Wilson, J. R. & Grey, S. M., "Reach requirements and job attitudes at laser scanner checkout systems", in Ergonomics, 1984 Vol. 27, No.12, pp. 1247-1266.
(5) For example, Development of an Ergonomic Cash Register Working Place for Self-Service Shops, International Federation of Commercial, Clerical, Professional and Technical Employees (FIET), Vienna, 1982.
(6) Usdaw, ADM - Report of Proceedings 1986.
(7) Forrester, K., Thorne, C. and Winterton, J., "Worker' Investigations into the Working Environment", in Industrial Relations Journal, winter 1984, Vol. 15.

THE RECORD TO DATE: ONE UNION'S EXPERIENCE

J. CHURCH

ASTMS, Whitehall Office
Dane O'Coys Road, Bishop's Stortford
Hertfordshire   CM2 3JN

This paper discusses the steps one union has taken to provide advice and guidance for its members on several important aspects of the working environment. It describes the needs of the union's members and the virtual absence of published guidance in a form which could be used by trade unionists. The paper questions the reasons for the low priority which has been given to these issues by the Government's regulatory body, the Health and Safety Executive, and proposes steps which should be taken to ensure that ergonomics is given a higher priority by both managements and Inspectors.

Introduction
Aspects of the working environment including lighting, temperature, ventilation, seating, noise and the design of the work station are given low priority by managements and by Inspectors. Trade union representatives are rarely consulted about new working environments or changes to existing working environments and they do not have comprehensible guidance upon which to formulate their arguments. Occupational ill-health arising from poor working environments remains unmeasured and the scale of the problem unknown.

The Union and the needs of its members
The union covers a very broad spectrum of workplaces including office work in insurance companies, voluntary bodies; laboratory work in the health service, universities and in public and private research institutions; industrial work throughout the industrial sector including off-shore oil exploration and production. Although many of the queries relate to specific workplace hazards such as work with toxic substances, human pathogens, flammable materials, a substantial number, particularly from the office workplaces, relate to lighting,

heating, ventilation, smoking, seating, VDUs, and other aspects of the work environment.

While lengthy discussions take place on the design and furnishing of directors' offices, scant attention is given to the 'general offices' within which the bulk of our members work. The advent of new technology has transformed these open plan offices into a jungle of trailing cables and extension leads, supplying a myriad of VDUs, copiers, and other modern equipment. Many offices were inadequately designed originally and are now overheated, underventilated, noisy and overcrowded.

Complaints to management are difficult to formulate because there are no legally enforceable standards and because, for the most part, the regulatory authorities show little interest in investigating such complaints. Trade union members need a bargaining strategy. They need to know what standards to aim for, how to formulate their case and how to produce evidence of the harmful effects of such poor environments.

One union's response

Over the past ten years we have produced guidance for our members on many of these issues: temperature, VDUs, occupational stress, humidifier fever and Legionnaire's disease, and women's health. Other major areas are currently being researched including lighting, ventilation and seating. We do not intend to provide academic works but rather to explain in non-expert terms the problem, the cause, its solution and steps that our members can take.

We are very concerned at the activity of the Health and Safety Executive and its apparent failure to appreciate that, while chemical accidents and machinery injuries do affect working people, far more people are affected all day and every day by poor working environments. We continue to press the HSE to employ more ergonomists and to develop its expertise in this area whilst at the same time funding research into these important areas.

Until effective occupational health services are active in all workplaces we believe that the extent of ill health arising from the working environment will remain unrecognised. We also believe that many employees do not report backaches, eyestrain, headaches, stress, swollen joints etc. because they know that nothing will be done to investigate the cause of these problems and to rectify them.

Proposals for the future

(i) a legal requirement for architects, planners etc. to consult with their consumers, i.e. employees;

(ii) a legal requirement for employers to consult with employees on the working environment and changes to it;

(iii) a commitment by the Health and Safety Executive to give higher priority to ergonomics;

(iv) more training of Inspectors in ergonomics;

(v) published standards on aspects of the working environment;

(vi) funding by the Health and Safety Executive of research;

(vii) a comprehensive Occupational Health Service for every workplace so that the work environment can be monitored, measured and improved.

**WORKER DESIGNERS:
A WORK ENVIRONMENT ACT FOR BRITAIN?**

D. FEICKERT

NUM
St James House, Vicar Lane
Sheffield S1 2EX

In Britain trade union safety representatives were given limited rights in law in 1978, following the Health and Safety at Work Act 1974. In some other countries, safety representatives have more extensive rights, including the right to take part in the design of buildings, machinery and work processes. New legislation is now required in Britain that will encourage 'worker designers' to become actively involved in designing the working environment.

INTRODUCTION
   Throughout the 1970s many countries, especially in Europe, improved their workplace health and safety legislation. In Britain, the Health and Safety at Work Act was passed in 1974. For the first time all workers were to be covered by health and safety legislation. Of equal importance was the provision within the 1974 Act for the appointment of Safety Representatives. The Safety Representatives and Safety Committees Regulations, drawn up subsequently by the Health and Safety Commission, took effect in 1978.
   This new legislation gave trade union representatives a limited range of legal rights - to inspect their workplaces, to paid time-off for training, to workplace facilities and access to information. Now, a decade later, it is appropriate to review not only the experience of activity within the legislative framework in Britain, but also to assess the relevance of the experience in other countries with more advanced legislation.

OTHER EUROPEAN EXPERIENCES
   British trade unionists have followed the development of health and safety activity and legislation in Scandinavian

countries with particular interest. The Swedish Workers Protection Act 1949 influenced trade union attitudes in Britain. (Williams, 1960) In 1974 the legal rights Swedish safety delegates had been given in 1931 were further strengthened. Amongst other rights they gained the legal right to stop dangerous work.

Following the 1974 amendments a second phase of improvements was initiated. This culminated in the Work Environment Act 1978. Under the 1978 Act new directives on the working environment can be made by the National Board for Occupational Safety and Health. These legally binding directives cover both traditional health and safety concerns and new ones with mental and social aspects such as stress and shiftwork. Of interest to a Symposium on Trade Unions and Ergonomics is the following provision of the Work Environment Act. Section one of chapter two requires that:

"Working conditions must be adapted to human physical and mental aptitudes. The aim must be for work to be arranged in such a way that the employee himself can influence his work situation." (Ministry of Labour 1983)

The Act recognises the variability of human capacities. It gives safety delegates the right to take part in the design and planning of new buildings, work processes, machinery and working methods and in planning alterations to existing ones. Building permits can only be granted when the Inspectorate has verified that safety delegates have taken part in the planning stage.

Safety delegates see this latter provision as a breakthrough:

"The most important one, I believe is that we're now allowed to take part in planning all new workplaces, buying new machines and so on....
....It's cheaper for companies to listen to safety delegates before making investment in new workplaces. If they build workplaces which don't fulfill working environment standards, they risk major costs for improving the environment afterwards." (Andersson, 1979)

A British trade union study team, visiting Sweden in 1981, found the effectiveness of the new legislation impressive. (Feickert, 1982) The group visited two workplaces built after 1978. A new bus depot at Bjorknas provided a workspace layout in the workshop that allowed vehicle fitters a choice of working position while a new bus factory at Boras was remarkable with its superior physical environment and its semi-autonomous group working.

Denmark and Norway also have Work Environment Acts. In Norway the Working Environment Act 1977 was part of a comprehensive package of legislative reforms affecting working life. (Gustavsen, 1982) This Act, like its Swedish counterpart, requires employers to plan the working environment in such a way that workers are not exposed to physical or mental stress.

Gardell and Gustavsen (1980) recognised that prior to the new legislation activity over safety issues was most effective in workplaces where job content and skill levels were high. On the other hand where work was highly rationalised workers experienced greater difficulties in tackling working environment problems. A major objective in establishing a new legal framework was to focus attention on the design of the working environment and to encourage worker initiatives.

In West Germany the emphasis on working environment issues is seen in the £36M research programme on the humanisation of work. This programme is similar in content to much of the research sponsored by the Swedish Work Environment Fund. This Fund is financed by a payroll levy of 0.155% on all employers and has £50M annually at its disposal. (ASF, 1982) Between one third and one half of the research allocation is spent on the study of ergonomic problems.

WORKER DESIGNERS

In Sweden special grants are available from the Work Environment Fund for worker-organised research. The philosophy behind this and the new legislation generally has been explained by Gustavsen (1982); if workplace activity is to be stimulated the evaluations flowing out of workers' experiences must be seen as legitimate. To speed up progress in work environment standards, Gardell and Gustavsen (1980) argue, the criteria of proof needs to be re-defined to allow for cruder data based on well-established experiences among workers.

In Britain the Health and Safety at Work Act 1974 and associated Safety Representatives Regulations also encouraged worker initiatives. In recent years a wide variety of workers' investigations have been made. Most of these investigations have been conducted with little or no funding.

Those worker studies dealing with design aspects of the working environment have confirmed yet again the widely held view among ergonomists that action is vital at the design stage. It is often very difficult to resolve ergonomic and work organisation problems after machinery and equipment has been put into production or certain forms of work organisation implemented.

This was the experience of Leeds busworkers who made a detailed investigation of many aspects of their working environment (T&GWU, 1981) The busworkers identified a range of ergonomic problems in the design of vehicle cabs and workload problems associated with One Person Operation and irregular shift working. The study team experienced great difficulty in negotiating any changes to their working environment in the absence of any legislative support.

By contrast a major study of the working environment of busworkers in Sweden produced a set of detailed recommendations which were then implemented and tested as part of the modernisation programme in Gothenburg's public transport system. (Gardell, 1980) The upgrading of the working environment was seen as the priority. The Swedish study arose from a request by the busworkers' trade union and was funded by the Work Environment Fund.

In West Germany and Scandinavian countries a great deal of effort is also being made to ensure that new technology more closely matches workers' needs. Research programmes in this area are often established within the context of work environment legislation. In Britain activity in this research area is limited and workers have rarely been involved in planning the introduction of new systems.

In the design of the MINOS computer control system for coal mines, for example, the National Coal Board (NCB) have ignored the evidence from within the industry that worker involvement in the design phase has led to better systems. (Burns, 1982) There has been no consultation about what priorities the design teams should adopt. This is in complete contrast to the development of the shearer loader, the standard coal cutting machine used in NCB mines. Over 50% of the NCB awards for innovations to the shearer during its development phase were made to miners. (Townsend, 1976)

In a recent study of computerised mining operations Best et al (1985) found that operators in computerised colliery control rooms considered their skills and autonomy to have been reduced. The study also concluded that factors in the design of the system, management and organisational interactions have together resulted in less satisfaction among operators of computerised control rooms compared with operators in traditional ones.

In order to optimise the new technology a key management policy issue was identified - the need to agree the exact role of control room operators. This fundamental issue is of obvious concern to workers and their representatives and is one in which they should be fully involved. Many mining jobs are being transformed as the industry becomes more capital intensive. With this increase in capital intensity, the NCB is also proposing major changes in working practices. Many of these proposals touch upon important working environment issues and have already produced much strenuous debate.

CONCLUSION

While the Health and Safety at Work Act 1974 and Safety Representatives Regulations 1978 represented a major legislative advance in Britain the time has come for a new initiative. There is a need to draw not only on British experience in this area but also to assess the experiences of other countries with more developed working environment legislation and programmes.

This experience suggests that any new legislation must be informed by new understandings gained from ergonomics and other work environment sciences. Secondly, after identifying the design of the working environment as a central issue, there must follow an equally clear recognition that workers and their representatives have a crucial role to play in the design process. Safety representatives must be given new rights. Improved training, including courses in ergonomics, must be made available to enable them to use these new rights effectively.

REFERENCES

Andersson, K., 1979 For Health and Safety, Labour Market Reforms in Sweden, edited by A. Larsson (Swedish Institute)

Best, C.F., Ferguson, C.A., Martin, R., Mason, S., Simpson, G.C., Talbot, C.F., 1985, The Human Aspects of Computer Based Monitoring and Control of Mining Operations, (Institute of Occupational Medicine)

Burns, A., Feickert, D., Newby, M., Winterton, J., 1982, An Interim Assessment of MINOS (University of Bradford)

Feickert, D., 1982, The Scandinavian Experience: Legislation and Research, Stress at Work Conference (University of Leeds)

Gardell, B., Aronsson, G., Barklof, K., 1980 The Working Environment for local public transport personnel, (University of Stockholm)

Gardell, B., Gustavsen, B., 1980 Work Environment Research and Social Change: current developments in Scandinavia, Journal of Occupational Behaviour 1, 3-17

Gustavsen, B., 1982, Direct Workers Participation in Matters of Work Safety and Health: Scandinavian Strategies and Experiences (Work Research Institute)

Ministry of Labour, 1983, The Swedish Work Environment Act and Work Environment Ordinance

Swedish Work Environment Fund, (ASF), 1982, Programme of Activities and Budget 1981/82 - 1983/84

T&GWU 9/12 Branch, 1981, Stress at Work: Final Report (Workers Education Association)

Townsend, J.F., 1976, Innovation in Coal Mining Machinery: The Anderton Shearer Loader (SPRU, University of Sussex)

Williams, J.L., 1960, Accidents and Ill Health at Work (Staples)

NEW TECHNOLOGY AND RE-DESIGNING WORK

J. WINTERTON

Working Environment Research Group
School of Industrial Technology
University of Bradford, Bradford   BD7 1DP

Issues arising out of new technology are ranked in a hierarchy reflecting workers' interests. Most trade union effort has been directed at minimizing the negative effects of issues low in the hierarchy, like health and safety, rather than promoting positive aspects like skill and autonomy. Workers' investigations and participatory ergonomics offer methodological innovations to raise the level of strategic union intervention.

TRADE UNIONS AND NEW TECHNOLOGY

From the earliest days of ergonomics, practitioners have accepted that the effects upon operatives of technological developments may not be apparent until some time after their introduction. However, the problems associated with microelectronics are quantitatively and qualitatively different from any previous technical changes. The introduction of new technology raises five issues for trade unionists: job contraction, job content, job control, health and safety, and earnings (Winterton & Winterton, 1985). Job contraction and earnings are usually the immediate issues but the qualitative changes wrought by new technology will persist long after these issues have been resolved and it is in this area that ergonomists can make the greatest contribution.
The qualitative issues may be ranked in a hierarchy rather like Maslow's (1954) hierarchy of "human needs". At the base are conditions that need to be fulfilled in order

to avoid the <u>negative</u> effects of new technology, while at the top are objectives, the attainment of which would constitute a <u>positive</u> advantage of new technology over existing work. Unlike Maslow's, this hierarchy is constructed according to the interests of workers, rather than management, so the ranking is also an expression of the degree of conflictuality between management and workers.

At the bottom of the hierarchy are conventional health and safety issues. The potential of microelectronics to reduce exposure to physical hazards is a major justification for automation but the <u>design</u> of work will determine the extent to which this happens. Moreover, there are new physical hazards of VDU operation which were quickly recognised by trade unionists and ergonomists (Hunting <u>et al</u>, 1981).

At the next level in the hierarchy are new psychological hazards that are liable to arise from the software or work organization aspects of new technology. Ergonomists (Smith, 1980) and trade unionists (TGWU, 1981) have tended to approach occupational stress from a health and safety perspective, attempting to uncover stressors in a particular work situation. However, Gardell (1980) notes dehumanization, deskilling, work intensification and decreased job control contribute to the high workload, low discretion working environment associated with a high incidence of stress-related illness. It may, therefore, be more fruitful to address directly the issues further up the hierarchy which are major sources of stress where new technology is involved.

Dehumanization is considered the next level in the hierarchy since, like health and safety, it is an unintended consequence of new technology; it serves the interests of neither party and is detrimental to increased productivity. Dehumanization arises where physical isolation prevents social interaction and where communication is primarily via a computer terminal. Robot assembly and word processors create dehumanized working environments.

Skill, is seen as a positive attribute of work; a means of self-expression and fulfilment. Deskilling must rank above dehumanization in terms of its degree of conflictuality since it represents a central management dynamic of technological change aimed at reducing labour costs. Ure (1835) wrote of nineteenth century automation that "skilled labour gets progressively superseded, and will, eventually, be replaced by mere overlookers of machines". Established production and maintenance skills become incorporated into the technology, but a smaller number of highly-skilled jobs associated with design and programming are introduced. With this polarization of work, ever more complex tasks become incorporated into system software and more software functions become incorporated into the hardware itself. Some unions have negotiated <u>socially-constructed</u> skill labels for work from which the <u>genuine</u> skill foundation has been removed, rather than attempting to maintain skills.

At the top of the hierarchy is the question of control over the labour process. Managements have used new technology to reduce the autonomy of individual workers in three ways. First, removing <u>skill</u> facilitates management control since unskilled operations are easier to direct. Second, workers may have less <u>discretion</u> over the way in which a job is performed and less control over the <u>pace of work</u>. Production management seeking further increases in labour productivity can use new technology to reduce job discretion and intensify the pace of work by continuously monitoring work activity. In recent years Management Information Systems (MIS) have emerged with software dedicated to the surveillance of workers. Higher performance targets may be established and natural breaks taken by workers eliminated. Increased management control makes it more difficult for workers to use informal workplace sanctions.

A NON-REACTIVE STRATEGY

The centre of trade union strategy on new technology has been the negotiation of agreements through existing institutions of collective bargaining. All the major trade unions have produced model technology agreements as part of their policy on new technology, but there is considerable disparity between model agreements and those actually negotiated (Manwaring, 1981). Moreover, two recent surveys found that most technological change takes place without any formal agreement (Williams and Moseley, 1982; LRD 1982). Even where technological change is negotiated, little importance is attached to the quality of working life.

The unions need to develop a non-reactive strategy in order to address the higher-ranking issues. Moreover, trade unionists need to reject the notion of the neutrality of technology and recognise it as a social product. Technological design choices reflect political decisions, conscious or otherwise, even though technology is neutral in the sense that it can be <u>designed to serve the interests of either party</u>. Ergonomists can play a major role in formulating an alternative approach to the design of new technology which enhances the working environment, retains skills and increases workers' autonomy.

Research in Stockholm and Leeds has demonstrated that

factors present in urban transport work, like irregular shift working and tight vehicle timetabling, have negative effects on workers' health. Information from such research can be used to build working environment <u>rules</u> into the system software in order to limit shift irregularity, the length of the working day and week, and unsocial hours. In this way new technology can help minimise occupational stress.

Alternative systems design is not confined to information systems, but is also possible in the organization of numerically controlled machine tools. Rosenbrock (1982) describes a project at UMIST to design a lathe as part of a flexible manufacturing system that builds upon the skill of operators. Cooley (1983) argues that human-centred alternatives can be as readily conceived with intellectual work as with manual, and outlines alternative approaches to computer-aided design. The potential of new technology to <u>enhance</u> workers' skills has only begun to be explored but offers exciting possibilities.

New technology can also be designed to facilitate increased worker participation and job autonomy (Winterton, 1984). Conventional centralized computer systems constrain workers' preferences, such as shift choices, that may have been accommodated informally under a manual system. The complexities of organizational information systems demand the involvement of all those affected at the design stage in the interests of efficiency. Worker involvement should extend into the operational life of the system.

Scandinavian experience suggests that autonomous workgroups are superior from a managerial perspective, as well as being preferred by workers, so managers should abandon the strategy of attempting to increase control in the application of technology, and seek to enhance workers' autonomy and responsibility. Managers need the cooperation of workers to make the new systems work; often they are modified on the shop floor. If managers continue to use technology against workers they will be faced with an alienated and resentful workforce. All the shopfloor ingenuity that has gone into the creation of a complex web of informal custom and practice rules will develop mechanisms to subvert the technology.

Both conceptual and methodological changes are necessary for ergonomists to facilitate the development of a non-reactive strategy on new technology. In conceptual terms, their concerns will have to continue to move away from the physical sciences and incorporate more of the social sciences. This enlargement of ergonomics should lead to an acceptance of the inherent conflict of interests within industry and a recognition of the political character of decision making. The views of workers will have to be seen as of at least equal importance to those of management and of technical specialists, ergonomists included. As Davis (1975) has argued: "no one has the moral right to design the work and work situation of another person. The role of the expert... is to help the worker design his own work situation to assist his own efficiency and job satisfaction needs".

The methodological changes are automatically necessitated by the conceptual developments. Workers themselves must be involved in the processes of redesigning work, both on the moral grounds argued by Davis and because only the end user of technology has the necessary experience. "Workers' investigations", where groups of activists using academic specialists as resources shape their own research (Forrester et al, 1984) provides one model, but the methodology of "participatory ergonomics" needs to be refined (Forrester, 1986). The involvement of trade unions has two additional advantages. First, the unions may become sufficiently committed to the approach to provide financial support for worker-orientated research. But equally important, the trade union at shop floor level provides an institutional channel and power base through which the results of joint research can be pursued. Ergonomists may discover they have more allies among union activists than among managers.

REFERENCES

Cooley, M., 1983, The New Technology: Social Impacts and Human-Centred Alternatives, TPG Occasional Paper 4, Open University.

Davis, L.E., & Cherns, A.B., (eds) 1975, The Quality of Working Life, London: Macmillan.

Forrester, K., 1986, Involving workers: participatory ergonomics and the trade unions, in Osborne, D.J., (ed) Contemporary Ergonomics 86, London: Taylor and Francis.

Forrester, K., Thorne, C., & Winterton, J., 1984 Worker's investigations into the working environment, Industrial Relations Journal, 15, 4, 28-37.

Gardell, B., 1980, Production Techniques and Working Conditions, Stockholm: Swedish Institute.

Hunting, W., Laublie, T., & Grandjean, E., 1981, Postural and visual loads at VDT workplaces Ergonomics, 24, 919-44;

LRD, 1982 Survey of new technology, Bargaining Report 22, Labour Research Department.

Manwaring, T., 1981, The trade union responses to new technology, *Industrial Relations Journal*, 12, 4, 7-27.

Maslow, A.M., 1954, *Motivations and Personality*, New York: Harper & Row.

Rosenbrock, H., 1982 Engineers, robots and people, AGM Society of Chemical Industry, London, May 1982.

Smith, M., 1980, Job stress in video display operations, in Grandjean, E., and Vigliani, E., (eds) *Ergonomic Aspects of VDTs*, London: Taylor and Francis.

TGWU, 1981, *Stress at Work: Final Report*, Transport and General Workers' Union 9/12 branch, Leeds.

Ure, A., 1835, *The Philosophy of Manufacturers*, London: Charles Knights.

Wilkinson, B., 1983, *The Shopfloor Politics of New Technology*, London, Heinemann.

Williams, R., & Moseley, R., 1982, Technology agreements: consensus control and technical change in the workplace, EEC/FAST Conference, The Transition to an Information Society, London, 28 January.

Winterton, J., 1984, Industrial democracy and new technology, *Industrial Tutor*, 4, 10, 5-11.

Winterton, J., & Winterton, R., 1985, *New Technology: The Bargaining Issues*, Occasional Papers in Industrial Relations, No. 7, Universities of Leeds and Nottingham.

RESTRUCTURING THE RELATIONSHIP:
THE ERGONOMIST AS A RESOURCE PERSON

K. FORRESTER

Department of Adult and Continuing Education
The University of Leeds
Leeds   LS2 9JT

This paper focuses on the relationship between the ergonomist and the research subjects. It suggests that research relationships, structured around the ergonomist as a 'resource' rather than as an 'expert', addresses a number of possible criticisms levelled against ergonomists by users of research.

> "Research knowledge, which confers enormous power, is too often held and maintained by an elite (some of them us) and utilised sometimes for, often against, people who are ignorant and maybe helpless against it".
>   J Roby Kidd (1981).

Introduction

Any analysis of published articles, literature and texts within the discipline of ergonomics clearly reveals the problem-oriented focus of research concerns and activities. Moreover, the variety of audiences addressed by ergonomists is bewildering, from the disabled, the consumer, the road-user and the domestic worker through to the more mainstream audience of those at work. It might appear, therefore, as surprising that this same literature displays few examples of the ergonomist critically reflecting on the relationship between this concern to overcome problems and the audience immediately affected. To state the problem from a different perspective, we can agree with Sell (1985) when he writes "Traditional ergonomics has relied very much on the view that the expert is the person who knows best whilst the person who has to carry out the job has to use the equipment as provided or the procedures as specified". The absence of involvement with those at the receiving end of the ergonomic research effort, is puzzling. It is likely to reduce, for example, the practical outcome of the research work. There is no simple correlation between research endeavours and ergonomic changes in the workplace, in the home or on the road. Secondly, many ergonomic projects have a hidden industrial relations agenda. The efficiency, safety, comfort and productivity of the total man-machine system will convey different, often conflicting, meanings to different groups within the workplace. Without due attention being paid to the research subjects' involvement in the project, it is possible for suggested beneficial ergonomic changes to be rejected on non-ergonomic grounds by these it the receiving end. Finally, it is puzzling that more emphasis has not focused on user involvement, because it might restrict and inhibit the very nature of the research design and research activity, as will be further detailed below.

In short, an ergonomic practice structured around assumptions of 'the expert' is a practice that is unlikely to realise the potential and usefulness of the discipline. This paper will argue that there exists an alternative conception to this traditional 'expert' role of the ergonomist ie. the ergonomist as a 'resource person'. For reasons of analysis, the differing conceptions of the ergonomist as an 'expert' or as a 'resource' will be seen as ideal-typical constructs. Neither approach to ergonomic research are mutually exclusive: rather they are significant distances apart but on the same continuum. For reasons of simplicity, this paper will assume that the field of concern is an industrial context.

The Ergonomist as a Resource Person

What, then, are some of the characteristics that contribute towards this alternative conception of ergonomic activity? The five points below, it will be argued, suggest partial answers to this question.

1. **The importance and legitimacy of the workers knowledge and experience:** Recognition of the accumulated experience and intimate knowledge of work processes available within a group of employees is an essential starting point for the model of ergonomic practice suggested in this paper. This recognition extends beyond the simple participation of employees in various forms of testing or surveys ie. beyond passive sources of data. It is a perspective acknowledging that employees have a constructive role to play in the formulation, development and outcome of particular research

activities. It, conversely, recognises that the ergonomist is not a busworker, or a shop worker or a hospital porter. Rather he or she is a highly trained and skilled practitioner in a particular discipline and can contribute towards resolving problems confronting a range of employees. Above all, such a conception of the ergonomists role recognises the importance of employees gaining "confidence in their own ability to solve problems" (Gustravsen, 1979) rather than having problems resolved on their behalf.

2. **The issue of accountability:** It flows from the point above that the issue of control and decision-making is at the forefront of research activity, rather than an area which is taken for granted or defined away. Collective decision making is the result of an environment and context shaped by mutual trust and respect by all participating parties. Negotiation and education are the mechanisms for the achievement of this collective and supportive research environment. Central to this renegotiation between the researcher and a group of employees is the dismantling of the "enormous power" (Roby Kidd, 1981) invested in and normally assumed by the 'outside' specialist/researcher.

3. **Research Methodology:** Over and above the normal complexities associated with research methodologies an alternative ergonomic practice has to confront additional difficulties. Research methods employ their own logic, theoretical commitments and methodological implications often resting on competing epistemologies. (Long, 1984; Gustavsen, 1979). A participatory research design will include, in consideration of the appropriate methodologies, a variety of 'non-research' factors such as accessibility to research tools by participating groups, the financial costs of particular research techniques and the possibility of generalising the research methods to other, wider concorned groups. Encouragement and support to workers in certain investigative areas, for example epidemiological studies, already exists (Fox <u>et al</u>, 1982; Forrester, 1986). The eventual choice of appropriate research methods will reflect then the blend of skills and expertise available within the project group and address directly the, often 'hidden', social and political assumptions underpinning much research activity.

4. **The research outcome:** A fifth characteristc in reformulating the role of the ergonomist is consideration of the results of the research work. As a professional research worker observed, when reflecting on a research project involving Scandinavian Iron & Metal Trade Unions, "In research projects, the term 'results' usually is understood to be the reports being written by research

workers .... In (our) project, it was now instead stated that as a 'result' of the project, we will understand actions carried out by the Iron & metal Workers Union, centrally or locally, as part of or initiated by the project" (Nygaard, 1979). There is an important concern, then, of overcoming the division between the development and application of knowledge. The 'reorientation' of the professional in such circumstances is a complex process requiring clarity of role and purpose within the research activity. To define the results of the research work as the actions flowing from a particular project significantly alters the traditional role of the professional 'outsider'.

5. **The Ergonomist as a resource person:** Finally, and as a summary of the points above, the research role of the ergonomist is better conceptualised as a 'resource' rather than as an 'expert'. A resource worker implies a radically changed relationship to the research subjects, someone who is in dialogue and acting as a partner in the area which is to be changed. The ergonomist as a resource person also implies a different perception of the research process itself ie. as a collaborative supportive process where resultant activity "is not only application of knowledge, it is (also) generation of knowledge". (Gustavsen, 1979) This unity of the development and application of knowledge, or research and action, implies a different level and commitment of responsibility by the ergonomist. In short, the ergonomist as a resource person implies a very different research activity necessitating a critical evaluation of all the primary research relationships.

Conclusion

As indicated earlier, the rival conceptions of the ergonomic research activity implied in roles of 'expert' or 'resource' are most usefully understood as part of a continuum. There are a variety of modifications possible around each ideal-type. The paper advocates and encourages a closer examination of the ergonomist as a resource person. The promise entailed in such a perspective includes the increased possibility of resolving the discrepancies between the environment-employee relationship. Above all, such a perspective begins to address the relationships of power and exploitation within the research activity identified by Roby Kidd in the introduction to this paper.

References

Forrester, K., 1986, Involving Workers: Participatory Ergonomics & The Trade Unions, *Contemporary Ergonomics 1986* edited by D. J. Oborne. (Taylor & Francis), pp. 59-63.

Fox, J., Gee, D., and Leon, D., 1982, *Cancer and Work: Making Sense of Workers' Experience.* Published by the City University Statistical Laboratory and General, Municipal and Boilermakers.

Gustavsen, B., 1979, Liberation of Work and the Role of Social Research, in *Work & Power: The Liberation of Work and Central of Political Power*, edited by Tom R. Burns, Lars Erik Karlsson and Veljkorus. (Sage Publications), p. 342.

Long, A. F., 1984, *Research into Health and Illness: Issues in Design, Analysis and Practice*, (Gower).

Nygaard, K., 1979, The Iron & Metal Project: Trade Union Participation, in *Computers Dividing Man and Work: Recent Scandinavian Research on Planning and Computers from a Trade Union Perspective*, edited by A. Sandberg. Swedish Centre for Working Life, Stockholm.

Roby Kidd, J., 1981, Research needs in Adult Education, *Studies in Adult Education*, Vol. 13, No. 1, p. 5.

Sell, R.G., 1985, It Takes Both Experts and Job Holders to Design Jobs, *Ergonomics International 85*, Edited by I. D. Brown, R. Goldsmith, K. Coombes and M. A. Sinclair (Taylor & Francis), p. 358.

Late Paper

CAPRICIOUS BEHAVIOUR AND HUMAN RELIABILITY

W.W. SUOJANEN

Department of Management
Georgia State University, University Plaza
Atlanta, GA 30303, USA

In ergonomics, human reliability in work organizations assumes rational behavior on the part of the worker, the supervisor, and the general management. Recent discoveries in neurophysiology and psychopharmacology suggest that the opposite may instead be the case. Michael Gazzaniga in The Social Brain argues that a considerable amount of human behavior is capricious in nature. In the United States, our research finds that about 80% of the workforce behaves irrationally a good deal of the time, which then results in high costs, low output, and below standard quality.

Traditional concepts of training in human reliability, work organization, and job design must be modified to accept the fact that capricious, rather than rational, behavior is often the fact in the American workplace.

Behavior Formation

During the past fifteen years or so, numerous articles have explored how the differences between the left and the right sides of the human brain relate to individual and organizational behavior. These articles argue that the left-new brain (of the right-handed person) is analytical and verbal and that the right-new brain (of a right-handed person) is intuitive and synthesizing.

A recent article argues that left-brain/right-brain differences are founded on myth and that in view of this, attempts to improve performance and training by relying on non-existent left-brain/right-brain differences are unlikely to be productive (Hines, 1985).

Hines views the human brain as a single unitary, linear, hierarchic organ which always functions rationally in accordance with the current assumptions of the behavioral and social sciences as these are embodied in ergonomics.

Michael Gazzaniga in the Social Brain differs with this notion of linear, unified conscious experience. He argues that human behavior has a modular-type organization and often engages in capricious thoughts and actions. By modularity, Gazzaniga means that the brain is organized into relatively independent functioning units that work in parallel. Given this approach, the mind is not an indivisble whole, operating in a single way to solve all problems. Rather, it consists of specific and identifiably different units (or modules) which breakdown and begin to work on information which enters the mind.

Gazzaniga's theory does however support the contention of Hines and those authors who claim that too much has been made of left-brain, right-brain differences. His findings, based on extensive research with neurologically impaired, split-brain patients, shows that the right and left-new brain organization is not necessarily the same for everyone. He contends that the location of right and left-new brain characteristics are far less significant than the basic premise of his theory - that specific brain systems handle specific tasks, therefore exhibiting a modular brain structure.

Gazzaniga (1985) explains the modular organization of the brain in order to illustrate how beliefs and behaviors are formulated. This is important because of the implications for ergonomics. Gazzaniga shows that the brain is organized into independent processing modules and that each of them generate specific behaviors. Such behaviors are then rationalized by special non-language processes in the "inerpreter" part of the left-new brain and reasons are developed to support them. These reasons or hypotheses are then reported to the left-new brain language center or "the talker" and verbalized.

Because Gazzaniga's work is essentially the study of belief and behavior formation, it presents new insight to the study of ergonomics. The special capacity of the left-new brain "interpreter" reveals how important the carrying out of behaviors is for the formation of many theories about the self. Gazzaniga contends that the dynamics that exist between our mind modules and our left-new brain interpreter module are responsible for the generation of human beliefs.

Unlike animals, the evolution of Homo Sapiens was marked by intense conflict between the new brain and the visceral or animal brain. The conflict as to which brain is master and which is servant in the control of behavior and of each other is now completely embodied in all concepts of mental health and mental illness. This conflict for control of behavior also determines the state of human consciousness.

The Decisive Mind

The human brain has three homo modules-the think left-new brain (TLNB), the feel left-new brain (FLNB), and the right-new brain (RNB). Simeons (1961) argues that the decisive mind is located, not in the new brain, but in the visceral brain (VB). Recent research data, including those of Gazzaniga, appear to support this hypothesis. Watts (1975), for example, puts it as follows:

> "The thalamus...controls the unplanned, non-thinking, reflex systems of brainstem and the voluntary, conscious, spontaneous actions of the individual...the thalamus supplies the coordination for thinking, memory, and precise somatic activity and shares control of visceral parasympathetic and dange sympathetic response with the anterior and posterior hypothalamic areas-a partnership between intellect and mood. At times, a healthy, mature thalamic system excerises the final decision. At other times, the hypothalamus makes the final decision in immaturity, in powerful moods, and in mood disorders."

According to Watts, there are two modules in the decisive mind. One of these, the thalamus, is linked to the new brain and governs our behaviors. When the thalamus is in charge, intellect governs emotion-this is the I/E state of consciousness. In this state of consciousness, the human being functions rationally, or as I have termed it, an open, adaptable, knowledgable human being. Using the acronym of highlighted letters to form the word OAK. According to Watts (1975):

> "The mood is more fundamental than thinking. On the evolutionary scale the mood appears in lower, simpler, and older animals. The lateral portion of the cerebrum, the thinking regions of the nervous system, increases in size and significance as the evolutionary scale is ascended. The lateral cerebrum system inhibits viscero-somatic brain circuits. This cortical influence brings wisdom and conditioned reflexes to bear on mood. A gross incongruity between mood and thinking is a hallmark of malphrenia (schizophrenia)."

What Watts terms as mood is that state in which the animal brain module rules the human brain module. This E/I state is the opposite of the I/E state of consciousness. When the hypothalamus is in control, chemistry determines behavior and the person behaves in an addictive, complusive, obsessive, really nutty way. The acronym ACORN is developed and this person is in an altered state of consciousness (ASOC) verses the OAK who operates in a normal state of consciousness (NSOC). The person who operates in an ASOC does so because of the euphoria it produces. Watts (1975) indicates that "Pleasant sensory input from visceral, skin, and cranial nerve receptors maintains a euphoric mood."

Wilder Penfield (1975), one of the great neurosurgeons of our time, published a book which greatly strengthens the Simeons thesis. Penfield also went beyond Simeons, Watts, and McClean and located the seat of consciousness in the visceral brain module:

> "...the indispensable substratum of consciousness lies outside the cerebral cortex, probably in the diencephalon (the higher brain stem)...the cerebral cortex, instead being the "top", the highest level of integration, was an elaboration level..."

If one accepts the Simeons-Watts-Penfield hypothesis, then the seat of consciousness of the human being is in the DM rather than in the cerebral modules. It is important to note that Watts and Penfield have located the decisive mind in the visceral brain (VB) module. Equally, the reader must remember that conflict between the new and the old-the human and animal brain-is a basic element of the human condition.

One summary points out that the human brain has two memory systems-one for the new brain and the other for the old brain:

> "...two forms of memory reside in different structure in the brain, may be under different biochemical control and may have separately evolved..." (Blakeslee, 1985)

According to Blakeslee, old brain memory consists of motor and cognitive procedural skills that are rarely forgotten and which are seldom knocked out "by an insult to the brain." Such memory may involve "changes in specific areas that generally do not communicate with the conscious mind."

New brain memories are factual and declarative. These higher memories are found in the factual left-new brain and the imaging right-new brain brain. New brain memory is extremely flexible, relying on new brain connections. Like the new brain, new memory arrived later in evolution.

The Modular Brain and States of Consciousness

Following Gazzaniga (1985), the TLNB may be viewed as giving expression to the normal state of consciousness (NSOC).

In a like manner, the VB mediates the unconscious, the FLNB, the subconscious, and the RNB, the preconscious. I would argue that the human being experiences many different altered states of consciousness (ASOC) in the course of a lifetime. This is nothing new—William James prophesized this more than a half century ago:

> "Our normal walking consciousness...is but a special type of consciousness, whilst all about it, parted from the silkiest of screens, there lie potential forms of consciousness entirely different, without suspecting their existence...and at a touch, they are all there in their completeness."

The real significance of cerebral asymmetry research is that it has served to explain the modularity of the human brain. Gazzaniga (1985) states:

> "...it is not important that the left brain does this or the right brain does that. But it is highly interesting that by studying patients with their cerebral hemispheres separated certain mental skills can be observed in isolation...It is a highly important point and find."

The modular brain, then, is organized into a confederation of relatively independent modules that work in parallel, rather than in a series. When engaged in processing memory about happenings, storing affective reactions about those events, and in responding to stimuli associated with a particular memory—"all of these activities are routinely carried out by cats, dogs, and by humans. These activities occur without language and with abandon." (Gazzaniga, 1985)

All of the modules of the human brain, except the TLNB, work in non-verbal ways. These other modules express themselves through overt actions and/or more covert behavior. Many of these actions are <u>capricious</u>. As Gazzaniga (1985) phrases it:

> "The realization that the mind has a modular organization suggests that some of our behavior should be accepted as capricious and <u>that a particular behavior might have no origins in our conscious thought processes</u>...humans resist the interpretation that such behaviors are capricious because <u>we seem to be endowed with endless capacity to generate hypotheses as to why we engage in any behavior.</u>"

The synonyms related to the word "capricious" include whimsical, implusive, moody, flighty, volatile, fanciful,

unstable, irresponisble, to name only a few.
The human being indulges in a number of capricious behaviors. Only one of its' modules possesses the power of speech-the left-new brain (LNB) of the right-handed person. According to Gazzaniga (1985), the brain's interpreter is also located in the LNB-the brain's talker, as follows:

> "There are unique neural systems present in the left hemisphere of (right-handed) humans that compel the brain system communicating with the external world to make sense out of the diverse behaviors humans produce. These beliefs take on a central importance to the person and as such can override the tugs produced by more mundanely delivered rewards or punishment for the behavior."

Thus, through this recent work, the human being is seen as more of a sociological entity than a single unified psychological entity. By realizing this fact and by understanding Gazzaniga's theory of modular organization of the brain, we may apply this knowledge to today's management theory.

Conclusion
The idea that human beings do not always behave rationally goes back a hundred years to one of the greatest works on behavior ever published, that of William James, Principals of Psychology in 1890. Gazzaniga now shows us that the works of James, Simeons, Penfield, Watts, and McClean (1973) support the concept that we must always make an allowance for a certain amount of capricious or ACORN behavior in designing organizations. Many large organizations are already utilizing Employee Assistance Programs (EAP) to cope with capricious or ACORN behavior in the workplace.

References
Bessinger, R. Carlton & Suojanen, Waino W., Management and The Brain, Business Publishing Division, Georgia State University, Atlanta, GA, 1983.
Blakeslee, Sandra, "Clues Hint at Brain's Two Memory Maps", The New York Times, February 19, 1985.
Gazzaniga, Michael S., The Social Brain: Networks of The Mind, Basic Books, New York, New York, 1985.
Hines, Terence, "Left Brain, Right Brain: Who's on First", Training and Development Journal, November, 1985, p. 34.
Simeons, A.T.W., Man's Presumptuous Brain, E.P. Dutton, New York, New York, 1961, p. 34.
Watts, George O., Dynamic Neuroscience: Its Application to Brain Disorder, Harper and Row, New York, New York, 1975, pp. 102-103, p.332.

AUTHOR INDEX

Akinmayowa, N.K. 298
Aldridge, M. 321
Astley, J.A. 135

Ball, P.W. 46
Barton, P.H. 129
Boyce, P. 1
Britten-Austin, H.G. 108
Brotherton, C. 321
Brouwer, W. 95
Brown, C. 333
Browne, R.M. 251
Buckley, P. 233

Cadman, J.S. 129
Chan, W.L. 140
Chapman, A.J. 239
Charleton, S. 321
Church, J. 356
Cooper, S.L. 275
Corlett, E.N. 197, 263
Cox, T. 328
Coyle, K. 83
Craig, A. 105
Crampin, T. 110

David, H. 227
Davis, A. 328
Davis, G.N. 179

Embrey, D.E. 39
Emerson, T.J. 108

Farmer, E.W. 77
Farrow, A. 209
Feickert, D. 359
Fenn, S. 233
Forrest, M.A. 239
Forrester, K. 371

Gallwey, T. 89
Gay, L.A. 280
Graves, R.J. 140, 162

Hallman, P.J. 315

Harvey, R.S. 146
Haslam, R.A. 275
Haslegrave, C.M. 197
Hockey, R. 103

Ing, J.J. 129

John, P. 269
Joyner, S. 346

Kanis, H. 156
King, N. 340
Kirwan, B. 70
Kleberg, I.G. 203
Kompier, M. 185

Leather, P. 321, 333
Life, M.A. 117
Long, J. 117, 245
Lowe, T.J. 179

Mabey, M.H. 162
Mawby, F.D. 150
Matzdorff, I. 167
Meijman, T. 185
Mohindra, N. 101
Mulders, H. 185

Norman, C.J. 150
Norris, B.J. 191

O'Boyle, P. 89

Parsons, K.C. 257, 275, 292
Payne, A.P. 292
Pethick, A.J. 140
Porter, J.M. 173, 191
Porter, M. 209
Powrie, S.E. 123, 215

Randle, I.P.M. 286
Reason, J. 21
Reising, J.M. 108

Ridd, J.E.   203
Robertson, S.   173
Rothengatter, T.   95
Russell, D.   352

Sheehy, N.P.   239
Simpson, M.R.   263
Sinclair, M.A.   221
Smith, A.   83
Smith, T.A.   257
Spencer, E.   101
Stammers, R.B.   106, 135, 251
Stone, P.T.   303
Street, P.J.   150
Suojanen, W.W.   376

Taylor, N.K.   263
Taylor, R.   101

Thomas, N.T.   280
Thorne, C.   352
Tracy, M.   197

Ussher, M.H.   77

Van Noord, F.   185
Van Wolffelaar, P.   95
Visick, D.   64

Waters, T.   37
Webber, G.M.B.   315
West, M.   340
Whalley, S.P.   52
Whitfield, D.   58
Wilson, J.R.   129
Winterton, J.   365
Wood, J.   309

SUBJECT INDEX

Applications
  Aircraft 108, 227
  Aircrew 108, 227
  Armed forces 110, 146
  Automated manufacturing systems 346
  Automotive industry 269, 333
  Business 321
  Coal industry 135, 162, 359
  Clothing industry 280
  Computer aided design 179, 215, 269
  Decision support systems 108, 123, 269
  Driving 95
  Electricity industry, 309
  Expert systems 257, 263, 269
  Fault diagnosis 64
  Fire services 150, 280
  Food industry 1
  Helicopters 110
  Hospitals 209
  Information technology 321
  Inspection 89
  Maintenance 197
  Management information systems 365
  Manual materials handling 140, 209, 286
  Manufacturing industry 140
  Medicine 83, 209, 298
    colds 83
    influenza 83
    respiratory diseases 83
    vibration white finger 298
  Metal working 298
  Metallurgical examination 89

Applications continued
  Nuclear power plants 37, 58, 70
  Nursing 209, 340
  Offices 1, 203
  Physically handicapped 303
  Postal services 129
  Process control 46, 52, 64, 135, 309
  Public information systems 179
  Radiography 89
  Sea diving 77
  Shipbuilding 328
  Shops and supermarkets 352
  Space technology 70
  Teleshopping 233
  Textile industry 340
  Truck manufacturing 333
  Urban transport 179, 185
  Viewdata 233

Absenteeism 140, 333
Accident causation 95
Age effects 1
  elderly users 95
Allocation of function 108, 117, 221, 227
Anthropometry 167, 197
Anxiety 77
Arousal 77
Assembly lines 140, 333
Autonomous work groups 365

Bar coders 352
Body core temperature 292
Body size 275
Bus design 179, 185

Cognitive performance 77, 83
Cognitive processes 21, 39, 70

Subject Index

Colour judgement 1
Conspicuity aids 315
Control design 46, 185
Control room design 309
Corridors 315
Cost benefits 140, 167

Data analysis methods 239
Decision making 21
Design methods 117, 123, 156, 179, 263, 269, 359, 371
Dialogue design 117, 215, 221, 233, 251
Display design 46, 64
Display format 64
Dynamic work capacity 286

Education in ergonomics 156
Employment 321
Entrepreneurship 321
Environmental design 257
Equipment design 146, 162, 209
Evacuation 315
Expert opinion 263, 371

Gender effect 321
Graphics interface 215, 251

Hand tool design 162
Health and safety 356
Heat stress 280
Human error 21, 37, 39, 46
Human performance 83, 101, 103, 108, 129
Human reliability 37, 39, 52, 58, 70, 376
Human-computer interface design 117, 215, 227, 233, 239, 245

Icons 215, 251
Individual differences 167, 197, 221, 275, 280, 321
Information clutter 64
Introduction of new technology 328, 340, 352, 365

Job analysis 227
Job attitudes 321, 328, 333, 340
Job characteristics 340
Job design 46, 365
Job organisation 365, 376

Keyboard design 129
Knowledge acquisition 257, 263
Knowledge base 21, 106, 257

Learning 101
Legislation 356, 359
Lighting 1, 303
  colour rendering 1, 309
  daylight factor 309
  design of illuminants 309
  emergency lighting 1, 315
  glare 303
  illuminance 303, 309, 315
  indirect lighting 309
  layout for illumination 309
  luminance contrasts 1, 303, 309

Management 328, 340, 346
Manikins 150
Manual control 95
Marketing ergonomics 110
Measurement techniques 103, 105, 108, 239, 315
Memory scanning 83
Mental models 245, 257
Mental workload 101, 103, 105, 106, 108
Menus 215
Mock-ups 309
Motivation 321
Motor performance 77, 129
Motorcycle design 173
Movement analysis 239
Musculo-skeletal disorders 185, 352

Non-verbal communication 239

Subject Index

Organisational design 70, 328, 333, 346
Oxygen consumption 286

Performance shaping factors 21, 52
Physical fitness 280
Postural comfort 129, 173, 203, 209
Posture 129, 140, 191, 197, 203, 209
Problem solving 21, 123
Product design 156
Productivity 333
Protective clothing 150, 280

Reach capabilities 197
Reaction time 83, 95
Reading 1
Reasoning 21
Risk assessment 70

Safety 58
Screen format 233
Seating 173, 185, 191, 203, 209
Selection 89
Simulation 37, 110, 117, 227, 275, 286
Skin temperature 298
Sociotechnical systems 333
Software reliability 70
Speech interface 117
Stairways 315
Standards 146, 167, 315
Strength capabilities 197
Subjective measures 105
System adaptability 221, 227
System design 46, 58, 101, 108
System reliability 52, 58
System usability 257

Task analysis 39, 110, 135, 227
Task luminance contrast 1, 303
Task taxonomy 135
Teamwork 346
Test rig 179
Thermal comfort 280
Thermoregulation 275, 292, 298
Time perception 83
Tracking 83
Trade unions 352, 356, 359, 365, 371
Training 46, 64, 89, 106, 110, 135, 321, 328, 346

User acceptance 150, 292
User assistance 64
User control 340
User experience 215
User involvement 156, 269, 340, 352, 359, 365, 371
User models 245
User needs 117, 221
User trials 179
Users' models 245, 257

Vehicle design 173, 179, 185, 191
Veiling reflections 1
Vibration 298
Video recording 140, 315
Vigilance 83
Visual discomfort 1
Visual lobe 89
Visual search 1

Walkways 309
Work analysis 239
Work measurement 140
Work procedures 46
Working conditions 356
Working posture 129, 140, 191, 197, 203, 209
Workplace design 46, 140, 185